ROUTLEDGE LIBRARY EDITIONS:
HISTORICAL SECURITY

Volume 6

# THE COMMANDERS

# THE COMMANDERS
Australian Military Leadership in the Twentieth Century

Edited by
D. M. HORNER

LONDON AND NEW YORK

First published in 1984 by George Allen & Unwin

This edition first published in 2021
by Routledge
2 Park Square, Milton Park, Abingdon, Oxon OX14 4RN

and by Routledge
52 Vanderbilt Avenue, New York, NY 10017

*Routledge is an imprint of the Taylor & Francis Group, an informa business*

© 1984 D. M. Horner

All rights reserved. No part of this book may be reprinted or reproduced or utilised in any form or by any electronic, mechanical, or other means, now known or hereafter invented, including photocopying and recording, or in any information storage or retrieval system, without permission in writing from the publishers.

*Trademark notice*: Product or corporate names may be trademarks or registered trademarks, and are used only for identification and explanation without intent to infringe.

*British Library Cataloguing in Publication Data*
A catalogue record for this book is available from the British Library

ISBN: 978-0-367-61963-3 (Set)
ISBN: 978-1-00-314390-1 (Set) (ebk)
ISBN: 978-0-367-63563-3 (Volume 6) (hbk)
ISBN: 978-1-00-311970-8 (Volume 6) (ebk)

ISBN: 978-0-367-63572-5 (pbk)

**Publisher's Note**
The publisher has gone to great lengths to ensure the quality of this reprint but points out that some imperfections in the original copies may be apparent.

**Disclaimer**
The publisher has made every effort to trace copyright holders and would welcome correspondence from those they have been unable to trace.

Major-General Sir William Bridges

Lieutenant-General Sir Leslie Morshead

Lieutenant-General Sir John Lavarack

Vice-Admiral Sir William Creswell

THE COMMANDERS

EDITED BY D.M. HORNER

# The Commanders

Australian military leadership
in the twentieth century

GEORGE ALLEN & UNWIN
Sydney London Boston

© D.M. Horner 1984
This book is copyright under the Berne Convention. No reproduction without permission. All rights reserved.

First published in 1984 by
George Allen & Unwin Australia Pty Ltd
8 Napier Street, North Sydney, NSW 2060 Australia

George Allen & Unwin (Publishers) Ltd
Park Lane, Hemel Hempstead, Herts HP2 4TE, England

Allen & Unwin Inc.
9 Winchester Terrace, Winchester, Mass 01890 USA

National Library of Australia
Cataloguing-in-Publication entry:

The Commanders.

Bibliography.
Includes index.
ISBN 0 86861 496 3.
ISBN 0 86861 504 8 (pbk.).

1. Military biography. 2. Generals — Australia — Biography. 3. Australia — History, Military. I. Horner, D. M. (David Murray), 1948-

355'.0092'2

Library of Congress Catalog Card Number: 83-73043

Set in 10/12 pt Times by Setrite Typesetters, Hong Kong
Printed in Australia by Macarthur Press Pty Ltd

# Contents

|   |   |   |
|---|---|---|
| Acknowledgements | | vii |
| Illustrations | | viii |
| Maps | | x |
| Contributors | | xi |
| Guide to Army Ranks and Command Structure | | xv |
| Approximate equivalent ranks | | xvii |
| Abbreviations | | xviii |
| 1 | Introduction  *D.M. Horner* | 1 |

**Part One: The First World War**

| | | |
|---|---|---|
| 2 | Major-General Sir William Bridges: Australia's First Field Commander  *Chris Coulthard-Clark* | 13 |
| 3 | General Sir Brudenell White: The Staff Officer as Commander  *Guy Verney* | 26 |
| 4 | Vice-Admiral Sir William Creswell: First Naval Member of the Australian Naval Board, 1911–19  *Stephen D. Webster* | 44 |
| 5 | General Sir Harry Chauvel: Australia's First Corps Commander  *A.J. Hill* | 60 |
| 6 | General Sir John Monash: Corps Commander on the Western Front  *P.A. Pedersen* | 85 |

**Part Two: The Second World War**

| | | |
|---|---|---|
| 7 | Lieutenant-General Sir John Lavarack: From Chief of the General Staff to Corps Commander  *A.B. Lodge* | 129 |
| 8 | Lieutenant-General Sir Vernon Sturdee: The Chief of the General Staff as Commander  *D.M. Horner* | 143 |
| 9 | Lieutenant-General Henry Gordon Bennett: A Model Major-General?  *A.B. Lodge* | 159 |
| 10 | Lieutenant-General Sir Leslie Morshead: Commander, 9th Australian Division  *A.J. Hill* | 175 |
| 11 | Field Marshal Sir Thomas Blamey: Commander-in-Chief, Australian Military Forces  *D.M. Horner* | 202 |
| 12 | Lieutenant-General Sir Sydney Rowell: Dismissal of a Corps Commander  *D.M. Horner* | 225 |

13  Lieutenant-General the Honourable
    Sir Edmund Herring: Joint and Allied Commander
    *Stuart Sayers* ............................................................. *244*
14  Major-General George Alan Vasey:
    Commander, 7th Australian Division
    *D.M. Horner* ............................................................. *263*

**Part Three: Post-Second World War**

15  Lieutenant-General Sir Horace Robertson:
    Commander-in-Chief British Commonwealth
    Occupation Force    *Ronald Hopkins* .................. *281*
16  Air Chief Marshal Sir Frederick Scherger:
    Chairman, Chiefs of Staff Committee
    *Harry Rayner* ............................................................ *298*
17  General Sir John Wilton:
    A Commander for his Time    *Ian McNeill* .......... *316*
    Notes ........................................................................... *335*
    Index ............................................................................ *351*

# Acknowledgements

This book would not have been possible without the enthusiasm, skill, and hard work of the other ten contributors. I thank them for their contributions. I am grateful to Dr Robert O'Neill, Director of the International Institute for Strategic Studies, London, for his help in the conceptual stages of the book.

I am indebted to the Council of the Australian War Memorial for a grant to cover administrative costs. In addition, most of the photographs and many of the maps are reproduced with permission of the Australian War Memorial. Other maps were provided by the draughting section at the Australian Army Command and Staff College, while the maps for Chapter 5, drawn by Wendy Gorton, were provided by Alec Hill. The National Library of Australia and the Imperial War Museum supplied some photographs and gave permission for their publication. Unfortunately, after an extensive search, I was unable to locate the copyright holder for Map 19. Finally, I thank John Iremonger of George Allen & Unwin for his enthusiastic support.

D.M. Horner

*Endpapers*: All portraits reproduced courtesy of the Australian War Memorial

*Front endpaper*: (Detail from) Florence Rodway *Major-General Sir William Bridges*; John Longstaff *General Sir Brudenell White*; Vice-Admiral Sir William Creswell; W.B. McInnes *General Sir Harry Chauvel*; John Longstaff *General Sir John Monash*; Ivor Hele *Lieutenant-General Sir John Lavarack*; Murray Griffin *Lieutenant-General Sir Vernon Sturdee*; James Quinn *Lieutenant-General Henry Gordon Bennett Back endpaper*: (Detail from) Ivor Hele *Lieutenant-General Sir Leslie Morshead*; Ivor Hele *Field Marshal Sir Thomas Blamey*; Ivor Hele *Lieutenant-General Sir Sydney Rowell*; William Dargie *Lieutenant-General the Honourable Sir Edmund Herring*; A.M.E. Bale *Major-General George Alan Vasey*; John Longstaff *Lieutenant-General Sir Horace Robertson*; Geoffrey Mainwaring *Air Chief Marshal Sir Frederick Scherger*; *General Sir John Wilton*

# Illustrations

1  Major-General Bridges   *14*
2  Sir George Reid, General Maxwell and Major-General Bridges   *20*
3  Artillery orders 1st Australian Division   *22*
4  Lieutenant-Colonel White at Anzac   *35*
5  Major-General White in Grevilliers, France   *38*
6  Vice-Admiral Creswell   *45*
7  Brigadier-General Chauvel in Monash Valley   *63*
8  Lieutenant-General Chauvel at Damascus   *83*
9  Major-General Monash   *97*
10  Headquarters 3rd Australian Division   *102*
11  Lieutenant-General Monash with his staff   *104*
12  William Hughes with General Monash   *118*
13  Breaking the Hindenburg Line   *123*
14  Major-General Lavarack, Brigadier Allen and Brigadier Stevens   *136*
15  Lieutenant-General Lavarack reviews French troops   *139*
16  Lieutenant-General Sturdee   *144*
17  General Sturdee, General Blamey and Percy Spender   *147*
18  General Sturdee and General ter Poorten   *149*
19  Major-General Bennett   *160*
20  General Bennett and Captain Collins   *164*
21  Major-General Morshead and Colonel Lloyd   *180*
22  Lieutenant-General Morshead and Winston Churchill   *190*
23  Lieutenant-General Morshead, Rear-Admiral Royal and Air Vice-Marshal Bostock   *199*
24  R.G. Menzies and General Blamey   *206*
25  F.M. Forde and General Blamey   *209*
26  General Blamey and General MacArthur   *212*
27  General Blamey with Generals Herring and Eichelberger   *217*
28  General Blamey with General Krueger   *221*
29  General Blamey   *223*
30  Lieutenant-General Rowell   *227*
31  General Rowell with Brigadier Rourke   *232*
32  Lieutenant-General Herring and General Blamey   *252*
33  Herring with Colonel Doe, Colonel Murray and General Vasey   *254*
34  Lieutenant-General Herring   *256*
35  Headquarters 7th Australian Division   *267*

*Illustrations*

36  Major-General Vasey    *271*
37  General Vasey, General Blamey and General Berryman    *274*
38  General Vasey with a soldier    *277*
39  Review, 1st Armoured Division    *290*
40  General Blamey, Brigadier Fergusson and General Robertson    *291*
41  Lieutenant-General Robertson with New Zealand Minister for Defence    *294*
42  Air Chief Marshal Scherger    *299*
43  Air Chief Marshal Scherger hands over to General Wilton    *314*
44  General Wilton    *317*

# Maps

1. Gallipoli. The Anzac Positions, May 1915   *62*
2. Quinn's Post, May–August 1915   *64*
3. The Canal Defences and the Turkish advance July–August 1916   *67*
4. The Battle of Romani   *69*
5. The Capture of Beersheba, 31 October 1917   *75*
6. The Battle for Es Salt, 30 April–4 May 1918   *77*
7. The Destruction of Army Group F, 19–25 September 1918   *80*
8. Monash Valley, May–August 1915   *93*
9. Gallipoli. The Anzac Positions, May 1915   *95*
10. 3rd Australian Division in Third Ypres, June–November 1917   *100*
11. Capture of Hamel, 4 July 1918   *107*
12. August–October 1918 Offensive   *112*
13. Area of 7th Australian Division operations in Syria, June–July 1941   *137*
14. The Japanese advance through the Indies and to Rabaul   *150*
15. The Conquest of Malaya, December 1941–January 1942   *170*
16. Cyrenaica 1941   *177*
17. Defence of Tobruk. Dispositions, afternoon 5 May 1941   *182*
18. 9th Australian Division in the first Alamein Battle, 10–27 July 1942   *187*
19. 9th Division attack at El Alamein, 23/24 October 1942   *192*
20. The Alamein Line on 23 October 1942   *194*
21. 9th Division attacks at El Alamein, 25–31 October 1942   *196*
22. The Boundaries of the South-West Pacific Area and the Extent of the Japanese Advance   *203*
23. Axes of advance, Papuan Campaign, outlined by Blamey in September 1942   *238*
24. Allied advance to Buna, Gona and Sanananda, October–November 1942.   *249*
25. The advance to Lae-Nadzab, 1–5 September 1943   *258*
26. Allied advance on Buna, Sanananda and Gona, 16–21 November 1942   *269*
27. The envelopment of the Huon Peninsula, September 1943–February 1944   *273*

# Contributors

*Chris Coulthard-Clark* works in the Strategic and International Policy (SIP) Division of the Department of Defence in Canberra. A graduate of the Royal Military College, Duntroon, he served in the Australian Army until 1979 and took his MA in 1980 at the Faculty of Military Studies of the University of New South Wales. His publications include *The Citizen General Staff. The Australian Intelligence Corps 1907-1914* (Canberra, 1976) and *A Heritage of Spirit. A Biography of Major-General Sir William Throsby Bridges, KCB, CMG* (Melbourne, 1979), as well as numerous newspaper and journal articles on military historical subjects. He is a contributor to the *Australian Dictionary of Biography* and also edited the autobiography of Air Commodore R.J. Brownell, *From Khaki to Blue* (Canberra, 1978). At present he is working on a history of the Royal Military College under a research grant from the Australian War Memorial.

*A.J. Hill*, MBE Ed, was a senior lecturer in the faculty of military studies of the University of New South Wales at the Royal Military College, Duntroon, from 1968 until he retired in 1979. Educated at Sydney Grammar School, the University of Sydney and Balliol College, Oxford, he was commissioned in the Citizen Military Forces in 1936. During the Second World War he commanded a company in the 9th Australian Division in the Defence of Tobruk, served on Morshead's staff at El Alamein and was brigade major of the 20th Brigade in New Guinea and Borneo. After the war he taught history and geography at Sydney Grammar School. He is the author of *Chauvel of the Light Horse* (Melbourne, 1978) and has written numerous articles on Australian soldiers for the *Australian Dictionary of Biography*.

*Major-General Ronald Hopkins*, CBE, US Legion of Merit, knew Sir Horace Robertson well and served with him on many occasions in a military career spanning almost 40 years. After graduating from the Royal Military College, Duntroon in 1917 he saw active service with the 6th Australian Light Horse in Palestine in 1918-19. He was seconded to the United Kingdom for armoured training in 1937-39; after the fall of France in 1940 he was largely concerned with the establishment of an armoured training organisation in Australia and the raising of the 1st Australian Armoured Division of which he became GS01. During the first New Guinea campaign he was Chief of Staff, 1st Australian Corps, and later was Land Forces adviser with the 7th Amphibious Force, USN.

He commanded the Australian Component of the British Commonwealth Occupation Force in Japan from 1946 to 1948. Other appointments included Commandant, Australian Staff College, Deputy Chief of the General Staff and Commandant, RMC Duntroon. After retirement, General Hopkins spent some years with Advertiser Newspapers Ltd, Adelaide. He was chief executive officer of the first Adelaide Festival of Arts. He is the author of *Australian Armour* (Canberra, 1978) and many journal articles.

*Major David Horner* is a regular army officer currently serving in the Department of Defence (Army Office), Canberra. He is a graduate of the Royal Military College, Duntroon, the University of New South Wales, and the Australian National University, where he was awarded the J.G. Crawford Prize, the university's most prestigious PhD research prize. He served in Vietnam as a platoon commander in 1971, has had regimental postings with the Royal Australian Regiment, was a Churchill Fellow in 1976, and is a graduate of the Australian Army Command and Staff College. He is the author of *Crisis of Command: Australian Generalship and the Japanese Threat, 1941-1943* (Canberra, 1978) and *High Command, Australia and Allied Strategy, 1939-1945* (Canberra and Sydney, 1982) and co-editor (with R.J.O'Neill) of *New Directions in Strategic Thinking* (London, 1981) and *Australian Defence Policy for the 1980s*, (Brisbane, 1982). He is currently working on a biography of Major-General George Vasey.

*Brett Lodge* is a postgraduate student in the Faculty of Military Studies of the University of New South Wales at Duntroon. Born at Bathurst, NSW in 1955, he was educated at St Stanislaus' College and the Royal Military College, Duntroon, which he entered in 1974. He resigned from the army in 1978 and for a short time worked as a journalist in Queensland. His article, 'A Share of Honour: Lavarack and Tobruk—April 1941' was published in the *Journal of the RUSI of Australia* in April 1982, while in February 1983 Mr Lodge presented a paper at the Australian War Memorial History Conference on Lieutenant-General H. Gordon Bennett's conduct of operations in Malaya in 1941-42. Work is continuing on a thesis examining the career of Lieutenant-General Sir John Lavarack and a biography of Lieutenant-General H. Gordon Bennett.

*Ian McNeill* was a regular army officer in the Royal Australian Infantry Corps from 1954 to 1982. Born in Melbourne, he was educated at Wagga Wagga High School, the Royal Military College, Duntroon, and the University of New England. He was an adviser in the Australian Army Training Team in Vietnam and subsequently joined the military history cell in the Department of Defence (Army Office), where he was responsible for research and recording the military aspects of Australia's involvement in Vietnam. He is the author of *The Team: Australian Army Advisers in Vietnam, 1962-1972* (Canberra, 1984) and has written articles on the

Australians in Vietnam. He left the army as a major and is currently working on a biography of General Sir John Wilton as part of a post-graduate program in history at the Australian National University.

*Major Peter Pedersen* is a regular army officer currently serving as a company commander in 5th/7th Battalion, The Royal Australian Regiment. He graduated from the Royal Military College, Duntroon, in 1974, and holds the degrees of Bachelor of Arts and Doctor of Philosophy from the University of New South Wales. He has written widely on the First World War and his study of Monash as a commander will be published in 1984. He has twice visited the Australian Imperial Force's battlefields on Gallipoli and the Western front.

*H.R. (Harry) Rayner*, MVO MBE, is a journalist with lengthy experience in Defence. A graduate of Queensland University, majoring in History and English literature, he has been editor of several provincial dailies and correspondent for metropolitan newspapers and the ABC. He served with the RAAF in the Second World War and at the end of the war was adjutant of No. 2 Medical Air Evacuation Transport Unit, on strength of Scherger's 1st Tactical Air Force, RAAF. In 1952 he joined Defence Public Relations, later becoming Director PR, Air, with the reserve rank of group captain. Seconded to the Prime Minister's Department as Australian press liaison officer to US President Johnson in 1966 and again in 1970 for the Royal Visit, he was appointed a Member of the Royal Victorian Order by the Queen. Soon after his retirement in 1979 he received the MBE for services to journalism. Foundation President and currently a Vice-President of the National Press Club, he has recently completed a full-scale biography of Air Chief Marshal Sir Frederick Scherger.

*Stuart Sayers*, journalist, became Literary Editor of the Melbourne newspaper, the *Age*, in 1964 after some twenty years as a reporter, feature writer, columnist and leader writer. Educated at State School, Ivanhoe Grammar and Geelong Grammar School, he served in the Second World War with the Royal Australian Navy in the waters north of Australia and surrounding New Guinea and in the Pacific. Except for three years in London with Australian Associated Press and Reuter, he has always worked for the *Age*. *By Courage and Faith*, a history of Carey Baptist Grammar School, his first book, was published in 1973. His biography of Lieutenant-General the Honourable Sir Edmund Herring, *Ned Herring*, written with the close cooperation of the subject and his wife, Dame Mary Herring, was published in 1980. He is a contributor to the *Australian Dictionary of Biography*, wrote the chapter on the Press in *Australia, New Zealand and the South Pacific: A Handbook*, and has published papers in the *Victorian Historical Journal*.

*Guy Verney* works in the Department of Defence, Canberra. He was educated at the University of Queensland, the Flinders University and the University of Sydney where he completed his PhD thesis on 'The Army High Command and Australian Defence Policy, 1901-1918'. In 1976 he was a lecturer in social studies at Kelvin Grove College of Advanced Education and in 1981 was a tutor in Australian history at the University of Sydney. He has published articles on Australian defence history, has contributed to the *Australian Dictionary of Biography*, and is currently working on a biography of the professional career of General Sir Brudenell White.

*Stephen D. Webster* is senior adviser to the Premier of Tasmania. American born, he completed his BA (Hons) at the United States Naval Academy and Miami University of Ohio in 1968. In 1970 he completed an MA in history as a Fulbright Scholar at Victoria University of Wellington, New Zealand, with a thesis entitled 'The Tasman Connection: New Zealand's Trade and Defence Relations with Australia, 1900-1914'. He was awarded his PhD from Monash University in 1976 for his thesis, 'Creswell, the Australian Navalist: A Career Biography of Vice Admiral Sir William R. Creswell, RN, RAN, 1852-1933'. Dr Webster tutored in Australian and United States history at Monash University from 1974 until 1978 and lectured in British imperial history from 1978 until 1980.

# Guide to Army Ranks and Command Structure

| Serial | Unit or formation in ascending order | Rank of officer commanding | Chief operational staff officer |
|---|---|---|---|
| (Each unit or formation contains two or more of the subordinate units plus specialist, support and administrative troops) | | | |
| 1 | infantry platoon; armoured, cavalry or light horse troop | lieutenant | |
| 2 | infantry company; artillery battery; armoured, cavalry or light horse squadron | major, sometimes a captain | |
| 3 | infantry battalion; artillery regiment (brigade in First World War); armoured, cavalry or light horse regiment | lieutenant-colonel | Adjutant is a captain. In more recent years there have been operations officers who are majors. |
| 4 | infantry or light horse brigade; divisional artillery | brigadier (colonel or brigadier-general in First World War) | Brigade major—normally a major. |
| 5 | division | major-general | General Staff Officer Grade 1 (GSO1)—colonel |
| 6 | corps | lieutenant-general | Brigadier, General Staff—brigadier |
| 7 | army | general or lieutenant-general | Major-General, General Staff—major-general |
| 8 | army group | field marshal or general | Chief of the General Staff—lieutenant-general |

A 'force' can be any size. Thus the Australian Imperial Force in the First World War, commanded by General Birdwood, reached seven divisions, but the GOC AIF exercised administrative control only. As GOC AIF Middle East in 1941 Lieutenant-General Blamey commanded three divisions and corps and administrative troops. As GOC AIF Malaya Major-General Bennett commanded little more than his own understrength division. When established in mid-1942 New Guinea Force amounted to about corps strength and was the same formation as the 1st Australian Corps. However, it soon expanded to become an army command, with 1st Australian Corps being separated to form a subordinate formation. The British Commonwealth Occupation Force was nominally a corps-sized formation, but also commanded airforce and naval base units. The Commander, Australian Force Vietnam, a major-general, commanded the army task force (brigade) in Phuoc Tuy Province, the logistic support group at Vung Tau, and the Training Team and RAAF and RAN units throughout the country.

# Approximate Equivalent Ranks

| Navy | Army | Air Force |
|---|---|---|
| sub-lieutenant | 2nd lieutenant<br>lieutenant | pilot/flying officer |
| lieutenant | captain | flight-lieutenant |
| lieutenant-commander | major | squadron leader |
| commander | lieutenant-colonel | wing commander |
| captain | colonel | group captain |
| commodore | brigadier | air commodore |
| rear-admiral | major-general | air vice-marshal |
| vice-admiral | lieutenant-general | air marshal |
| admiral | general | air chief marshal |
| admiral of the fleet | field marshal | marshal of the air force |

# Abbreviations

| | |
|---|---|
| AA & QMG | Assistant Adjutant and Quartermaster-General |
| ABDA | American, British, Dutch, Australian (Command) |
| ADC | Aide-de-Camp |
| AGPS | Australian Government Publishing Service |
| AHQ | Army Headquarters |
| AIC | Australian Intelligence Corps |
| AIF | Australian Imperial Force |
| AMF | Australian Military Forces |
| ANZAC | Australian and New Zealand Army Corps |
| ANZUS | Australia, New Zealand and United States (Treaty) |
| AOC | Air Officer Commanding |
| AVM | Air Vice-Marshal |
| AWM | Australian War Memorial |
| BCFESR | British Commonwealth Far East Strategic Reserve |
| BCOF | British Commonwealth Occupation Force |
| BEF | British Expeditionary Force |
| BGGS | Brigadier-General, General Staff |
| BGHA | Brigadier-General, Heavy Artillery |
| BGRA | Brigadier-General, Royal Artillery |
| BGS | Brigadier, General Staff |
| BM | Brigade Major |
| BMRA | Brigade Major, Royal Artillery |
| C-in-C | Commander-in-Chief |
| CAS | Chief of the Air Staff |
| CB | Companion of the Order of the Bath |
| CBE | Companion of the Order of the British Empire |
| CCSC | Chairman, Chiefs of Staff Committee |
| CG | Commanding General |
| CGS | Chief of the General Staff |
| CID | Committee of Imperial Defence |
| CIGS | Chief of the Imperial General Staff |
| CMF | Citizen Military Forces |
| CMG | Companion of the Order of St Michael and St George |
| CO | Commanding Officer |
| CRA | Commander, Royal Artillery (of a division) |
| DA & QMG | Deputy Adjutant and Quartermaster General |
| DCAS | Deputy Chief of the Air Staff |
| DCGS | Deputy Chief of the General Staff |
| DMO | Director of Military Operations |

| | |
|---|---|
| DMO & I | Director of Military Operations and Intelligence |
| DSC | Distinguished Service Cross |
| DSO | Distinguished Service Order |
| G1 | Chief operational staff officer (of a division). See GSO1 |
| GCMG | Knight Grand Cross of the Order of St Michael and St George |
| GHQ | General Headquarters |
| GOC | General Officer Commanding |
| GOC-in-C | General Officer Commanding-in-Chief |
| GSO1 (GSO3) | General Staff Officer, Grade 1 (Grade 3) |
| GSO1 (Int) | General Staff Officer, Grade 1 (Intelligence) |
| HQ | Headquarters |
| IG | Inspector-General |
| JCOSA | Joint Chiefs of Staff in Australia |
| JIB | Joint Intelligence Bureau |
| JIO | Joint Intelligence Organization |
| KBE | Knight Commander of the Order of the British Empire |
| KC | King's Counsel |
| KCB | Knight Commander of the Order of the Bath |
| KCMG | Knight Commander of the Order of St Michael and St George |
| LGA | Lieutenant-General in charge of Administration |
| LHQ | Land Headquarters |
| MBE | Member of the Order of the British Empire |
| MC | Military Cross |
| NCO | Non-commissioned officer |
| NGF | New Guinea Force |
| NLA | National Library of Australia |
| NZ & A | New Zealand and Australia |
| NZEF | New Zealand Expeditionary Force |
| NZMR | New Zealand Mounted Rifles |
| P&O | Peninsular and Oriental Steam Navigation Company |
| QMG | Quartermaster-General |
| RAAF | Royal Australian Air Force |
| RAF | Royal Air Force |
| RAN | Royal Australian Navy |
| RANR | Royal Australian Naval Reserve |
| RMC | Royal Military College, Duntroon |
| RN | Royal Navy |
| RSM | Regimental Sergeant Major |
| SCAP | Supreme Commander for the Allied Powers |
| SEATO | South East Asia Treaty Organization |
| TAF | Tactical Air Force |
| TEWT | Tactical exercise without troops |
| VC | Victoria Cross |

# Introduction

## D.M. HORNER

The purpose of this book is to analyse the performances of selected Australian senior military commanders under the stress of action and policy-making. Although the chapters deal with issues as varied as the commanders they discuss, there are some common questions which appear throughout the book. These questions are: how far can it be said that there is an Australian style of command; how has the nature of military command changed in the twentieth century; how well have Australian officers been prepared for high command; how well have Australian commanders handled the problems of coalition war and cooperation with their Allies; and to what extent have political direction, organisation structures and entrenched defence policy impacted upon the performance of commanders in the field? By coalition war is meant war conducted by two or more Allied countries, involving not only cooperation on the battlefield, but also the development of combined or inter-locking strategies.

There have been few true analyses of the performances of senior Australian military commanders, and most of those available have been written in recent years. The Australian official histories have been limited by the concern of the official historians to provide a multi-strata picture of the Australian war effort with emphasis on the actions of the private soldier. In the opinion of Professor Encel they are democratic war histories:

> The great strategic decisions, the historic battles, the political conflicts, the personal rivalries of generals and admirals in search of glory, are no more than a distinct back drop for hundreds of personal dramas taken from regimental diaries. It is the kind of history that would have commended itself to Tolstoy.[1]

This comment is, perhaps, a little unfair to Gavin Long and Paul Hasluck, but the point is made.

Sir Basil Liddell Hart believed that if military history was to be of practical value, it 'should be a study of the psychological reaction of the commanders, with merely a background of events to throw their thoughts, impressions and decisions into clear relief'.[2] In other words, the heart of the military problem is the personality of the commander, for it shapes the battle, and in turn, the battle affects the psychological state and effectiveness of the commander. 'The melting-point of warfare,' as Barbara Tuchman put it, is 'the temperament of the individual commander'.[3]

Australians have an unfortunate habit of cutting down the 'tall poppies', unless of course the tall poppy is a racehorse or a gangster. Even outstanding cricketers eventually become the victims of excessive egalitarianism. But the two world wars made such a deep impact on Australian society that the more outstanding military leaders could not be ignored. As Field Marshal Helmuth von Moltke observed, 'the military commander is the fate of the nation'. Marshal Foch added: 'Great results in war are due to the commander. History is therefore right in making generals responsible for victories—in which case they are glorified; and for defeats—in which case they are disgraced.'[4] As the memory of war fades, however, so too does the remembrance of a nation's great commanders.

The reader is entitled to ask why particular commanders have been included and others excluded from this book. The book is, of course, merely a selection with the aim of giving insight into Australian commanders. The list does not pretend to be comprehensive. The first criterion was that the commander should have exercised command at a senior level in war. But what is command, what is a senior level, and what is war? (A guide to ranks and command structures can be found on pp. xv–xvii.) On the face of it, a number of commanders selected appear to have failed the entrance test. General Sir Brudenell White spent all of the First World War as a senior staff officer, but as Guy Verney argues, he exercised many of the prerogatives of command, and had an important impact on Australian operations and military development. Vice-Admiral Sir William Creswell, as First Naval Member of the Australian Naval Board during the First World War, had little opportunity to command Australian naval forces in war. However, I have included the chapter on Creswell by Stephen Webster because it demonstrates the problems of senior policy-making faced by Australia's top naval officer during that war.

In the section on the Second World War, I have written a chapter on Lieutenant-General Sir Vernon Sturdee, Chief of the General Staff from August 1940 to September 1942, in which I argue that for the period December 1941 to March 1942 he was, in effect, commander-in-chief for the defence of Australia.

In the third section covering the post Second World War period the chapter on Lieutenant-General Sir Horace Robertson by Ronald Hopkins is unique in that it deals with the performance of a commander in a situation short of war. As commander of the British Commonwealth Occupation Force Robertson handled a range of difficult problems with skill and style, and viewed against his command in battle at more junior levels in both world wars, raises the question of how he would have performed if given the opportunity of senior command in war. The chapters on Air Chief Marshal Sir Frederick Scherger by Harry Rayner, and on General Sir John Wilton by Ian McNeil, reveal the problems of senior command and policy-making in the undeclared wars and low-level conflicts of the post-war period.

Since this book is about senior commanders, it was decided to exclude those who had not commanded at, at least, the level of an army division or its naval or air equivalent. Thus commanders like Brigadier-General Pompey Elliott from the First World War, or Brigadiers Dougherty and Windeyer (both later to be major-generals and to be knighted) from the Second World War, had to be excluded. General Robertson and Air Chief Marshal Scherger each commanded brigade-sized formations in the Second World War, but they both went on to more senior commands after the war. Similarly, General Wilton commanded a brigade in the Korean war, but as Chief of the General Staff and as Chairman, Chiefs of Staff, exercised senior direction in the Vietnam war. Vice-Admiral Sir John Collins was Commodore Commanding China Force in 1942 and in 1944 and 1945 was Commodore Commanding the Australian Squadron in important battles in the Pacific. As Australia's senior naval commander during the Second World War an exception might have been made and he could have been included, but fortunately we do have his autobiography, *As Luck Would Have It*.

Notwithstanding the criteria, the selection of the commanders might still be open to some comment. Major-General Sir William Bridges commanded the 1st Australian Division in battle for only three weeks. As Chris Coulthard-Clark points out in his chapter, Bridges was very much an imperial, rather than an Australian general, but it is a commentary on Australia's development as a nation that Britain may not have tolerated anything else at the beginning of the First World War.

There can be little argument with the inclusion of chapters on General Sir Harry Chauvel by Alec Hill and on General Sir John Monash by Peter Pedersen. Chauvel was the first Australian to command a corps, the Desert Mounted Corps in Palestine in 1917-18, and Monash commanded the Australian Corps during the important offensive that broke the Germans in 1918. In view of Monash's reputation as Australia's greatest soldier, his chapter is longer, and examines his development as a commander from the period well before the beginning of the First World War. Perhaps also there was scope for a chapter on one or more of Australia's First World War divisional commanders: Legge, M'Cay, Hobbs, Holmes, Sinclair-MacLagan, Glasgow, Rosenthal and Gellibrand.[5] Although half of these men commanded their divisions for little more than six months, most of them were very competent and there is work here for future historians.

For the Second World War three prominent corps commanders have been omitted: Lieutenant-General Sir Iven Mackay, who commanded the 6th Division in the Middle East in 1940 and 1941, and then New Guinea Force during 1943; Lieutenant-General Sir Stanley Savige, the commander of the 3rd Division at Salamaua in 1943 and of the 2nd Australian Corps on Bougainville in 1945; and Lieutenant-General Sir Frank Berryman, commander of the 2nd and then the 1st Corps as it advanced from Sattleberg along the New Guinea coast at the beginning of 1944. Two of these generals have been the subjects of full-length biographies.

Ivan Chapman's book *Iven G. Mackay, Citizen and Soldier*, is a balanced account and provides a good insight into Mackay's performance. W.B. Russell's biography of Savige, *There Goes a Man*, is in my view somewhat flattering. Both generals could have warranted inclusion in this book if space and an available author could have been found. There could well have been a chapter on Berryman. As Blamey's senior staff officer for much of the last three years of the war he had an important impact on Australian operations and strategy. Commanding Berryforce, the *ad hoc* force formed to deal with the French counter-attack at Merdjayoun in Syria in 1941, and a corps in New Guinea, he showed courage and leadership of a high order.

The other corps commanders have been covered. In my chapter on Field Marshal Sir Thomas Blamey I have focused on his period as Commander-in-Chief of the Australian Military Forces, rather than on his command of the 1st Australian Corps, or of the AIF in the Middle East. Brett Lodge has written on Lieutenant-General Sir John Lavarack, the commander of the 1st Australian Corps in Syria; I have discussed the dismissal of Lieutenant-General Sir Sydney Rowell from his command in New Guinea in 1942; and Stuart Sayers has examined the performance of Lieutenant-General Sir Edmund Herring, commanding the 1st Australian Corps and New Guinea Force in 1942-43. Alec Hill has concentrated on Lieutenant-General Sir Leslie Morshead's command of the 9th Australian Division at Tobruk and El Alamein, rather than his later command of the 1st Australian Corps in New Guinea and Borneo. There is no chapter on Lieutenant-General Sir John Northcott who commanded the 2nd Corps in Australia for part of 1942. Northcott then became Chief of the General Staff under Blamey until the end of the war, and to his great regret did not command in action, although he was Acting Commander-in-Chief during a quiet period while Blamey was overseas in April and May 1944.

There are two chapters on divisional commanders. I have written on that grand regular soldier, Major-General George Vasey, commander of the 7th Division in New Guinea, and Brett Lodge has examined the performance of the controversial citizen soldier, Lieutenant-General H. Gordon Bennett in Malaya. Major-General Sir George Wootten, commander of the 9th Australian Division from 1943 to 1945, could well have warranted a chapter, although whether in preference to Vasey is arguable. The efforts of Major-General Tubby Allen, first in Syria and then commanding the 7th Division on the Kokoda Track, and of Major-General Cyril Clowes, at Milne Bay, were worthy of note, but both received no further chance in action after those campaigns. Generals Milford, Stevens and Bridgeford fought competent but less spectacular battles.

In the post-war period chapters could have been included on: General Sir Thomas Daly, brigade commander in Korea and then Chief of the General Staff during the Vietnam war; General Sir Arthur MacDonald, Commander, Australian Force Vietnam during the important Tet offensive of 1968 and later Chief of Defence Force Staff; and Lieutenant-General Sir Donald Dunstan, deputy commander of the

Australian Task Force during the Tet offensive in 1968, Commander, Australian Force Vietnam in 1971, and later Chief of the General Staff. Sufficient material is not yet available for adequate accounts to be written, but it is to be hoped that the official historian for the Vietnam war will not disregard the roles of these important commanders.

The lack of chapters on important naval and air commanders does not denote a biased selection, but rather the reality of Australian military history. No Australian naval officers reached positions of senior command in the First World War, and in the Second World War only Commodores Collins and Farncomb commanded the Australian Squadron in action. Admiral Sir John Crace commanded the Australian Squadron during the Battle of the Coral Sea, but although he was born in Australia he was a Royal Navy officer detached to the RAN and he returned to Britain soon after the battle. The three Chiefs of Naval Staff during the Second World War were all British officers.

The paucity of senior Australian airforce commanders during the Second World War demonstrated the unfortunate aspects of the Empire Air Training Scheme by which Australia provided aircrew for RAAF units in Europe but received few opportunities to fill senior command positions. The Chief of the Air Staff (CAS) from early 1940 until early 1942 was a British officer, and after that the airforce was wracked by the feud between the CAS, Air Marshal Sir George Jones, and the commander of the operational forces, Air Vice-Marshal William Bostock. As Air Officer Commanding RAAF Command Bostock was responsible to the American General George Kenney (commander of the Allied Air Forces in the South West Pacific Area) for operations, but he was responsible to Jones for administration. The struggle for power, caused by an unsatisfactory command relationship, soon developed into a bitter personal argument that was damaging to the effectiveness of the Air Force. The problems of command in the RAAF during the Second World War are worthy of close study, but this book is not the place.

Air Commodore J.E. Hewitt's command of No. 9 Operational Group was of limited duration before he was removed from command. His own account is published in *Adversity in Success*. Perhaps a chapter on Air Marshal Sir Richard Williams could have been included. Williams's senior active command was as a lieutenant-colonel commanding an RAF Wing (which included Australians) in Palestine in 1918. He was Australian CAS for most of the inter-war period but did not hold the position during the Second World War, when he was for a while Air Member for Organization and Equipment and later RAAF representative in Washington. While his autobiography, *These Are Facts*, does not always live up to its title, it is a good starting point for an examination of RAAF operations and administration.

The greatest restriction on the selection of commanders has been the availability of authors. Yet the mere fact that this book has proved possible at all is a tribute to the increasing acceptance of military history as a legitimate area for academic study in Australia. Nevertheless, many of the new military historians are concerned with sociological issues; only a small number are concerned with aspects of strategy and combat and of the men who decide strategy and win and lose battles. It is to be hoped that this book stimulates further study in these areas.

The chapters in this book have been written by individual authors with their own points of view. Chris Coulthard-Clark has hardened his view of Bridges since he published *A Heritage of Spirit* in 1979, but some critics might believe that his account is still too favourable. Perhaps Brett Lodge has been a little severe on Bennett, but his evidence is impressive. For my own part, I have changed my position slightly on the dismissal of Rowell since I published *Crisis of Command* in 1978. I have been accused by different critics of being both too soft and too hard on Blamey. While I have tried to be balanced in my chapter on Blamey some might argue that I should not have included aspects of his private life. Others surely will claim that I have not made enough of this aspect. In my view Stuart Sayers has not been quite critical enough of Herring. I have proposed a modified view in my book *High Command*, but it may well be that Sayers is right—certainly, many officers who fought with Herring agree with him, and at least one, I know, believes that Herring was the greatest man he ever knew. Similarly, I believe that Hopkins has played down Robertson's ambition and lack of tact, but he knew Red Robbie well and I never met him.

But whatever the views each author might have on his commander, before examining them it is worth reflecting on the requirements of a good commander. Field Marshal Lord Wavell, in *Generals and Generalship*, discerned various qualities which he saw as being essential to a good commander. The first essential, he believed, was 'the quality of robustness, the ability to stand the shocks of war'. Montgomery called it 'toughness'. Napoleon had the same idea: 'The first qualification of a general-in-chief is to possess a cool head, so that things may appear to him in their true proportions and as they really are. He should not suffer himself to be unduly affected by good or bad news.'[6]

A commander may be subjected to two types of pressures, those imposed by enemy action, and those imposed by other forces, for example interference from superior commanders or politicians. Both these pressures play upon the psychological well-being of the commander. Colonel J.F.C. Fuller wrote that 'the enemy does not attack [the commander] physically, but mentally; for the enemy attacks his ideas, his reasons, his plan. The physical pressure directed against his men reacts on him through compelling him to change his plan, and changes in his plan react on his men by creating a mental confusion which weakens their morale.'[7] Monash, Morshead and Vasey resisted the enemy's pressure and exerted pressure back on their opponents.

On the other hand, the pressures from areas other than the enemy are more insidious and can be even more damaging. The superior who shows a lack of confidence in his subordinate, by implication, and often directly, reminds him that if he makes even an apparent mistake, he will be disgraced and fall into oblivion. Lieutenant-General Sir Francis Tuker noted that commanders venture themselves in command in battle. He says, 'ventured' because the commander stakes his reputation, his career, all that he is and may be in his profession, on the issue of a fight in which his instinct must derive so much that is unknown and his science weigh all that is imponderable: above all, he stakes the coin he values most—the trust that his officers and men have in him.[8]

Field Marshal Dill maintained that 'in war you must either trust your general or sack him',[9] and if neither is done then an atmosphere is created which may lead to disaster and which would certainly not promote the bold stroke that invariably wins a great victory.

As a commander increases in rank, he is more open to criticism from all sides. Furthermore, the closer the fighting comes to the homeland, the more intense is the criticism from civilians. Political leaders, fearful of their own positions, tend to interfere with superior commanders who then apply pressure to their subordinates. Blamey and Rowell were to discover this harsh reality in 1942. As Field Marshal Montgomery has written, 'the crucible of war will determine the fine metal of which a general is made'.[10] Sturdee threatened the government with resignation if it did not recall the AIF to Australia in 1942. Rowell, too, preferred to stand firm as a matter of principle, although he might have done better to have bent like a willow before the storm, as did Herring who thus maintained his position and carried out his task. Blamey's position was the most precarious; his survival was a tribute to his sagacity and ruthlessness. Few commanders in this book were completely insulated from political pressure, and their success was dependent as much on handling this influence as upon their performance in the face of the enemy.

Almost on a par with the requirement for 'robustness' Wavell stressed the aspect of administration. The principles of strategy and tactics are simple, but it is another matter to marshal the resources to execute the plan. It was easy for MacArthur to give Rowell a brigade to capture Kokoda. It was another matter to transport the brigade to New Guinea, feed and supply it along the Kokoda Track and put it into the line in a condition fit to fight. White and Monash excelled at administration. Ironically, while Robertson prided himself on his dashing leadership in battle, he won his greatest tributes as an administrator. Of course, command and administration go hand in hand, and a commander cannot exercise his command without a knowledge of the mechanics of war—the principles and practice of logistics.

The personal factor in war should never be overlooked. The attributes required of a commander in the days of Wellington and Napoleon, such as courage, leadership and a personal presence on the battlefield are just as relevant today. While the commander should no longer personally have to lead his troops into battle, he should at least have a detailed knowledge of conditions in the forward areas and he must be seen by his troops. A criticism of Monash was that he was rarely seen in the forward areas. By contrast, Vasey's leadership of the 7th Division was outstanding.

Fuller saw three pillars of generalship: courage, creative intelligence and physical fitness, and observed that these were the attributes of youth rather than of middle age. Taking the ages of one hundred generals from Xenophon (401 BC) to Moltke (1866) he discovered that on the average the period of most efficient generalship was between the years of 30 and 49, and that the peak was reached between 35 and 45.[11] Sir Basil Liddell Hart listed a number of additional qualities, including mental initiative, a strong and positive personality, an understanding of human nature, a capacity for ruthless determination, an ability to express his thoughts clearly, the power to grasp the picture instantly, and a combination of exceptional ability and commonsense. His supreme requirements were creative intelligence and moral courage.[12] Field Marshal Slim added the quality of integrity.

It is against a background of such premises that this book examines the Australian commanders. It must be remembered, however, that there were very few commanders with completely independent commands, and it is arguable whether it is only with a completely independent command that a commander can realise his full potential. Clausewitz wrote:

> For each station, from the lowest upwards, to render distinguished services in War, there must be a particular genius. But the title of genius, history and the judgement of posterity only confer, in general, on those minds which have shone in the highest rank, that of Commander-in-Chief. The reason is that here, in the point of fact, the demand on the reasoning and intellectual powers generally is much greater.[13]

Nevertheless, it is still valid to examine the performances of the Australian commanders for they were faced with a wide array of distinctive problems. These problems included the maintenance of the Australian forces as distinctive national forces, the burden of command in terrain for which they were largely unprepared, command structures which were improvised and unrehearsed, inexperienced politicians with little understanding of strategy, and the requirement to operate under, or at best to cooperate with, an allied senior commander.

Another underlying strand in any study of Australian commanders is the rivalry between the regular and the amateur soldier. The rivalry was not so marked in the First World War since almost all middle-rank officers were citizen soldiers—they had either been members of the pre-war militia, or had joined the army on the outbreak of war. Nonetheless, as Peter Pedersen shows in his chapter, Monash had firm views that the citizen soldier was better prepared for high command in war.

After the First World War, many citizen officers believed that it was their role to command Australian formations in a future war, and that Duntroon officers, members of the Staff Corps, were suitable only to fill staff appointments. They believed that Duntroon officers had little capacity to command troops. In fact, Duntroon graduates had few opportunities to exercise command in the inter-war period, and many wasted, and finally lost, their energy and enthusiasm on routine administration. Underpaid and ignored, they were often humiliated by being overtaken in rank by younger citizen officers.

At the beginning of the Second World War, Duntroon graduates such as Rowell, Robertson, Berryman, Vasey, Clowes and Bridgeford, who had seen action in the First World War, were not content to remain staff officers to citizen officers who had had no more operational experience and less formal training. The rivalry between the Staff Corps and the militia officers reached its peak in the period 1939-41, and then gave way to a rivalry between the voluntarily enlisted Australian Imperial Force (AIF), and the conscripted militia. The government's failure to resolve satisfactorily the problem of having two armies, the AIF and the militia, affected Australia's military effort during the Second World War, and had ramifications through the Vietnam war and to the present time. From 1942 to 1945 the Staff Corps-militia rivalry could be maintained only at the highest levels, and generally appointments were determined more by an individual's performance than by his pre-war professional background.[14]

Although the regular officers filled only a small proportion of the command positions, by the end of the war they were well placed to assume complete control at all levels of the Army. Their ascendancy was helped by the Chifley government's decision to form a Regular Army and to maintain an Australian contribution to the British Commonwealth Occupation Force in Japan. In these forces all command and staff appointments went to regular officers or AIF officers who had chosen to remain in the Regular Army. Furthermore, the nature of the Australian Military Forces changed from one based primarily upon citizen soldiers to one based upon regulars.

It was natural that the large and influential group of ex-wartime citizen officers should be critical of this situation, but they overlooked the fact that Australia's fine military reputation from the war was achieved not because of the smallness of the pre-war Regular Army, but despite it. The senior regular officers, men such as Rowell, Chief of the General Staff, 1950-54, and Berryman, GOC of Eastern Command, were determined that their subordinates should not lose their wartime ranks nor that they should be humiliated by a government with a short memory, as they themselves had been 25 years earlier.

This book shows commanders operating in a wide range of situations, each bringing its own problems and pressures. Clearly, commanders must be prepared for changing conditions. Perhaps commanders of the future will have to be prepared not only for more rapid and intensive warfare, but also for a higher level of political involvement. This book does not provide the answers for the future, but it might be a jumping-off point for further thought. It is worth considering, too, whether Barbara Tuchman was right when she wrote in 1972 that she thought that generalship was undergoing 'as a result of developments of the past twenty-five years, a radical transformation which may make irrelevant much of what we now know about it'.[15]

The biographies of Australia's military commanders taken together provide a sound basis for further study of the history of Australia's armed forces. Military history is concerned with the solution of problems, and the commanders are the men chiefly concerned with solving the problems. While the historian, the final master of the battlefield, sets out to re-examine these problems and to provide some hindsight judgment, it is well to remember that it is the little things which go wrong that compound the problems and determine the psychological effectiveness of the commander. As Liddell Hart said, 'it matters little what the situation was at any particular point or moment; all that matters is what the commander thought it was.'[16]

# Part One
The First World War

# Major-General Sir William Bridges: Australia's First Field Commander

## CHRIS COULTHARD-CLARK

When it was announced on 7 August 1914 that Australia would send troops to assist Britain in the great conflict that had just erupted across Europe, the Minister for Defence in Melbourne also announced that 'the duty of organising and taking the expeditionary force to England had been confided to Brigadier-General Bridges, who would be relieved of his duties as Inspector-General of the Commonwealth military forces'.[1] William Throsby Bridges (1861-1915) thus became the first Australian to fill the role of national commander, and additionally became the first local officer to hold a field command of formation size in battle,[2] leading the 1st Australian Division of the Australian Imperial Force (AIF)—as the expeditionary force was known—during the Allied landings on the Gallipoli peninsula in April 1915. His period in command of his division on Gallipoli was shortlived however, as on 15 May, barely three weeks after landing, he was mortally injured by a sniper's bullet and died of his wound three days later.

Bridges was succeeded both as General Officer Commanding (GOC), AIF and GOC 1st Division by another local officer, Colonel Gordon Legge, who was despatched from his post as Chief of the General Staff at Army headquarters in Melbourne to take up command. Legge's tenure of these dual appointments was brief, as in late June, a month after reaching Gallipoli, he was appointed to raise the 2nd Australian Division back in Egypt. He returned to Gallipoli with his new command in September and continued to lead it on the Western Front throughout 1916 until ordered to return to Australia in January 1917. His powers in respect of the AIF were, however, delegated in September 1915 to Sir William Birdwood, the British general commanding the Australian and New Zealand Army Corps (ANZAC) which included the AIF. Birdwood was formally appointed GOC AIF, a year later and retained administrative control of the AIF for the remainder of the war and until September 1920.[3] That a British general was commander of Australia's national forces for all but the first year of the war has naturally overshadowed the fact that two local officers first held this post, providing an additional point of interest to any examination of the performance of the first Australian to command a formation in the field.

*Bridges as national commander*

Bridges owed his position at the pinnacle of Australia's defence

1 Major-General Sir William Throsby Bridges KCB, CMG (AWM Negative No. A2687)

establishment in 1914 to his proven capacity as an administrator. His career had largely conformed to the narrow pattern, common to virtually all officers of the small permanent cadre, of filling the administrative and instructional needs of the country's militia army in the comparatively undemanding context of peacetime conditions. In his case the appointments that had come his way had been more high-powered than most, the previous ten years having seen him in the posts of Chief of Intelligence on the Military Board administering the army, Australia's first Chief of the General Staff, Australian representative on the Imperial General Staff in London, and—up until his appointment as Inspector-General, at the age of 53, just two months before the war began—

founding commandant of the Royal Military College at Duntroon. His competence in these positions had established his reputation as the nation's pre-eminent soldier. It was therefore only logical that the task of raising the expeditionary force promised by the Government to the British authorities should have been settled on him, yet that Bridges should also have been given command of it was perhaps something less than a foregone conclusion.

Under the Defence Scheme that then existed, the role of the Inspector-General in war was to assume control of forces within Australia, there being no provision for the contingency of troops being sent outside the Commonwealth's territorial limits and, accordingly, no clearly defined command arrangements for such an undertaking. Bridges' claim to an appointment of command rested not solely on his seniority as the highest-ranking officer in the peacetime organisational structure (apart from the Director-General of Medical Services), but on the extraordinary circumstances that pertained at headquarters when news of the outbreak of hostilities in Europe arrived. The Military Board had only recently been reconstructed, with the newly appointed members taking up their duties on 1 August. The Permanent Head of the Defence Department was absent from Australia and the new Chief of the General Staff, Colonel Legge, was still at sea on the return trip from England where he had been Australian representative on the Imperial General Staff. As a wartime report on Defence administration put matters: 'The Department was fortunate... in having the services of... Bridges', although he, too, was initially absent from Melbourne on a tour of duty and had to be recalled from Brisbane.[4]

It can be argued that, of Australia's permanent officers, Bridges was the best qualified through experience and background knowledge to handle the administrative and policy questions likely to arise in connexion with the AIF. The range of his appointments had provided him with exposure to issues underpinning government defence policy and a detailed understanding of the workings of the country's defence machinery which was matched by very few other officers, possibly only Legge. But this could not disguise a lack of experience of regimental command, or of command in the field under active service conditions. He had served in the Boer War for five months in 1899-1900 but even this had been in the capacity of a Special Service Officer attached to British units for instructional purposes. As a choice for an operational field command he was, therefore, largely an unknown quantity, leading a contemporary to wonder in December 1914 'how the best soldier in Peace... [would] acquit himself in War'.[5]

Doubts as to Bridges' suitability for the task of leading the expeditionary force in combat centre on two aspects of his personal make-up. Of prime importance was the question of his qualities of leadership. Because his career had been preponderantly staff-oriented, it remains doubtful that he had ever come to terms with the democratic and independent spirit of his fellow Australians within the military setting. Bridges most probably lacked intimate knowledge of the characteristics of the men who voluntarily submitted themselves to service in the pre-war militia forces, let alone of the wider spectrum of Australian men entering uniform on the commencement of war, with little knowledge and even less liking of military discipline.

To a possible lack of empathy with the men of his command must be added certain facets of Bridges' personality. From first-hand knowledge C.E.W. Bean, the official historian, thought him an intensely shy man, a characteristic which can have only compounded any inadequacies Bridges himself felt regarding his career preparation for a position of senior command. This shyness, moreover, was combined with an aloof and intellectual manner. In one of several telling passages in the Official History, Bean remarked that 'Bridges was a profound student, and his colleagues soon began to look upon him as a learned soldier'.[6]

In essence, Bridges' academic demeanour and turn of mind seemed to fit him for the army's more cerebral posts, such as Chief of Intelligence or superintendent of the nation's principal military educational institution, and he was widely respected for his knowledge of military matters. It even caused him to be described on his epitaph as 'erudite', although that part of his writings which has survived does not indicate he was learned in a scholarly sense. These were, however, not qualities which were generally held to typify a successful military commander, British or otherwise. As Bean wrote, Bridges was not thought of 'as primarily a soldier, and in the opinion of those who knew him best he was likely to prove a purely academic commander. He had shown none of the qualities which commonly mark the leader'.[7] Bridges' strengths as an administrator and staff officer were, therefore, suspect credentials for his selection as a field commander.

It was this aspect to which Sir Brudenell White, Bridges' chief of staff on Gallipoli, later referred when he said that Bridges was fortunate that his first command was a division, for a brigade 'is really the last command in which an officer comes into intimate contact with his men'.[8] In similar vein, General Sir Ian Hamilton, Inspector-General of Overseas Forces in the British Army and later to command the Dardanelles operation in which the AIF took part, had sought to convince the Australian Defence Minister in 1914 that Bridges was not suitable for appointment as Inspector-General of the Commonwealth's military forces, arguing that, while he might make an excellent chief of staff, he 'had not...the magnetism and charm on the one side, or the strength of character on the other, which would carry the sympathy, concurrence, or have great weight with the troops'.[9]

Uncertainty regarding his capacity to be an effective battlefield commander appears to have been shared not least by Bridges himself, as evidenced by his proposal that command of the AIF should be offered to a British officer, Lieutenant-General Sir Edward Hutton. There are several alternative and plausible explanations for his action in this matter, including personal modesty on Bridges' part. Possibly he anticipated that the AIF would be headed by a British general in line with command arrangements for a similar force being raised across the Tasman, and he wished to secure the services of someone with previous knowledge of Australian troops and conditions. Hutton had twice served in Australia, as Commandant in New South Wales during 1893–96 and as GOC the Commonwealth's forces in 1902–04, while in South Africa he had commanded a mounted infantry brigade comprising contingents from the Australian colonies; he would, therefore, have seemed a very good choice to Bridges. But Birdwood may be credited with some insight into Bridges' character when he expressed the opinion in 1916, that he believed Bridges had asked for Hutton to command because he did not then have 'complete confidence' in his own ability to do so.[10] The Australian government decided, however, that it desired a local commander for the force and promoted Bridges to the rank of major-general on 15 August when confirming him as head of the AIF.

Bridges' initial assumption that command would go to a British regular would have been typical of him, and highlights the second aspect of his personal make-up which can be seen as throwing into question his suitability for the role in which he was cast. For while he was the officer best equipped to maintain the AIF's national identity and administrative independence from the larger British formations with which it was expected to operate, he was probably the officer with least inclination to place Australian national interests above those of the imperial authorities whenever areas of conflict might arise. As one historian recently observed, 'The life of Bridges had prepared him to be a soldier of the Empire rather than of Australia',[11] and his outlook was such that it probably occurred to few people to regard him as more than nominally Australian.

Bridges could point to substantial Australian connexions in the person of his mother, a member of the landed Throsby family of Moss Vale, New South Wales, but the fact remains that he was born in Scotland and his father was an officer in the Royal Navy. Educated in England and Canada, he entered the Canadian military college at Kingston in 1877, intending to train for a commission in the British Army,[12] but he forsook this ambition two years later, after his parents moved to Australia and settled at Moss Vale. His entry into the local forces was prompted by the occasion of the first imperial adventure by an Australian colonial government in 1885, when New South Wales sent troops to assist Britain in the reconquest of the Sudan.

His subsequent career provides many illustrations of the fact that his loyalty to his country of adoption was accompanied by an extended loyalty to the Empire, which was considerably stronger. In 1898 and again in 1902 he performed secret intelligence-gathering missions to islands in the Pacific for the War Office in London, although he would have known these to have been arranged through the senior British officers on loan to New South Wales and the Commonwealth at the time, without the apparent knowledge of the local governments. He found preferment on the Commonwealth army's headquarters staff during the three-year term of command of Hutton, who offended the nationalist sensibilities of many local politicians by his initiatives to give effect to War Office schemes involving Australia's military resources for imperial defence purposes.

In these circumstances it was particularly significant that Bridges put forward Hutton's name in 1914, for he never lost his esteem for his former chief and continued to press Hutton's ideas and aims long after the GOC's departure. It was Bridges' opposition to nationalist aspirations for an Australian blue-water (ocean-going) navy which caused Prime Minister Alfred Deakin to pointedly describe him in 1906 as 'the ablest of our Imperial officers...[but] imperfectly in sympathy with some of our aims'.[13]

Occasionally Bridges was to be found venting differences to London's policies, such as during a lecture he delivered to the British Army's staff college early in 1910 on the 'crimes the War Office has committed against Australia'.[14] These were, however, differences expressed still within the conviction of an essential commonality of British and Australian strategic interests, a point underscored by the strong possibility that Bridges was, in the same period, indirectly and confidentially corresponding with Field Marshal Lord Kitchener on matters relating to imperial defence.[15]

It was not simply that Bridges, and many other local officers, viewed the standards of professionalism of the British Army as the epitome of military virtue—the British martial tradition was, after all, the only one that Australia had known and the Commonwealth's forces still conformed to British organisational principles, tactical doctrine and equipment practices—it was more a matter that Bridges identified psychologically with the community of thought embodied by military imperialists at the War Office.

It was this imperial outlook which Bridges brought to the task of raising the AIF for overseas service. Although he was an active participant in the exchange of cables between Melbourne and London which decided the force's divisional composition, on 20 August Bridges remarked in a letter to his light horse brigadier that this composition was 'not quite my idea of what we should send but I naturally accepted the command'.[16] Most probably this comment referred to the initial proposal he found on his return to Melbourne for Australia to provide two-thirds of a division with the remainder to come from New Zealand. It may, however, also suggest that Bridges supported the War Office preference for several brigades lacking a formation headquarters, even though he carried through the Australian government's wishes in the matter.

Less ambiguous is the part he played in devising the force's title. Bean is unequivocal that 'Australian Imperial Force' was the title of Bridges' own choosing, because of its shortness and suitability of its initials when abbreviated, and also because the general refused to countenance using the word 'expeditionary' on the grounds that Australia was sending a fighting force that would not include all the units and services necessary to complete the self-contained organisation of a true expeditionary force.[17] As one historian has pointed out, however, Bridges' choice bore a remarkable similarity to the 'Imperial Australian Force' proposed by the War Office in 1902 as a dominion reserve upon which Britain might call in the event of a European war; it was, therefore, most probably a conscious attempt to ensure the imperial dimension was not forgotten.[18]

Bridges' anxiety on this score is fully revealed after he and the AIF arrived in Egypt in December 1914, in his anguished appeals to the Governor-General, with whom he maintained a personal correspondence, to help get the Australians to France; in view of the class of men in the ranks he feared an 'Imperial disaster' if the force did not get to the main theatre of operations before the war ended[19]—meaning presumably that Australians' attachment to Britain might be weakened by exclusion from the 'mother' country's defence in Europe. This was an attitude which established beyond doubt his devotion to the cause of Empire but it did not necessarily imply commitment to the preservation of the national identity of the force entrusted to him. It has been suggested, for instance, that Bridges rejected the status of an expeditionary force for his AIF because he feared the creation of potential obstacles to successfully 'welding' it into the British system.[20]

The desire to preserve the illusion of a single imperial army was manifested in a conscious effort to ensure that the AIF was structured along lines identical to British practice. In raising the force, Bridges decided to provide three field engineer companies for his division as against the British Army's allocation of two, on the basis that British thinking was known to favour the change. Similarly, anticipating a change in the number of guns allotted to a British artillery battery, Bridges reduced the number for his division from six to four—although in this instance a change did not eventuate in the British Army and the AIF consequently sailed with a one-third deficiency in organic field artillery support compared with the firepower of an equivalent British formation. Once in Egypt, Bridges also altered the AIF's organisation to comply with War Office wishes that infantry battalions consist of only four companies instead of the eight with which these had been raised.[21] Such changes were not strictly necessary to attain the capacity to operate with British forces.

Another illustration of Bridges' attitude seems to have concerned the matter of establishing an intermediate base for the Australians overseas. He evidently hoped the AIF would be absorbed into the British administrative system and was only persuaded to set up the base after realising that the British authorities in Egypt had no intention of accepting responsibility for dominion troops. But he refused to regard the base's operations as in any way important and initially allotted the bare minimum of resources necessary for its functioning. At the same time he showed no inclination to stand by his officers in battles to ward off the depredations of British officers seeking to extend their authority over the Australian medical units and reinforcements in the country.

The same attitude was probably also an element in Bridges' actions after December 1914 in referring major administrative decisions affecting the AIF to the British Army corps commander, Birdwood.

2  Sir George Reid's visit to Mena Camp, Egypt, December 1914. From left: Sir George Reid, Australian High Commissioner in London, Lieutenant-General Sir John Maxwell, commanding British troops Egypt, and Major-General Bridges (AWM Negative No. G 1600)

Admittedly Bridges was in a difficult position in terms of the military chain of command, for the Australian government had settled onto him powers in respect of appointing, promoting or removing officers, making internal organisational changes and control of reinforcements,[22] which cut across areas that would normally have been within Birdwood's ambit. The position may have been difficult for Birdwood to accept when the AIF became part of his command, yet the desire of the Australian government to retain administrative independence for its national force, to ensure its identity was preserved, was also understandable. Bridges undoubtedly owed his appointment to just such a concern, and there was a clear expectation that he would act as guardian of his government's interests in matters affecting the AIF.

The problem was that for a man of Bridges' sensibilities there was no distinction to be drawn in 1914–15 between Australia's national interests and Britain's—the two were inextricably bound together as one in the Empire. There was no precedent for the concept of a distinctive Australian style of command, and therefore no question arose of Bridges exercising his command of the AIF in a peculiarly Australian way. As one historian has noted, it was only by 1918—as a consequence of the war—that the term 'Australian' was even vested with a sense of separateness from the idea of being British.[23] What was important to perhaps the majority of the Australian population who saw the war as an imperial crusade, and particularly to the men of the AIF, was that they should gain acceptance as 'worthy sons of the Empire' by matching the high ideals and standards of the 'mother' country.[24]

Bridges was in this sense a supporter of Empire rather than an advocate of Australian nationalism, and yet, in being such, it is possible he contributed more to the achievement of a national identity for the AIF than is realised. On 25 August 1914 the British Army Council asked to be informed who had been given command of Australia's expeditionary force, and on being told, did not demur at Bridges' appointment.[25] The implication of this incident may well have been that had the name not been acceptable to London some sort of move to substitute a British general in his place would have followed. Among the several reasons advanced by R.A. Preston as to why Australian commanders of the AIF were not replaced, as happened with the Canadian army, the factor that Bridges was 'well known and highly regarded at the War Office'—in other words considered 'sound' by the imperial authorities—must be counted as high on the list.[26] It seems probable, therefore, that only because the first commander of the AIF bore a striking similarity to any other British general did the Australian division have a local officer at its head when the troops went ashore at Gallipoli to give birth to the 'Digger' tradition.

## Bridges as divisional commander

During the nine months in which he commanded the 1st Division before his death, Bridges' performance largely confirmed expectations as to his style of command. He was not a man who always got the best from subordinates, proving himself on frequent occasions to be an exacting and exasperating chief for whom to work. In Egypt he sapped the confidence of his aide-de-camp to the point where the officer requested to be returned to his battalion, and even the divisional intelligence officer, Major T.A. Blamey, said of him, 'Never have I admired and disliked a man so much'.[27]

His diffidence led him to project a rigid and wholly unsympathetic image to both officers and men on Gallipoli. Examples exist of Bridges kicking a man found asleep at his post, telling a man who implored him to seek cover when Bridges needlessly exposed himself to Turkish positions on reconnaissance to 'be damned', angrily reminding a group of men he found sheltering from enemy fire that they were Australians and calling them into the open to re-form, and doing much the same thing on another occasion and causing several men to be hit while so doing. He also expected the major who had been secretary of the headquarters officers' mess in Egypt to continue to act in this capacity for the small headquarters mess on Gallipoli and endure Bridges' complaints about the meals, a situation which produced quite unnecessary friction.[28] The grim reserve which he evinced in his relations with those about him stands in strong contrast to the informality which became part of the 'Digger' legend.

Yet Bridges did show acceptance of the fact that generalship was more than an intellectual pursuit. By living simply and sharing the same hard-

3  Artillery orders on 1st Australian division manoeuvres in Egypt, 5 March 1915. Seated are Major-General W.T. Bridges (nearest to camera) and Lieutenant-Colonel C.B.B. White. Behind them are Colonel J.T.T. Hobbs, commander of the 1st Division artillery, and his brigade major, Major S.M. Anderson (obscured). (AWM Negative No. G 1620)

ships and dangers that the men endured, he displayed elements of leadership that had not been expected of him. As Bean observed, it was the example of the Staff that ensured 'at Anzac there never grew between it and the fighting troops the gulf which sometimes under other conditions separated them'.[29] While it was true that on Gallipoli it was not possible for a senior commander to place his headquarters at a comfortable distance behind the firing-line out of harm's way, Bridges demonstrated this was not his inclination in any event. Daily visits through the trenches during which he routinely risked injury or death were a feature of his style of command. By the time of his wounding on one such tour of the line, Bridges was winning widening regard from his men for his coolness and courage. Indifference to danger was a standard displayed by all commanders on Gallipoli, however, and Bridges was not alone in observing it—Birdwood, for example, was no less assiduous in his studied disregard of risks—nor was he providing the lead in this, for, as Bean notes, both generals were simply observing a standard which was set by the rank and file of the force.[30] With Bridges, moreover, there was an element of wanton recklessness manifested in his refusal to shelter from incoming shells, or reconnoitring Turkish positions in dangerously exposed conditions, which stands in contrast to the quiet, cautious and calculating bravery later held to characterise the AIF's style of operations.

In terms of Bridges' tactical direction of his division, the judgment which can be made on the basis of his three weeks on Gallipoli suggests he was a resourceful, forceful and determined commander. His visits to the line, for example, were not undertaken simply for the purpose of being seen or 'showing the flag' but as a means of ensuring he knew what was happening and to exert his personal pressure on local commanders. He drove his artillery staff to locate suitable gun positions in the impossible terrain from which they could join in the battle, and also adopted quite unorthodox tactics to support his infantry by bringing up artillery into the front line (over the objections of his artillery commander). He showed, too, the courage to back his brigadiers and refuse to implement an ill-conceived corps plan of attack on 1 May.

In other respects though, Bridges presented little evidence of the independent spirit for which the men of his command became famous. His willingness to acquiesce to proposals and operational concepts devised by the British regulars planning the Dardanelles campaign produced a memorable clash with White, the indispensable chief of staff of the Australian division. Although in private correspondence Bridges lamented the loss of surprise for the operation, he does not appear to have questioned the outline plan originally proposed by General Sir Ian Hamilton's staff, unlike White who criticised the inherent risks so trenchantly that Bridges threatened to leave him behind. Perhaps a partial explanation for Bridges' apparent mental subservience can be found in the fact that he seems to have been at least partially misled over aspects of the plan.

The broad concept of the landing undertaken by the ANZAC in the vicinity of Gaba Tepe on 25 April flowed not from the brain of Bridges but from Birdwood. It was the senior commander who decided on a pre-dawn assault without preliminary bombardment of the enemy-held shore, in a deliberate attempt to achieve surprise. One British historian has suggested that Birdwood's idea of hurling his 'fiery Anzacs' at the Turk, having impressed on all the vital importance of speed and enterprise, commendably left room for initiative which was absent from plans for the British landing to the south at Cape Helles.[31] Even so, it is evident that herein lay the seeds of the debacle that occurred through the inadequate control procedures during the amphibious phase and the piecemeal commitment of troops to the fighting.

These deficiencies produced the spectacle of Bridges and his headquarters sitting on the beach during the fighting of the first day with little idea of what was occurring in the ridges beyond, and with scant control over the tide of battle other than through the apportioning of reinforcements to sectors of the battlefield. Accepting that, in modern military terms, the Gallipoli landing was a very crude operation, there is relatively little that can be said of Bridges' performance that reflects directly on his capacity as a commander, particularly anything that might distinguish him from British commanders similarly engaged. It is, however, worthy of note that Bridges entrusted the initial phase of the landing operation to the 3rd Brigade under Colonel E.G. Sinclair-MacLagan. There were several things that probably attracted Bridges to this choice for the divisional Covering Force, but chief among the deciding factors would have been that MacLagan was a British officer well-known to him through the latter having been Director of Drill at Duntroon during Bridges' term there. It could be suggested by some that Bridges was indulging his reverence of British Regulars in preference to the local officers commanding the other two brigades of his force.

Additionally, it warrants consideration that in suggesting evacuation of the ANZAC from the beach on the evening of the landing, Bridges had allowed himself to become demoralised or panicky by the palpable insecurity of the beach-head in the face of an anticipated Turkish counterattack. At a conference in Bridges' headquarters when the proposal was put to Birdwood, there were reportedly heated scenes in which Bridges' efforts to establish the exact situation off the beach were criticised as inadequate by Birdwood's chief of staff.[32] Whatever the truth of this accusation—and it is worthy of mention that the British commander of the New Zealanders, who had also come ashore that day and seen the situation that prevailed, was party to Bridges' recommendation—it is possible that Bridges' inexperience may have shown at this stage. Certainly both he and his staff made errors in the weeks that followed which were the result of British Army practices inexpertly applied, perhaps the most grievous of which was the inadvertent attack launched by the 4th Battalion on 26 April.

One matter for which Bridges later, but erroneously, received credit was the similarity in execution between the landing of his division and that of British forces under Abercromby at Aboukir over a century before, a coincidence given additional point by the fact that Bridges, acting on a remark by Hutton, had studied the 1801 operation and walked the battlefield in March 1915 during the AIF's stay in Egypt. In a letter to his former mentor Bridges had written, 'I hope to copy him very shortly'.[33] Hutton later pointed out that in both cases the troops were ordered to land with weapons unloaded and assault the enemy immediately with the bayonet, declaring this to be proof of 'the inspiration of... [Bridges'] genius as a leader in adopting the principles laid down by... Abercromby'.[34]

Superficially at least, this might be taken as an explanation for the loose arrangements instituted by Bridges for the landing of his men, as unquestionably his divisional Operation Order contained instructions relating to a requirement for rifles not to be loaded and these were repeated in the orders of the 3rd Brigade.[35] But closer examination reveals factors other than a conscious desire to emulate a past British feat of arms to have been the determinants for this policy.

Achievement of surprise dictated a need to avoid accidental discharge of weapons; there was also a risk of casualties to own troops in a pre-dawn landing through the men being unable properly to distinguish friendly and enemy locations in the dark. The latter fear led to a specific order to the 3rd Brigade that there should be no fire returned against enemy defenders until daybreak, an instruction which proved fully justified through not being observed.[36]

If one looks for qualities or aspects of Bridges' style which might be held as representative or indicative of an Australian commander, it must be concluded that he had shown no distinctively Australian approach to solving the problems of military command. It is perhaps more correct to describe Bridges as typifying the British style of command of the period; hence he was the point of departure for any evolving Australian mould. Of course, it was virtually inevitable, given all that had gone before in

1914, that the first Australian general to command in the field should have performed remarkably like his British counterparts. Indeed, there are grounds on which to ponder whether the imperial authorities would have tolerated anything else.

# 3 General Sir Brudenell White: The Staff Officer as Commander

GUY VERNEY

An examination of the career of Brudenell White (1876-1940) highlights the relationship between a staff officer and his field commander and how a staff officer may influence strategic, tactical and administrative decision-making. White's fine professional reputation and outstanding record, especially in the First AIF, raise the question of whether or not as the staff officer he was in fact the commander. The official historian, C.E.W. Bean, for example, noted in his diary on 18 June 1918 that General Sir Henry Wilson, Chief of the Imperial General Staff, had told Prime Minister Hughes that White had acted as Birdwood's brains all along. In his book, *Two Men I Knew*, Bean wrote about White in glowing terms and stated that in staff circles during the First World War he was regarded as 'something between the Ten Commandants and the laws of first-class cricket'. White had the personality and ability to work as 'chief assistant to a succession of leaders' and in Bean's opinion was the 'tactical and administrative commander' under Birdwood in all except 'name'.[1]

In his biography of Monash, Geoffrey Serle wrote that Bean idealised Brudenell White to the extent that he was his hero. A more sober judgment on White was made in 1940 by Field Marshal Lord Birdwood, who was his commander for most of the First World War. Birdwood paid tribute to the quiet way in which Brudenell White was responsible for the achievements of the 1st ANZAC Corps and to his devotion to the Australian soldier. Above all, he saw White as the exemplary staff officer who gave his views to his leader 'fearlessly and unhesitatingly' and then carried out the decisions of the leader although they may have been contrary to his. In assessing White, Birdwood wrote that his staff officer believed Australia's future was linked intimately to the 'Old Country' and used the words, 'a daughter in her mother's house but mistress of her own'. Prime Minister Hughes thought that Birdwood himself relied on the 'social art' as well as his letter writing—instead of his intellect—for his achievements. Birdwood's strength lay in his contact with the members of the AIF and the maintenance of morale. A sharp comment was made by Fihelly, Acting Premier of Queensland, about Birdwood during his post-war tour of Australia: 'He [Birdwood] is just one of those conceited grown-up puppies who imagines that the Australian Forces were made for his benefit and he used Pearce and Hughes on every occasion—as they used him...'[2] The conclusion can be drawn that Birdwood incurred a debt to White for taking a role as his tactical and administrative commander.

Another opinion of White was expressed by the war correspondent, F.M. Cutlack, who thought that there was something 'wanting in White' because he allowed Birdwood to 'overshadow him' during the war. Cutlack was referring to the question of the command of the AIF in 1918 in which White refused to permit a group of journalists and war artists to lobby for his appointment to the position of commander of the Australian Corps. Will Dyson, a war artist, held a similar view about White and thought that he displayed weakness about his position and career prospects in the AIF. The comments by Cutlack and Dyson implied that White was unequal to the leonine struggle for the highest honour in the AIF. However, they missed White's great ability to balance his personal ambition and his duty to Australia and the British Empire with the requirements of staff work and the pressure of war.[3]

In 1918, Brigadier-General Gellibrand, a friend of White's, superseded Brigadier-General Elliott, despite their equivalent seniority, and was appointed commander of the 3rd Division, AIF. Elliott was piqued by Gellibrand's success and wrote in a private memorandum that White was biased against him and that one had to toady to him to succeed in the AIF. Elliott thought that White favoured British officers inculcated with the 'Public School Spirit', especially Gellibrand who was 'well lacquered by his experience in the British Army'.[4]

The different assessments of White raise a number of questions (such as the relationship between Bean and White; and White's idea of imperial and Australian defence), which need to be addressed in a biography. The questions addressed here are White's impact on the Australian Army, and the degree to which, as a staff officer, he in fact exercised command.

*Early staff training*

White's performance as an outstanding professional staff officer in the First World War was based on the solid career that he built in the Australian Army from 1901 to 1914.[5] His military training and administration were marked by a determination to succeed, a capacity to work steadily, and a studious approach to military questions.

The appointment of White as ADC to Major-General Sir Edward Hutton, the first commandant of the Commonwealth Military Forces, in 1902 was significant in his career. He admired Hutton and thought of him as 'one of the finest and greatest soldiers I have ever known'.[6] White's diaries refer to Hutton's charm, and to dinners with Sir Edward and Lady Hutton. The friendship between Hutton and White spanned 21 years, and touched upon the questions of Australian and imperial defence. In particular, Hutton wanted to make Australian troops available automatically in the event of a military crisis involving Great Britain in Europe.[7]

As a matter of policy, Hutton wanted Australian officers to undertake higher military training as a means of raising military standards and increasing their knowledge of imperial defence. He was impressed by the cooperation of colonial and imperial military units during the Boer War, and saw the potential of a common imperial system for training officers to facilitate joint operations between the members of the British Empire in the event of a crisis. In nominating Brudenell White for the Staff College, Camberley, England, Hutton wrote: 'He has excellent abilities, has had exceptional experience as my A.D.C., and on many occasions as Private Secretary, and possesses many qualities which mark him out for a valuable Staff Officer'.[8] White was the first Australian to attend the Staff College and vindicated Hutton's faith in him.

For White, the Staff College experience contributed to the development of his military abilities and career. He completed the course successfully and marked himself for higher promotion in the Australian Army, gained a broad knowledge of military subjects, the British Army and imperial defence policy, and made personal contacts with British officers. As well, his attendance at the Staff College established training at a higher defence college in England or India as an essential rung on the career ladder in the Australian permanent army.

White's monthly reports on staff training at Camberley, which were sent to the Adjutant-General in Melbourne, showed that he was diligent and capable of applying himself to the different problems set by the instructional staff. The subjects taught at Camberley covered military history, strategy and tactics, and gave White the opportunity to learn about current military thinking in Britain. Above all, staff training broadened his 'soldiering experience'. He examined the process of entraining and detraining military supplies (White was associated with the War Railway Council in Australia as Director of Military Operations, 1912–14), the problem of landing an expeditionary force on a beach with Japanese experiences from the Russo–Japanese war as examples, the strategic value of terrain for the defences of the Belgium frontier and the north-east corner of Egypt, imperial strategy, the distribution and mobilisation of the British fleet in war, and theories about the employment of artillery, cavalry and infantry. Staff College courses toured the battlefields of Europe as part of the training syllabus.

Apart from academic training and staff papers, an important feature of White's training was the knowledge gathered about the British Army. His reports are replete with references to structure and organisation, information about staff tours, the artillery training centre at Okehampton, a scheme for reorganising the British Army volunteers on a county basis, and cavalry manoeuvres with the 5th Royal Irish Lancers. At a personal level, White mixed with British officers and made a lasting friendship with John Gellibrand, later a major-general in the AIF and founder of Legacy.[9]

The question of command in war also received attention, and in his report for March 1907 he wrote, 'That initiative in our Commanders is an essential, but that headquarters should be able to give clear and concise instructions as to the object to be attained in order that initiative may be correctly applied'. He regarded initiative as important, especially as future battles might 'run the danger of becoming separate struggles for the gaining of localities'. As well, he saw tactical intelligence of the highest order as vital to the selection of 'the decisive point' for attack. He also realised that cavalry was limited as a means for gaining information about the decisive point for attack in the face of likely future developments in armies which included mixed company forces of each arm. White wrote: 'Such cavalry in order to achieve anything may have to be supported by a force of all arms'.[10]

The Staff College experience also reinforced White's belief and commitment to the British Empire. In a paper entitled 'Observations', he wrote: 'I would ask you to build upon a firmer rock than this, something which will make room for the growth of an Imperial patriotism, and preserve the loyalty to a common Crown which I assert fearlessly does now exist'. He saw three strands in the British Empire namely, trade, defence, and political consolidation, which were linked by the blue line of the Admiralty, that is, protection by the Royal Navy. On defence, White believed that inculcation of the 'Imperial spirit' was a first step towards settling the question of imperial defence. He wondered if the Colonial Office and the Committee of Imperial Defence had the requisite wisdom 'to show that an Australian navy, in Australian control, can only be a source of weakness to the Imperial navy', and 'a very expensive means of inefficient defence to Australia'. It must be said, however, that White's view about Australian naval defence was set in the context of his broader vision of imperial unity:

> What our Empire wants is some elastic system of consolidation which fosters the spirit which well brought up children have for their parents and the family circle. All alliance is the most inelastic bond and once made would be rarely renewed by Colonies freed from parental influence and yearly becoming aware of their manhood.[11]

He wanted Australia to play her role in imperial defence and foster the links with the Empire at large.[12]

At the end of the course the commandant, Major-General Wilson, later Chief of the Imperial General Staff, wrote in his report on White:

> The officer started the course here with a lack of soldiering experience as compared with his companions. He is modest and unassuming, but possesses considerable ability and power of application. The results have been eminently satisfactory and he promises to be a valuable Staff Officer. His work now reaches a uniformly high standard. He can express himself well and clearly on paper. His opinion carries weight with the other students, and he deserves great credit for passing out so high. Popular and good horseman. He will do well either in the field or in an office. I recommend him for an appointment which will lead to employment on the General Staff.

As a result of his staff training, White was marked as a future high-ranking career officer of the Australian Army. He returned to Australia in 1908 and in the same year, travelled again to England for an appointment as General Staff Officer, 3rd Grade, at Army Headquarters in London for two years. From 1908 to 1911, he broadened his knowledge of the British Army and imperial strategy and then returned to Australia in 1912 as Director of Military Operations.[13]

*Apprentice policy-maker, 1912-14*

With the benefit of light from long afterwards, it seems probable that the most significant and revealing part of White's career was in the years between 1912 and 1914. In August 1911, one month before his thirty-fifth birthday, he was recalled from the War Office by Brigadier-General J.M. Gordon, Chief of the General Staff (CGS), Commonwealth Section, Imperial General Staff. The central task facing Gordon was the formulation of a general scheme of defence for Australia. He chose White who, he said, had established a solid reputation as a staff officer of considerable ability, temper and tact to assist him with the task.[14] As Director of Military Operations (DMO) from 1912 to 1914 Major White became directly involved in the preparation of strategic plans for the defence of Australia, and at the outbreak of the First World War was Acting Chief of the General Staff.

White worked directly for Gordon and carried out tasks which flowed from the initiatives taken by the CGS in strategic policy. Gordon took advantage of the efforts by Andrew Fisher and the Reform government in New Zealand to forge closer defence and trade links in 1912. Pearce, the Defence Minister, followed Gordon's suggestion and asked the New Zealand government to give General Godley, the New Zealand CGS, authorisation to discuss 'any proposed schemes of mutual assistance' when he visited Australia in November 1912. The discussions between Godley and Gordon on 18 November 1912 examined cooperation between Australia and New Zealand in the Empire and, more specifically, in the Pacific. On the question of a military contribution to the Empire, they agreed that the creation 'of an Australian unit was desirable to avoid the despatch of "fragmented organisations"'. Another avenue of cooperation discussed by the two officers was the conduct of operations in the East Indian Archipelago or the Pacific. Australia had responsibility for all foreign and other possessions west of the 170th degree longitude while the remainder was allotted to New Zealand.[15]

White's chief, Gordon, believed Australia had most to fear from Japan and believed that it was unwise to rely on British warships despatched from the North Sea or Mediterranean in the event of war. The logic of Gordon's fears about Japan was made clear from information sent from Colonel Legge, the Dominion Representative on the Imperial General Staff, to White in a letter on 25 July 1913. Legge calculated that Japan possessed the ability and capacity to send three divisions to Australia in under four weeks from the date of mobilisation, and he added that, 'the Japs could, if they chose, do it without giving us even indirect information of more than 7 to 14 days'.[16]

At the same time that the General Staff was assessing possible Japanese intentions in the Pacific during a European conflict Brudenell

White was instructed by Gordon and Pearce to draw up the composition of an Australian division for despatch overseas in time of war. According to Bean, White was asked by Pearce to keep the plan secret. On 14 November 1913, Major White presented Senator Edward Millen, the new Defence Minister in the Liberal government under Joseph Cook, with a plan for the employment of Australian troops overseas.[17] The plan included a division of 18 000 troops with 12 000 from Australia and 6000 from New Zealand. There is no record extant that Millen agreed to the scheme of cooperation presented by White. However, the plan of an Australasian division as an expeditionary force drawn by White provided Australia with the option of assisting with the defence of the Empire in Europe. As events showed later, it was the only contingency plan held by the military for action in the event of war in Europe.

The outstanding task for the Operations Directorate after the preparation of the local and general schemes for the defence of Australia was the completion of the War Book, which was designed to expedite Australian and imperial defence preparations in the event of war. By the outbreak of the First World War, the War Book was complete and Australia's military preparations for war were at an advanced stage. Due to the endeavours of White as DMO, Australia had military plans for defence in the Pacific and for participation in an imperial expeditionary force. His performance as DMO showed that he had ability to organise large numbers of men, units and equipment. Furthermore, he indicated a sharp understanding of how to translate strategic advice into operational plans. He was a supreme organiser.

The climax of White's involvement in policy-making from 1912 to 1914 came with the outbreak of the First World War, which he described as sudden, 'like a tropical storm'. At the time, he was Acting CGS and bore the immediate responsibility for advising the Australian government on military matters. On 11 August, White urged Millen to sanction military steps in accordance with the message from London on 30 July that the dominions should 'adopt precautionary stage'. Millen was reluctant to act without the Prime Minister's consent, especially as the telegram from London had been deciphered incorrectly. White called on Sir William Irvine, the Attorney-General, and urged him to persuade Millen to adopt military precautions. On 2 August, following Irvine's advice, Millen agreed to sanction precautions for the defence of Australia.[18]

The outbreak of war immediately raised the question of dominion support for the Empire. Again, White bore the responsibility for advising the Cabinet on the subject, and he was more than equal to the task. The decision to offer military aid to Britain was made by the Australian Cabinet at a meeting in Melbourne on 3 August 1914. The Prime Minister, Joseph Cook, asked White if plans existed for the despatch of Australian troops overseas. In reply, White told Cook and the Cabinet that a study of military cooperation with New Zealand in the Pacific, and the Empire, was undertaken in 1912. He believed that a force of 12 000 men could be raised and sent abroad within six weeks.

When asked about the size of the proposed force, White replied that in comparison with the South African War, 'the figure was within our resources'.[19] However, Cook wanted a larger force to match the Canadian offer of 30 000 men, and to demonstrate that Australia was carrying her imperial responsibilities. The Prime Minister then asked White if it was possible to raise a force of 20 000 men and the time needed to equip and despatch them overseas. White thought that the Australian nation was able to complete the equipping and brief training of a 20 000 size force, but he could not guarantee the completion of such a project in six weeks. The Australian Cabinet decided to offer a force of 20 000 men and pay for the expense of sending and maintaining them abroad. White wrote the cablegram which Cook sent to the British government offering 'to despatch an expeditionary force of 20 000 men of any suggested composition to any destination desired...to be at complete disposal of Home Government'.

*White and Bridges*

White's appointment as chief of staff of the 1st Australian Division, AIF, with Bridges as his commander, brought together a unique combination of two men with different personalities but bound by a common approach to the problems of Australian defence. Their personal association before the First World War spanned the development of the Australian Army from Federation. They were influenced by Major-General Hutton, GOC Commonwealth Military Forces from 1902 to 1904, to the extent that White described him as 'the founder of the Military spirit of Australia and its patron saint' and Bridges wanted him to command the AIF. Like Hutton, Bridges and White wanted to raise the standard of professionalism in the Australian Army. Throughout their careers they never allowed the imperial dimension of Australian defence to be forgotten. In contrast, officers such as Major-Generals Sir John Hoad and James Legge held a different view about the deference due to strategic advice from Whitehall.

The differences between Bridges and White and Hoad and Legge about Australian defence were reflected in the close personal relations of the Hutton protégés. In 1908, White applauded the appointment of Bridges as the first CGS and told him that Hoad, his rival and a political general in the narrowest sense, had not grasped the idea of a general staff. White's close personal association with Bridges led him to discuss the personal ambitions of Hoad, and he advised Bridges to let Hoad, if he became CGS, 'work out his own salvation' and wait for the results of his first term of office. He told Bridges that 'he was praying hard that the Labor ministry may remain in office'.[20] The creation of the Fusion Party in Parliament in 1909 favoured Hoad's aspirations and he was appointed the second CGS.

White's association with Bridges as his commander covered the creation of the AIF, its training in Egypt and operations at Gallipoli. From August 1914 until Bridges' death on 18 May 1915, the story of

White is the story of Bridges.[21] Both men believed that the AIF had to include members from the citizen army and soldiers who had served in the militia or in the Boer War. White, in particular, wanted to develop the AIF along the lines of the local military organisation but he realised that the citizen army was too inexperienced and new to play such a role. However, Bridges and White insisted that the units of the AIF be recruited on a local and state-wide basis.

The creation of the AIF also brought forward the question about the identity of the AIF and the protection of its interests. White, like Bridges, wished to prevent the dismemberment of the AIF among British units which had occurred in the Boer War. At Bridges' request, he drafted regulations which gave the AIF autonomy and a channel of direct communication with the Australian government. The common imperial banner under which the Australians fought did not exclude development of the idea that they were dominion units in their own right. To preserve the Australian national interest in the AIF, the senior AIF leaders, at the direction of the Australian government, established a machinery for its administrative self-government.

The efficient despatch of the AIF from Australia to Egypt in four months was a result of the smooth working relationship between White and Bridges. In Egypt, White supported Bridges in his efforts to train and administer the AIF. He made a practice of moving among the units in the desert every day and lifted as many administrative tasks as he could from Bridges' shoulders.

The only hint of any strain in the relationship between White and Bridges arose over the plan to employ the AIF in the naval attack against the Dardanelles. The central question was the suitability of landing the ANZAC Corps at Cape Helles or at Gaba Tepe. Bridges and Birdwood, the ANZAC Corps commander, agreed with General Sir Ian Hamilton, Commander-in-Chief, Mediterranean Expeditionary Force, that Cape Helles presented the best chance for success because the fleet could provide support for both flanks of the Army. White, however, thought that a force of 150 000 men was necessary for such an operation and preferred an attack on Gaba Tepe. Furthermore, he argued that a large force needed room to manoeuvre and there was little spare at Cape Helles. Bridges' reaction to White's criticism of the Cape Helles plan resulted in a comment to Birdwood: 'I have told White that he can stay behind if he doesn't like it'. Their difference over the merit of the Cape Helles plan was resolved when the ANZAC Corps was given orders to land at Gaba Tepe. In fact Bridges' high regard for White had not suffered over their disagreement about the proposed Cape Helles operation, and before the operation he wrote to Mrs White: 'no one could have given me more help and assistance than your husband has given. He has been more than my right hand and has smoothed over innumerable difficulties. I am indeed very grateful to him.'[22] Bridges realised that White was indispensable to his command of the 1st Division.

The brief association of White and Bridges for twenty-one days at Gallipoli saw their smooth working relationship on the battlefield. On 25 April White accompanied Bridges during his tour of Australian positions and noted how his commander was reckless in exposing himself to enemy fire. Both he and Bridges worked frantically that day to grasp the results of the fighting, which, as a result of the pre-dawn attack and poor system of communication, was 'touch and go'. Their most difficult and important task was to decide where and how to reinforce battalions without jeopardising the hard-won tactical advantages of the fighting. By dusk, Bridges realised that the Australians had lost their position on 400 Plateau, and the left and right ANZAC flanks were far from secure. The landing had failed to secure its objective and White, along with Major-General Godley, the New Zealand commander, argued that Hamilton should be informed and given advice to withdraw the troops. Bridges accepted the advice from White and Godley and asked Birdwood to visit ANZAC Headquarters to discuss the situation. Birdwood agreed to put the position before Hamilton but he refused to make preliminary arrangements pending a decision to withdraw. In *Two Men I Knew*, Bean recorded that Bridges gave White a stiff reply when he asked what arrangements to make for the staff in the event of a withdrawal, stating that, 'We will stay here to the last'. The decision not to withdraw was made by Hamilton who was told that an evacuation was impossible in the time available.

Until Bridges' death, White accompanied his chief on tours of inspection, the latter still showing scant regard for his safety by standing in the open. Even the ANZAC Headquarters was frequently hit by shrapnel and on one occasion White was struck while he stood at lunch. White was with Bridges when he was shot on 15 May.

Bridges' death ended a unique partnership between commander and chief staff officer in the AIF. The success of their working relationship rested on their common views about Australian and imperial defence, the need to professionalise the Australian Army, and their link with General Hutton. White wrote to Hutton after Bridges' death and revealed how he saw his former commander:

> Lacking as he may have the human qualities of a great leader he undoubtedly proved himself not only the organising and administrative master but a general capable of leading troops in the field. His calm judgement and imperturbability in times of stress were most inspiring.[23]

In terms of the relationship between a commander and his staff officer, White was the perfect complement to Bridges on the battlefield.

### White and Birdwood

An assessment of White as the tactical and administrative commander of the AIF under Birdwood[24] must be made against his eulogistic treatment by C.E.W. Bean, the official war historian, in his diaries and in his book, *Two Men I Knew*. Bean outlined what he saw as White's noble character

4 Lieutenant-Colonel C.B.B. White, GSO1 1st Australian Division, in his dugout at Anzac, May 1915. (AWM Negative No. C 1815)

and the major decisions that he made in the development of the AIF. Birdwood underlined White's 'complete selflessness' and stated that he was responsible for many achievements of the ANZAC Corps. White, in his opinion, was the key operator who facilitated the coordination, administration and military operations of the Australian troops in the field.

White's success as Birdwood's chief staff officer from the end of 1915 to the end of 1918 was built on his ability to appreciate the strengths of his commander and their similar views about the greatness of the British Empire. White was convinced that Australia's future lay with Britain in the Empire. Birdwood was a commander who enjoyed contact with his troops and in this role he exemplified the military link between Australia and the Empire on the battlefield. For the AIF, White viewed Birdwood's high profile as 'far too valuable to lose'. From Birdwood's viewpoint, White was the ideal complement to his style of command, and in a letter to Hutton on 30 September 1916, he set down his assessment of White:

You so rightly describe White, for whom I have the greatest affection, and whom I regard as one of the best officers it has ever been my good fortune to have with me. It is so very much the old head on young shoulders, a well-balanced mind—much knowledge and the activity of youth. My only regret about him is that he is not commanding a division for which he is thoroughly suited. I have, however, thought it out very carefully, and come to the conclusion that he really is of more value to the State in his present position than he would be in command of a division... He is still very young and I think can afford to wait, though I often have a feeling of regret that he is perhaps being kept back on account of his usefulness as a staff officer...[25]

The performance of White as staff officer to Birdwood shows how he was able to influence profoundly the tactical and administrative decisions affecting the AIF. At a tactical level, White's greatest achievement came as Brigadier-General, General Staff of the ANZAC Corps, when he worked out the plan for the evacuation of Australian troops from Gallipoli in December 1915. From start to finish, the plan was 'a model of precision and clear thinking' in which the Turks were conditioned to short periods of silence from the Australian trenches. White planned to deceive the Turks by withdrawing the troops gradually and maintaining an appearance of normal activities in the trenches. On the evenings of 18 and 19 December the last 20 000 troops were scheduled for evacuation. The detailed plans of the evacuation included the display of periscopes in trenches, the firing of rifles as men left their positions, the rigging of tins of water to the firing mechanism to create the impression that troops were still in the trenches, and boot mufflers. The evacuation from Gallipoli was successful and no lives were lost.

As the senior Australian officer on Birdwood's headquarters, White never hesitated to criticise senior British officers on the Western Front when he believed that Australian lives were at risk, and he gained support from Birdwood for his actions. Convinced of the need for professionalism in military planning, he was never overawed by higher ranking officers when he thought that their ideas or plans were wrong. In July 1916, British and Australian troops attacked the Sugar-loaf salient south of Armentieres between Fromelles and Aubens. The commander of the British 11th Corps decided to mount an infantry attack against the Aubens ridge to straighten out the bulge in the enemy line. He decided to use the 61st British Division and the 5th Division AIF and proceeded with the operation despite the recent failure of a similar attack. The 5th Division lost 5533 men in the attack on 19 July. Brudenell White was bitterly critical of the lack of planning and preparation by the British Corps. On another occasion, he spoke plainly to Haig who was critical of Major-General Legge and the 2nd Division AIF in July 1916 when it attacked Pozieres. White told Haig that General Gough had pushed Legge into action without any regard for adequate preparations and that the development of the battle was very different to the information provided by British staff officers.

Because of Birdwood's dual responsibility as Commander of the 1st ANZAC Corps and Commander of the AIF, White was promoted to major-general (on 1 January 1917) and therefore had even greater authority than was usually exercised by a chief of staff. Birdwood was almost invariably away from his headquarters inspecting troops, and White planned and wrote the orders for every operation conducted by the ANZAC Corps. Some commanders might have been happy to give such authority to their chief of staff, but most would have seen it as an abrogation of their responsibility—perhaps even a dereliction of duty. White's ability to support Birdwood on tactical matters was made patently clear later in the war by an entry in General Rawlinson's diary in May 1918:

> I visited the corps and discussed with Birdie and White the plans for the Somme offensive which Foch is very keen about—Birdie is not much good at making plans but White is excellent and I have told him to write a paper on the basis of the 5th or 6th Div. They move to the V Army on 30 by which time Monash will be well in the saddle I hope....[26]

Rawlinson's comments only understated the way in which White had assisted Birdwood in his command.

At an administrative level, White was, in effect, the commander of the AIF. After the evacuation from Gallipoli he was chosen by Birdwood to draw up a scheme for the remaking of the Australian divisions in Egypt. Birdwood wanted to create an Australian army of two corps by splitting existing battalions in half and bringing them to strength with reinforcements. The War Office did not approve the idea of an army but sanctioned the establishment of four Australian divisions. Before White left for France in 1916, he planned the Australian Reinforcement Training Centre, and took the precaution of warning Colonel Sellheim, commander of the centre, that General Murray, the Commander-in-Chief Egypt, held different views about the training of Australians. In fact Murray did not respect the arrangement worked out by White and Birdwood, and Sellheim discovered that a British staff was training Australian troops.

On the Western Front, White had not only to cope with operational plans drawn by British officers but also the difficulties of administering the 1st ANZAC Corps in France. By November 1916 the preparation of trenches was overdue and soldiers suffered from frozen feet, lack of food and shelter. When White arrived at the front he laid down plans for the movement of troops, food, and equipment to relieve the pressure of fighting under winter conditions. His reputation at the time was 'a man who could get things done'. The result by the end of the second half of 1917 were 'Anzac Light Railways', the remaking and extension of roads and the construction of Nissen huts to provide shelter for troops near the front.

In personnel matters, Birdwood gave White duties which he, as commander, should have taken upon himself. For example, the prospect of changes in the command of AIF divisions in 1918, following the proposed appointment of Monash, brought problems over the issue of Australian identity at the divisional level. When Brigadier-General H.E. Elliott was informed by White that Gellibrand would supersede him and

5 Brigadier-General C.H. Foote, Chief Engineer, and Major-General C.B.B. White, chief of staff of the 1st ANZAC Corps, walking down a duckwalk in front huts at Grevilliers, France, 22 May, 1917. Major Smith, Assistant Provost Marshal, is in the doorway. (AWM Negative No. E616)

take command of a division, despite their equivalent seniority, he asked if, under the Defence Act, he might appeal to Pearce, the Defence Minister. White told Elliott that he was not sure what Birdwood might do, but he would most likely insist that he retire from his command pending the minister's decision. In a letter to Elliott on 22 May 1918, White told him that Gellibrand was an Australian, and not a British officer as he mistakenly thought, and that the appeal to political intervention from Pearce was a hasty action. As noted earlier, Elliott believed that White was biased against him and that he favoured officers schooled in the British tradition.

For White, the growth of the inward nationalism of the AIF, which brought forth a demand for the appointment of an Australian officer as commander of the Australian Corps and the AIF, was an opportunity to realise his potential as a commander. In early 1918 both Charles Bean and the journalist, Keith Murdoch, organised a campaign to promote his appointment but their efforts came to nought. White was a natural candidate for the command of the AIF, but he was not prepared to allow the high command of the AIF to be wracked by internecine rivalry between his supporters and supporters of General Monash. More to the point, Field Marshal Haig was not prepared to contemplate White as commander of the Australian Corps. In July 1917 he had incurred Haig's displeasure when Haig had remarked to White that he should be commanding a corps. White had replied that General Birdwood's position in the AIF was invaluable and that he was essential for the Australian fighting command.

As well, White's decision to remain as Birdwood's chief of staff after Gallipoli denied him vital experience as a divisional commander which was usually a prerequisite for the highest field command posts. The upshot was that Monash received command of the Australian Corps in May 1918 and Birdwood retained command of the AIF as well as receiving command of the 5th Army. White went with Birdwood as his chief of staff, thereby maintaining his control over the administration of the AIF.

Under Birdwood, Brudenell White exercised the powers and influence of a tactical and administrative commander in all but name. His influence was not only a result of Birdwood's style of command and his ability to work with his commander as a staff officer, but also his grasp of tactics, administration and organisation. He played a vital role in the development of the Australian Corps as a distinctive fighting force and laid the foundations for a tradition which has continued to this day. The irony was that his indispensability as chief of staff to the commander of the AIF in the end denied him the appointment as commander of the Australian Corps.

*Chief of the General Staff*

As Chief of the General Staff of the Australian Army from 1920 to 1923 White brought a wide range of knowledge and experience to administering and organising a peacetime army at a time when financial resources for defence were decreasing. His aim was to build a citizen army capable of defending Australia in the Pacific and playing a responsible role in imperial defence. The performance of the AIF in the First World War had convinced him that the citizen army concept was ideal for Australia: 'We can, however, as a result of the war, safely conclude that a citizen army can be made an efficient instrument whereby to maintain national policy'.[27] White admired the ability of the Australian nation in the First world War to respond to 'sacrifice and trial'.

The strategic considerations for the defence of Australia in the Post-First World War period were discussed by two committees of which White was a member. Pearce, the Defence Minister, asked White, Major-Generals J.M. M'Cay and J.G. Legge and G. Swinbourne, a businessman, to recommend ways of reducing defence expenditure and modifying the civilian forces' training, as well as advising on the establishment of an air service. The main recommendation was the continuation of compulsory military training and the development of a citizen force consisting of six infantry and two mounted divisions (approximately 180 000 men). Recalling their recent experience in war, the army officers wanted longer periods of training for the citizen forces to avoid additional training on the outbreak of war. Their concern about Australia's strategic circumstances was revealed in their desire to train 180 000 troops in five years. The committee also endorsed the need for an air service.

The second committee of which White was a member met in January 1920 to discuss the military defence of Australia.[28] The other members of the committee were Generals Chauvel, Monash, M'Cay, Hobbs and Legge. A notable feature of the report was the concentration on Australian security in the Pacific. Japan was seen as the only potential and probable enemy, especially as Japanese military capabilities were extensive. According to the committee, Japan had the ability to move 100 000 soldiers in one convoy. Australia, therefore, needed a large army to counter a possible invasion. The senior army officers wondered if Japan would wait until the Anglo-Japanese alliance expired in 1921 and stressed that a force of 180 000 was necessary to meet a Japanese invasion. They recommended that the government amend the Defence Act to allow compulsory military trainees to serve overseas.

White's major task when he became CGS in June 1920 was the training of a citizen army. In accordance with the Committee's suggestion of building a citizen army, he decided to use the higher commanders and staffs of the AIF and supported the introduction of a 70-day training camp for 18-year-olds under compulsory military training. In October 1920, the Military Board asked the Military Districts to prepare plans for mobile field forces to operate in the field and meet an invading force. The units in the District plans were to take AIF designations, and the peacetime organisation of the army was modelled on the AIF division. By May 1921, the plans were complete and accepted government policy.

The challenge facing White from 1920 to 1923 was the preservation of a nucleus for the expansion of the army in war. There was pressure from some parliamentarians to abolish compulsory military training while others wanted economy in the defence estimates. In the 1919-20 estimates, the reduction was less than expected, but this was due to the unsettled situation in the Pacific. However, the government statement on defence policy in September 1920 reduced the size of the army recommended by the senior officers from 180 000 to 130 000 men and the proposal for 70-days' training of 18-year-olds was shelved.

After the Washington Conference, the government decided to reduce defence spending to the minimum level and White responded by retaining the nucleus of commands, divisions, brigades, battalions and companies. The problem was put clearly in a letter from Field Marshal Lord Birdwood to White on 15 November 1922: 'In your case, your original nucleus was so small, that I am always apprehensive that drastic cuttings down may reduce you to a state of inefficiency and hardly leave you with an organisation on which you could recreate should occasion arise'.[29] Thus White decided to maintain the war organisation by cutting 'horizontally and not vertically'. Training under compulsory military training was also cut to ten days, and each trainee served for two years.

Despite the 'onslaughts of the axe' and the depression which he felt about the defence retrenchments, White's sound judgment about a citizen defence scheme was borne out in the Second World War. Menzies proposed a system of compulsory training in the citizen forces along the lines which White had advocated while CGS. Furthermore, the National

Service Act of 1951 established a training system for citizen soldiers similar to White's schemes.

In June 1923 White was called to reorganise the Federal Public Service and retired from the army. However, he faced the policy arena again in the Second World War, at the age of 63, when he was asked to take the position of CGS after General Squires died in March 1940. According to Bean, he said, 'I could wish to have been spared this task,...but one must do as one is bid...'[30] On 20 March he was promoted to full general, the first time for an Australian in wartime, and brought his accumulated knowledge, experience and wisdom to the duties of First Member of the Military Board. The wheel had come full circle since the beginning of the First World War when he was Acting CGS.

White's immediate concern was the protection of Australian interests in the Second AIF which would fight under British or Allied commanders. He knew that the AIF had to be led by a strong personality capable of preventing its dismemberment among British and Allied forces. For the solution of the identity and protection of Australia's interest in the AIF, White drew upon his experience of the First AIF. A charter of the Second AIF was drawn up based upon the document drafted by White for the same purpose in the First World War. The central principle was that the AIF was an Australian force under its own commander, who was responsible to his own government. The operational control of the AIF was under the commander-in-chief in the theatre where it served. White also recommended Blamey as commander of the new AIF corps. In his view Blamey had the ability and force of character to make British politicians and generals appreciate Australian needs. Furthermore, Blamey had the requisite experience as his successor as chief of staff in the Australian Corps in 1918. Blamey was subsequently appointed as commander of the Corps.[31]

In May 1940 White advised the government of the need to concentrate on assisting Great Britain and her allies 'rather than on local defence'. The Military Board in that month recommended the establishment of the 8th Division AIF and the War Cabinet suggested the suspension of militia training for six months. White also urged, when the War Cabinet met after Dunkirk, that the 7th Division AIF and Blamey should proceed overseas. According to Bean, White valued the psychological impact of Australian support for the British Army in Europe. Bean also described a War Cabinet meeting on 4 June 1940 at which White was called upon to give an opinion about the date of the next German offensive against France. White thought the Germans would need six to nine months to prepare. The German offensive against France on 5 June shook the 'confidence of Ministers in his judgement'. According to Bean, members of the Cabinet could conclude from his advice that he was approaching the problem in terms of 1918.[32] Seven days later, White recommended the raising of a fourth division for the AIF.

With reference to Japan, he believed that Japanese targets would be Singapore, British bases and the Royal Navy. In his view, Japan had sufficient sea power to force the Empire to sue for peace. He thought that the United States Navy would deter the Japanese from invading New Caledonia and the Dutch East Indies, and the possibility of the Royal Navy assisting Australia could not be disregarded. He thought that Japan would not invade Australia if she were successful in taking Singapore and ousting the Royal Navy from Eastern waters. The defences of northern Australia needed a battle fleet for protection and there was no point in garrisoning Darwin. When the British government asked the Australian government on 25 June 1940 to send a division and two air squadrons to Malaya, White argued that such a commitment would run the risk of operating in unknown terrain and the division might be lost. On 6 August the Joint Planning Committee advised the chiefs of staff to make plans against invasion using the forces available in Australia. It is not known how White would have reacted to this recommendation, for on 13 August 1940 he was tragically killed in an air crash near Canberra Airport.

White served as an important policy-maker and executive in critical periods in Australia's military history and showed that he was more than equal to the task of advising government at the highest level. His staff work conformed to the highest standards and his judgment and sense of proportion in professional military matters were highly regarded by his fellow officers.

*The most complete general staff officer*

Despite the fact that Brudenell White served as CGS for over three years, including a period of some five months during the early days of the Second World War, he achieved his greatest impact on the Australian Army during the First World War. Although the Australian Corps won its final and momentous victories under Sir John Monash, White had helped forge the weapon—perhaps the finest body of Australian soldiers ever to serve in the field. Aspects such as careful planning, attention to detail, concern for saving Australian lives and strict professionalism, insisted upon by White, have become hallmarks of Australian military operations, though they are often less obvious aspects than the well-known courage and dash of Australian soldiers. A mere staff officer could not have developed and inculcated these characteristics in the Australian Army.

Brudenell White was no mere staff officer. Highly professional, shy and loyal to his commanders, he never attempted to usurp the role of commander. Yet through his ability and clear-sightedness he stamped his methods and ideas upon the formations of which he was chief of staff. In doing so, he demonstrated many of the qualities of a commander, and under Birdwood he undertook many of the responsibilities of a commander. The men of the 1st AIF and indeed the Australian Army have good cause to be thankful that for so much of the First World War White was at the helm.

Sir Alfred Kemsley, a staff officer on HQ Australian Corps during the

First World War, and a senior officer on Blamey's staff in the Second World War, recalled his impression of White:

> Probably the most complete General Staff Officer of all times and least known because of his retired nature. Did all possible to keep journalists away from frontal areas because of their inadequacy of battle knowledge. Helped give Birdwood his high status because of loyalty and the free hand Birdwood permitted him. Kind, encouraged juniors, gently transferred those of lesser success, untiring, loyal.

There is much truth in Menzies' view that White was 'the most scholarly and technically talented soldier in Australian history'.[33]

# Vice-Admiral Sir William Creswell: First Naval Member of the Australian Naval Board, 1911-19

STEPHEN D. WEBSTER

As First Naval Member of the Australian Naval Board Rear-Admiral Sir William Rooke Creswell (1852-1933) played a key role in the direction of local naval operations and policy during the First World War. Creswell's long and colourful career—beginning in the Royal Navy of Queen Victoria, then in Australia's motley colonial naval forces, and finally as the senior officer of the Australian Navy—mark him as the single most important figure in the gestation, early development and first testing of the Royal Australian Navy. By virtue of his long, continuous involvement in the debate surrounding Australian naval defence no other figure, politician or naval officer, played such an influential part.

*Early experience*

Creswell first came on the Australian colonial naval scene in Adelaide in 1885 at 33 years of age. After a brief but eventful career in the Royal Navy fighting pirates and slavers, Creswell, frustrated by minimal promotional chances, had resigned and shipped out for Australia with a younger brother in 1879.[1] Thereafter, he had lived the life of adventurer in Queensland and the Northern Territory. He was a long-distance drover, pioneering stock routes with men such as Charles Armstrong.[2] Just how the commandant of the Colonial Naval Forces in South Australia, Commander John Walcot, learned of the nautical drover's presence and whereabouts in the North is unknown. Nevertheless, when he did he invited Creswell to join his sole effective weapon, Her Majesty's Colonial Naval Ship, *Protector*.

Creswell took up duties as senior lieutenant aboard the 920 ton (935 tonne) *Protector* in October 1885. It was not long before he turned his mind to a defence strategy for South Australia. The reason was one which would motivate him time and time again; colonial politicians, due to the exigencies of public economy, lost sight of strategic truths and began to favour less costly, land-based military defence. It was a fateful decision when Creswell first put pen to paper in 1887 to set out his view of just how South Australian defence should develop. His arguments, in a series of private and public papers, were to change little, in terms of

6 Vice-Admiral Sir William Creswell KCMG, KBE (AWM Negative No. J 3055)

strategy, over the succeeding 22 years. Above all, Australia needed to develop its own naval defences to meet the only likely form of attack it faced—ocean-borne bombardments and coastal raids.³

While he set out his arguments the Australian colonies moved towards Federation. Creswell succeeded Walcot as commandant and then moved north to become Queensland's naval chief, from which position he was released to command the *Protector* when it was deployed with the China Fleet during the Boxer Rebellion. After the formation of the Commonwealth in 1901 he worked his way into the position of professional head of the federated naval forces. The post seemed much more than it really was. The new Federal government had inherited a collection of ageing vessels of doubtful fighting capacity from the colonies, and the majority of regulars and reserves who manned them were not much fitter. Yet Creswell remained obdurate. Australia had to have its own defence which should be largely naval.

After Federation Creswell's main opponents ceased to be Australian politicians. In their place were the Lords of the Admiralty who had only recently grasped the axiom of one great navy, at long last rejecting the strategic heresies of defensive, scattered naval forces.⁴ This rediscovery of the essence of the strength of the Navy of Nelson spelt doom for naval aspirations on the fringe of Empire—aspirations typified by the urging of Creswell.

Throughout the early 1900s Creswell and his dilapidated forces served as an embarrassing reminder to a succession of Defence Ministers and Prime Ministers, short of funds and swayed by Royal Navy officers, of a nationalistic urge that would not disappear. Creswell played upon the situation superbly, becoming one of the most skilled lobbyists of the early Federal era.⁵

Just as his tireless efforts seemed about to be rewarded his old employer and current foe cut the ground from beneath him. At the Imperial Defence Conference of 1909 the Admiralty did an about face by recognising dominion naval aspirations as legitimate. The self-governing dominions in the Pacific were invited to participate in building and manning a powerful Pacific fleet.⁶ Australia was to have its own Fleet Unit built around an all big-gun battle-cruiser. For a lesser man the dramatic shift might have proved too much, but not for Creswell. If the Admiralty could stand on its head so could he. In the long term Creswell proved the more agile. Australia proceeded with its Fleet Unit with Creswell still its senior naval officer and adviser.

By 1913 when the Australian Fleet Unit was nearing completion it became clear that the Pacific fleet was not going to eventuate. The reasons were to be found back in the North Sea where the Anglo-German naval rivalry had reached new heights. The Royal Navy was locked in a grim battle to maintain a clear superiority over Germany in numbers of modern battleships. It was a fight that strained the finances of Britain while making the development of a large Pacific fleet an impossibility.

Out of necessity Britain became increasingly dependent upon Japan through the Anglo-Japanese Alliance for defence of the Empire in the East.⁶ Unfortunately, imperial authorities had not informed Australia or

New Zealand of its change of plans after the Imperial Defence Conference of 1909 and the shock realisations in 1913 of the strength of the German fleet. The reaction in Australia was strongly indignant. Creswell, by then First Naval Member of the Australian Naval Board, was in a position to say, 'I told you so'. Development of Australian naval self-defence, regardless of details of form and type, was vindicated. So, too, was his ongoing suspicion of the Admiralty's intentions.

As Europe drifted towards war the future of imperial naval defence remained clouded. Australia had its Fleet Unit, which was the nucleus of the Royal Australian Navy, but there were no other British units in the Pacific.

Fortunately, Australia had taken steps to acquire a separate blueprint for development of its Fleet Unit and manpower. Late in 1910 and early 1911 Admiral Sir Reginald Henderson had toured Australia in order to make recommendations upon the future of naval development. He had been accompanied throughout by an old shipmate, Creswell.[7] Given what amounted to a free hand, Henderson and his specialist staff produced an ambitious plan for the fledgling Australian Navy. Australia was to develop by stages a fleet of considerable size and strength. By 1933 there were to be eight battle-cruisers and numerous smaller fighting vessels working out of a web of bases in both the Pacific and Indian Oceans.[8]

While Creswell's influence on the Henderson Report is arguable his commitment to it was unquestionable. It became the base document for Australian naval defence over the next few years. To ensure its implementation Henderson recommended that Creswell become the government's senior naval adviser as First Naval Member of the newly established Naval Board. The Board was clearly patterned on that of the Admiralty, the role and duties of the First Naval Member being similar to those of the First Sea Lord.

The other determining factor in the shape of the pre-war Navy was the settlement reached at the Imperial Conference of 1911, where it was agreed that there would be the closest cooperation between British and Australian forces. Interchangeability was encouraged so that the new Australian force would provide adequate career opportunities for officers and men. One of the more substantial files kept at the Navy Office bore the title 'Australian Unit of the Imperial Navy'.[9]

The newly knighted and promoted Creswell approved these developments. Australia would maintain full control of its force in peace, but after transfer in war the Royal Australian Navy would 'form an integral part of the British Fleet and will remain under the control of the British Admiralty during the continuance of war'.[10] From hindsight it can be said that this last provision spelt trouble for the Naval Board and for Creswell in war.

*The Naval Board*

The new Naval Board was hardly established before it fell upon troubled times. The reasons were twofold: first, there was a feud amongst members. Joining the Board in late 1912 as Second Naval Member was Captain Constantine Hughes-Onslow recently retired from the Royal Navy. It was to prove an unfortunate appointment as Hughes-Onslow possessed a difficult, aggressive personality which exacerbated several key differences on the Board.

Second, great pressures were put upon the Board by the failure of the proposed Pacific fleet to materialise, throwing into question the validity of the recommendations of the Henderson Report. Hughes-Onslow sided with the Third Naval Member, Captain Clarkson, and urged replacement of the Henderson recommendations with a new plan which featured an early form of forward defence in waters north of Australia. Hughes-Onslow's recommendations were clearly based upon a strategic assessment that saw Japan, not Germany, as the real threat. To meet it Australia needed to concentrate naval development in the north near Darwin. Base building and ship deployment in the south would have to take second place.[11]

These recommendations were the source of a long and increasingly acrimonious debate among Board members including two Ministers of Defence.[12] In view of the failure of the Pacific fleet to eventuate Creswell saw some merit in these changes, but he was not prepared to abandon at once the Henderson recommendations. In this view he was joined by the Secretary to the Board and voting member, Paymaster-in-Chief H.W.E. Manisty, RN, who also had been a member of Admiral Henderson's staff during his tour of Australia. The feud that blew up over this issue in 1913 seriously affected the working of the Australian Naval Board. In the end, when matters became public, Hughes-Onslow was dismissed and Creswell's future was clouded.

Much of the good work carried out in this period of turmoil came out of Creswell's own area of responsibility. He had as his assistant, Commander Walter Thring, who, like Creswell, had left the Royal Navy due to minimal chances of promotion. It was a lucky stroke for Australia to pick an officer of such outstanding ability. Creswell and Thring combined smoothly and effectively to provide the RAN with detailed plans for mobilisation in the event of a war threat. Thring, who had expertise in intelligence work, carried on with Creswell's initiatives in this area by establishing a rudimentary, but effective information-gathering centre at the Navy Office in Melbourne.

It is important to understand this immediate background of Australian naval development if the role and achievement of Creswell as an Australian wartime naval leader are to be assessed accurately. Both coolness of relations on the Board and the unresolved future of Australian naval development were to remain major issues throughout most of the war.

*Operations in the Pacific*

Creswell was in his sixty-third year when war broke out in August 1914. He was in excellent condition both physically and mentally. It was just as well that he went to war in fine fettle because the next four years were to be trying ones. Not only did he carry the burden of command, but also the strain and grief of personal losses in the later years of war. In addition, he was to endure strong criticism on more than one occasion—criticism in large part unjustified.

Creswell was on leave in Brisbane late in July, his first respite from the disagreements and recriminations lingering in the aftermath of the Hughes-Onslow episode. It was Commander Thring who, even before the receipt of an imperial warning message on 31 July, recalled Creswell and took the initial steps to place the Royal Australian Navy on a war footing.

The bulk of Australia was caught almost totally unprepared by the crisis in Europe. As war approached the nation was in the midst of a Federal election campaign. The warships of the Fleet Unit were dispersed on manoeuvres. Although momentarily caught off guard, Australia had a substantial force in being. Ready to meet the test were nearly 9000 officers and men manning and supporting the warships. Given his long career in support of the development of Australian naval strength it was fitting that Creswell should draft the cable officially transferring control of the Royal Australian Navy to the Admiralty.[13]

Under the agreement reached between the Imperial and Australian governments in 1911 the RAN passed under the control of the Admiralty in war. In war the role set out for the Australian Naval Board *vis à vis* the Admiralty was one of direct subordinate. For the duration, the Naval Board became Commander-in-Chief Home Ports in Australia.

In August 1914, the First Lord of the Admiralty, Winston Churchill, instructed the various naval branches to expect a war lasting one year of which the greatest effort would be concentrated in the first six months.[14] Though this forecast was sadly wrong for the global conflict it had some applicability to events in the Australian and Pacific regions. For Creswell and his war staff at the Navy Office the war was to be one of two distinct phases.

The first, shorter phase was dominated by the drive to eliminate the German presence in the Pacific, and the bulk of this task was achieved during the first six months of war. In January 1915 the Naval Board entered a second phase in which the main units of the RAN were called to duty half-way across the world in the North Sea. From then onwards the Naval Board, while experiencing momentary threats in Australian waters, for the most part concentrated on supplying the war effort while planning for the post-war future.

During the first phase Creswell virtually lived within the confines of the War Room at the Navy Office. He built up a small, efficient staff headed by Thring and dedicated to the early destruction of von Spee's German East Asia Squadron, the most potent enemy force in the Pacific.

Like his counterpart in the Royal Navy, the First Sea Lord Admiral Sir John Fisher, Creswell had always shown a willingness to embrace technological advances in warfare. He had been an early supporter of the locomotive torpedo and oil-fired, turbine-driven warships. He had also supported the establishment of a powerful wireless telegraphy network in

Australia. At his instruction the Navy Office was fitted out as a main transmission and receiving station in 1912-13.

This wireless telegraphy capability was to pay dividends in the early days of the war. Creswell's operators strained their ears listening for signals from the *Scharnhorst* and *Gneisenau*, von Spee's most powerful cruisers. Due to the interception of German signals the war orders for the Fleet Unit commanded by Rear-Admiral Patey were changed. Instead of sailing for the China Station the *Australia, Sydney* and *Melbourne* were ordered to New Guinea waters, thought to be von Spee's destination.

The story of that fruitless search and the subsequent alleged misuse of the Australian Fleet Unit has been told in the Official History.[15] The specific role of Creswell and his staff is worth examining here. From the beginning, Creswell, in line with the theory of Alfred Thayer Mahan (author of *The Influence of Seapower upon History*) sought to achieve the strategic goal of seeking out and destroying the enemy's main fighting force. In his view the Australian and imperial naval forces in the Pacific had to remain free to bring to battle the most powerful German units. For years he had anticipated that the enemy, whether it be Japan, Russia, France or Germany would pursue a limited, but nevertheless paralysing commerce war in waters crucial to Australian trade. As events unfolded von Spee did order a limited commerce war through the guns of the *Emden*.[16] Before she was beached on North Keeling Island in November 1914 the *Emden* sank or captured 25 Allied steamers and two warships, raided Penang and Madras and paralysed Britain's Far Eastern seatrade. Creswell was anxious to protect coastal and overseas commerce, but he instinctively ordered his priorities strategically. He was confident that the Admiralty would direct Patey to seek out the enemy. It is not within the scope of this appraisal to analyse the aims—aims unknown to Creswell and his Defence Minister—of the Committee of Imperial Defence in recommending that the Fleet Unit be used to escort expeditionary forces to the German colonial possessions of Samoa and New Guinea. It is, however, relevant to any assessment of Creswell as a war leader to note the effect of this decision upon him.

When he learned in mid-August that the main units of the RAN were to be used in a defensive, escort capacity his initial reaction was one of mild astonishment. Subsequently, when the operations dragged on, with von Spee slipping through the Navy's net, he became openly incredulous. It was one of the supreme ironies of Creswell's war career that he, not the Admiralty, became the upholder of strategic orthodoxy, and he became increasingly agitated as von Spee stealthily steamed across the Pacific towards those fateful encounters at Coronel and Falkland Islands. The Governor-General, Sir Ronald Munro-Ferguson, never a great admirer of Creswell or the Navy Office, made clear the general acceptance of Creswell's views in government circles:

> It is thought that a capital mistake was made in sending an expedition to Samoa before the wireless N.E. of Australia was dealt with, and the enemy chased from the sea; also that for some days past H.M.A. Ships in the Bismarck Archipelago have been wasted.[17]

To the end of his days Creswell remained convinced that had Patey been set free to chase von Spee across the Pacific the tragedy of Coronel might not have happened. He did, however, draw one grimly satisfactory conclusion from this disturbing sequence of events. Writing to the New Zealand Defence Minister, Colonel James Allen, he extolled the virtues of the Australian naval initiative of creating its own Fleet Unit:

> The *Good Hope* and *Monmouth* disaster [Royal Navy ships sunk by von Spee at Coronel] is due to the old idea that anything does that is to be used a long way off. They were old ships only just commissioned and commissioned with Fleet Reserve men. No time to shake down [and] get into gunnery efficiency and they had to meet two splendid modern cruisers that had been two or three years in commission...
>
> We should have had a pretty cheerful time in Australia and New Zealand under the old regime [of total dependence on imperial naval defence] with one armoured cruiser and a few of the unspeakably useless P class [cruisers]. That most undoubtedly would have been our state of defence had we not most providentially had our little fleet—we should have enjoyed a coastal Belgium state of affairs in shipping losses, bombardment and levies of cash in the pleasant German way.[18]

Creswell's disquiet over the direction of the war in the Pacific did not feed solely on the failure to bring von Spee to battle in mid-Pacific. Those diversions, the expeditions to the Pacific islands under German control, also resulted in deepening concern. While New Zealand was given the task of occupying German Samoa, Australia had been asked by the Imperial government to take over German New Guinea and those tiny, phosphate-rich specks that were part of island groups such as the Marshalls and the Carolines.

Though Creswell had been inclined to strike first at the Marshalls and Carolines, the initial Australian effort was aimed at New Guinea, the seat of German colonial administration in the Pacific. As it was to be a combined naval and military expedition the Navy Office was heavily involved in the planning during August.[19] Though Patey was under direct orders from the Admiralty, the Naval Board was in charge of providing naval components of the force.

Since the *Australia* and *Melbourne* were required to escort the New Zealand military force to Samoa early in August, Creswell and Thring had a breathing space in which to organise the naval side of the expedition to New Guinea. Even in the early days of the war men and ships were at a premium, but they were able to obtain 500 men from the Royal Australian Naval Reserve (RANR) and the P&O liner *Berrima* which was fitted out as an auxiliary cruiser at Cockatoo dockyard. This naval force, combined with that of the army, joined up with Patey's escorts to occupy Rabaul on 11 September.

Though the surrender of German New Guinea was achieved, the occupations of the Carolines, Marshalls, Pellews and Marianas (Angaur and Yap in particular) did not eventuate. Behind this failure lay the second chord of concern that exercised the mind of Creswell for the duration. Faced with the New Guinea expedition, Australian coastal protection and preparation for the first convoy to the Middle East, there were simply not enough war vessels available to undertake an additional expedition to Angaur and Yap. Once New Guinea was secured the matter was again brought to the fore on 15 October in a cable from the Colonial Office:

> Japanese Government state that in course of searching Western Pacific islands for enemy's ships and bases squadron called Yap on 7th October and landed Marines to investigate Wireless Telegraphy and cable stations there. They found that both had been repaired [HMS *Minotaur* had bombarded the Yap station on 12 August] and used by Germans and since destroyed again. They have temporarily occupied it but they are ready to hand it over to an Australian force. On account of strategic importance Island must be occupied by some force. Your Ministers will remember that it was originally intended that they should send force to occupy Yap and they will no doubt agree that it is desirable to relieve Japanese as quickly as possible of the task of holding the island.[20]

This information struck the Navy Office like a thunderbolt. As suggested, Creswell had maintained an open mind over the pre-war reports of Hughes-Onslow which cited Japan as the most likely enemy and emphasised the strategic importance to Australia of northern waters. Since the entry of Japan into the war on the Allied side on 23 August, the Naval Board had heard little of the movements of the Japanese Navy. The news that the Japanese were holding a position just north of the Equator was unsettling.

The reaction was swift with Creswell and Thring co-authoring a Board memorandum to the Defence Minister, Senator Pearce, recommending the use of one of Patey's cruisers to carry the expeditionary force to the islands. Pearce agreed, but Admiral Patey was reluctant to part with the ship. The Navy Office then looked for alternatives including captured German naval auxiliaries. By mid-November all logistic problems were overcome and a firm sailing date of 26 November set for the expedition designated Tropical Force.

Tropical Force never sailed to Yap and the other islands north of the Equator. Just two days before departure the first indications of trouble reached Australia in a Colonial Office cable:

> it would be discourteous and disadvantageous to the Japanese if we turned them out of Angaur when they are helping us in every way with their Fleet throughout the Pacific and convoy[ing] Australian contingents. Japanese are now erecting a wireless station on Angaur which they wish to use in connection with Fleet movements.
>
> I hope therefore that your Government will see that instructions are given to Australian ships not to call at Angaur or to interfere with its present occupation by Japan. This is of course without prejudice to permanent arrangements which will have to be made after the war when we come to settle terms of peace.[21]

Creswell and Senator Pearce were determined to cling to a faint thread of hope. Since the cable had confined comment to the occupation of Angaur perhaps the Australian force could proceed to other islands in the area. The Colonial Office was cabled. The reply from London finalised the matter. Any Australian expedition should not 'proceed to any islands North of Equator'.[22]

Creswell's reaction to this puzzling turnabout underlined his thinking about Australia's defence future. He ordered his staff to document the British claim to the German possessions north of the Equator.[23] This file, entitled 'Captured German Islands Australian Claim' grew steadily during the war. Subsequent information gathered was not encouraging. It gradually became clear that the Japanese occupation was taking on an air of permanence. Creswell realised that by this single stroke Australia's post-war security had become more uncertain and defence planning more difficult.

As operations began to wind down in the Pacific and Indian Oceans Creswell could reflect with satisfaction upon the Australian naval performance. He had seen preparations for transfer from peacetime to a war footing searchingly tested and found equal to the task. The Naval Board was working well in its new role of direct responsibility to the Admiralty and the German presence in the Pacific had been virtually eliminated. The Naval Board's relationship with government—first the Cook government and, after 17 September 1914, the Labor government of Andrew Fisher—was the best it had been since the Hughes-Onslow fiasco.

The working relationship between Creswell and the Defence Minister, Senator Pearce, was particularly effective. Pearce and Creswell held a common view of the importance of continued Australian naval development. It was a view which had been tested before and after the outbreak of war. On both occasions senior naval adviser and minister concluded that their beliefs had been totally confirmed. They shared, too, a mutual respect for each other's abilities and experience. They were both self-made men who spoke their minds with directness. Together they had managed to restore the effectiveness of the Naval Board during the first six months of war.

*Wartime policy-making*

The splitting of Pearce's portfolio early in 1915 was a further indicator that the naval war was entering a second phase. The winding down of the active war in the Pacific allowed the government to lighten the burden carried by Pearce in the Defence portfolio. The Navy became a separate ministry, and J.A. Jensen, who had been assisting Pearce, was appointed as the first Minister for the Navy. It was to prove a disastrous appointment both for the government and for the Navy.

With the passing of the active phase Creswell turned his attention once again to the future of naval development in Australia. During 1915 the Naval Board re-examined Australian naval policy in light of the experiences of the first six months of war. The need for this re-examination was made urgent by the government's desire to hold early talks with the Imperial government on post-war defences. Though these discussions did not take place until 1918 the Naval Board reached important conclusions during 1915 which were incorporated in a series of regularly updated strategic reports to the government. These reports argued that the Australian initiative to develop its own naval defence capability had been vindicated by the war experience. So too, had the immediate pre-war concerns about the future defence of the Pacific. The Empire's ally, Japan, was, given the unsettling experience of Angaur and Yap, confirmed as the real threat to Australia. To defend against this threat required an updated form of forward naval defence as espoused by Hughes-Onslow in 1913. Australia had to rely upon naval strength and agility in defending along a line running from Singapore across to New Guinea and down to New Zealand.

Creswell strongly supported this view of the post-war future. To him the Anglo-Japanese Alliance could only be viewed as a stopgap providing Australia and New Zealand with a breathing space in which to build up forward naval defences. He also supported the view that regional, as opposed to global naval defence had to be the basis on which Australia built its future. The old Admiralty concept that 'the Seas Are One' had been tested and found wanting.

These reports made a considerable impression upon the governments of Fisher and Hughes. In the absence of an effective body to coordinate strategic defence assessments and recommendations the work of the Naval Board influenced senior ministers considerably,[24] and the reports were the most extensive available on the changed defence situation. Their impact was substantiated when the long awaited discussions took place between the Imperial and Australian governments nearly two-and-a-half years later. The Imperial War Conference in 1918 demonstrated that Australian and imperial defence thought were on divergent paths. Though Creswell did not attend the Conference his views were frequently impressed upon the minds of the Australian representatives, Prime Minister Hughes and Navy Minister, Joseph Cook. Accompanying Cook as his personal secretary was Lieutenant-Commander J.G. Latham, whom Creswell had appointed to the Navy Office for special duties some months earlier.[25] Latham soon shared the Board's views upon the Japanese threat and the future development of a formal naval defence. In London and later at the Versailles Peace Conference Latham persistently bombarded Cook and the Prime Minister with the salient points of the Naval Board's view.

At the Conference the Imperial government put forward a post-war proposal for a single imperial navy in which the various dominions would be partners. The proposed administrative structure indicated the nature of the partnership. At the top there was to be a Board of Imperial Admiralty to exercise overall control 'both in peace and war'.[26] Directly under it would be a group of local naval boards responsible for all local shore establishments such as docks and training schools. All ships within the area of a local naval board would come under the control of a commander-in-chief who would report directly to the Admiralty. The proposal concluded with a tentative suggestion that each dominion be represented in London at least once a year by its navy minister.[27]

It was a reversion to the stance of 1902 when the 'blue-water' strategy was advanced as the only consideration for naval policy. Gone were those 'other considerations', principally dominion national sentiment, which when recognised in 1907 and 1909 had led ultimately to the Royal Australian Navy. It was a turnabout that neither Australian politician nor naval officer would accept, and predictably, back in Melbourne Creswell rejected the proposals. In direct reply to the strategic arguments put forward by Imperial officers Creswell observed:

> The Admiralty view has not the complete advantages which it claims for it. And although Admiral Haworth-Booth [Australian Naval Representative in London] is quite correct saying that we in Australia 'have a real natural aspiration' for developing our Naval capabilities it also has true strategy behind it. True strategy calls for centres of creative or producing Naval power at distant points in the Empire rather than perpetuating the grave disadvantages of depending on one central point situated possibly half the world's circumference from where it might be needed.[28]

Creswell thus played an influential part in the determination of a post-war, Australian-controlled, naval defence.

While Creswell and the Naval Board were successful in keeping an Australian-oriented naval defence outlook to the fore they encountered considerable difficulties in other areas from 1915 to the end of the war. These difficulties were of such magnitude as to call into question the competence of the Board and the First Naval Member. At the core of the problems which threatened to engulf the Navy Office in 1918 were the peculiar wartime status of the Board, the millstone of ministerial mediocrity and the unhealed wounds left by the Hughes-Onslow affair.

As previously discussed, the Naval Board, under the agreement reached between the Australian and Imperial governments, operated as a Commander-in-Chief Home Ports in war. The Board was thus placed in the position of reporting directly to the Admiralty in London while remaining a creature of the Australian government. As the official historian, A.W. Jose, observed: 'Stated on paper, this dual control obviously afforded unlimited opportunities for friction; and even the pettiest sort of friction might easily have imperilled the whole arrangement'.[29]

Though Jose concluded that there was, in practice, very little friction there was a good deal of evidence to the contrary. It was true that matters did not erupt during the days of action in the first six months of war, but there were problems from thereon. The problems were not between the Naval Board and the Admiralty, but between the Board and officials in Australia. From its inception in 1911 Creswell had supported the agreements on control. He reasoned that such a close connexion was essential if the Australian Fleet Unit were to function as a part of the Royal Navy in war. But from 1915 onwards the Naval Board came up against a formidable opponent in Sir Ronald Munro-Ferguson, the Governor-General from 1914 until 1920, who did not accept this unusual situation.

Backed sporadically by the Prime Minister, Munro-Ferguson repeatedly objected to direct communications between the Board and the Admiralty.

At various times the continual questioning of the propriety of direct lines of communication contributed substantially to the Board's difficulties. For example, the sinking of the merchant vessel, *Cumberland*, off Gabo Island in 1917 led to a confrontation over the matter. At the time concern was mounting in government circles over the risk posed by the actions of the Industrial Workers of the World. Information provided to the Naval Board indicated that the explosion which disabled the *Cumberland* had occurred internally. This led the Board, in consultation with the Admiralty, to take steps to protect wharf and dock areas from saboteurs.

As these actions were being taken, Munro-Ferguson's official secretary, Major George Steward, who was also acting as head of a counter-espionage organisation, had received information indicating that the explosion was external.[30] The Governor-General complained bitterly to Hughes about direct communications between the Board and the Admiralty while Australian officials were not kept informed. In fact, the Board had kept Major Steward informed of their information; it was Steward who did not inform the Naval Board of his contradictory findings.[31] As a result the Board continued to work on the internal explosion theory, leading Munro-Ferguson to question its competence, and, indirectly, to weaken its standing within government circles.

Munro-Ferguson's unfavourable conclusions on the Naval Board's performance during the *Cumberland* incident were at least partially based upon a personal confrontation with that body and on the reports of his official secretary who had his own axe to grind. Even though the path that led to his critical conclusions was not lined with objectivity, they were, in a general sense, substantially correct. But the source of the problems of the Board lay much less at the feet of Creswell and the other professional members than the Governor-General's comments suggested.

Earlier, in July 1915 when the Royal Australian Navy became a separate ministry, Munro-Ferguson had accurately identified the root of subsequent troubles. Commenting then upon the restructuring of defence administration he labelled the newly appointed Navy Minister as 'quite incompetent',[32] an assessment fully borne out by subsequent events. Jens August Jensen came to the Navy portfolio after spending nearly a year assisting Pearce. In that capacity he had been present at a number of Naval Board meetings and was acquainted with the personnel and administrative procedures.

In spite of this familiarity with the Navy Office, or perhaps as a result of it, Jensen as minister progressively pursued a course destined to reduce the effectiveness of the Board and thus justify many of the critical comments of Munro-Ferguson. Jensen took no real lead to ensure cohesion among the professional members of the Board. During his period as minister the old differences between Creswell and Clarkson resurfaced, and Jensen did nothing to resolve them. On the contrary, he deliberately chose to exploit the differences, with the net result that the Board went into a sharp decline as a decision-making body.

An example of Jensen's negative performance occurred when the Navy Department was considering the purchase of an old steam vessel, the *Emerald*. When inspected by the Acting Director of Naval Works in March 1916, the vessel was found to be unsuitable. The matter was considered closed, but two months later the vessel was offered to the Navy Department on hire. At the direction and in the company of Jensen the Acting Director of Naval Works again inspected the vessel, and this time reported that with substantial renovations the *Emerald* could be passed for naval work. On the following day Jensen entered into an agreement for purchase.[33]

It was not until after the agreement had been signed that Creswell or any other professional member of the Board learned of the reversal. When Creswell found out he immediately registered a protest with Jensen. He knew that the *Emerald*—even considering the great need for patrol vessels—was entirely unsatisfactory for virtually any type of naval work. In fact, he was so concerned over the implications of the minister's action that his protest took the extreme form of an offer to resign. Though his resignation was not accepted, the purchase of the *Emerald* proved a costly blunder for the Navy Department.

There was another, more extravagant example in 1916. Jensen again resorted to the tactic of relying upon the qualified support of a junior officer for justification to go ahead with a purchase. In this case he authorised the purchase of the Shaw Wireless Works in Sydney. Even though Creswell and Clarkson were not on good personal terms Jensen's high-handedness resulted in concerted action from the naval members. In a Board meeting of 29 July 1916 they protested to the minister over the purchase. Jensen replied that 'he did not want to interfere with the spheres of members or with the Board's powers as exercised in the past. At the same time he reserved the right to adopt a course even if the Board did not agree with it.'[34] The all-too-brief minutes of that meeting did not indicate whether Creswell or any other member pointed out to Jensen that he had not, in fact, even consulted the Board about this action.

The belated showdown between Jensen and his naval members in 1916 was the penultimate appearance of the minister at the Navy Board. From that confrontation until he was moved to Trade and Customs in 1917 the Naval Board met a total of fourteen times. Of those meetings Jensen was present at only one.[35] The Board's decline was thus accelerated by ministerial neglect.

The shift away from decision-making through the medium of the Board was matched by a corresponding increase in the compartmentalisation of the professional members. Creswell turned more and more to the specific responsibilities of the First Naval Member, that is plans, intelligence, operations and naval development. By 1917 the Naval Board had devolved into a body much different from its original form.

The troubled state of affairs at the Navy Office became public knowledge in 1918 when the Royal Commission on Navy and Defence Administration turned its attention from the considerable problems of the Defence Department to those of the Navy Department. The Commissioners diagnosed the causes of the ills of the Navy Office and the Navy Board as: first, the Naval Board had failed to fulfil the role of controlling body of the Navy Department, due largely to the actions of the minister in charge; second, the long-running conflict between Creswell and Clarkson had further prejudiced the chances of successful operation; and third, the failure of the Federal government to appoint a finance member to the Board, as recommended in the Henderson Report, had permitted wasteful use of public funds. A full restructure of naval administration was recommended.

One other recommendation directly affected the First Naval Member. Early in the hearings the Commission learned that Creswell was past the mandatory retirement age for a Rear-Admiral in the Royal Navy.[36] For this reason and Creswell's own expressed desire to retire the report recommended replacement. In the early stages of the enquiry Creswell had betrayed little discomfort at coming under examination. He was confident that he had acquitted himself well.[37] The prospect of retirement did not trouble him. The strains of heavy wartime responsibility, the internal difficulties of the Board and the crushing personal loss of two sons in 1917 had all taken their toll. He realised that he could not stay on to implement plans for the post-war Navy.[38] The time for a change was close at hand and in characteristic fashion he faced it squarely.

In the aftermath of the Royal Commission's report, Jensen, who had been transferred to Trade and Customs in 1917, was dismissed after refusing to step down voluntarily. Creswell had to endure a run of critical comment in the press. Much of it was unfair, given his attempts to resist the high-handed actions of his minister and his willingness to get on with his own responsibilities under difficult circumstances.

Characteristically, Creswell got back to work at the Navy Office while Jensen embarked upon a fruitless pilgrimage to London to plead his case to Hughes, who was attending the Versailles Conference. Creswell concerned himself in the early months of 1919 with preparations for the forthcoming visit of Lord Jellicoe to examine the naval defence of Australasia. As in 1911 with the visit of Admiral Henderson, it was Creswell who welcomed Jellicoe to Australia, but this time the arrival of a naval expert signalled an end not a beginning. He relinquished office on 14 August 1919.

The problems experienced by Creswell and the Naval Board were largely the result of ministerial incompetence. The poor personal relations between Creswell and Clarkson ran a distant second. But these negative aspects were only one side of Creswell's wartime career. Balanced against them was the important work done by Creswell and his staff within the realm of the First Naval Member. The establishment of an effective intelligence network, the most effective within the Australian sphere, was one of the major achievements of his period of office. The utilisation of the data collected and the experience gained were channelled into the planning of post-war defence. Creswell's commitment to making the best use of technological advances was another positive mark.

The final word upon Creswell's wartime performance lay with his stance upon the future relationship of the Royal Australian Navy and the Royal Navy. With admirable consistency Creswell opposed the Admiralty's recommendations of reviving the concept of one great navy. The future of imperial and Australian defence, in his view, was in the direction of cooperation among a group of independent navies. The passage of time has underlined the validity of Creswell's vision.

# General Sir Harry Chauvel: Australia's First Corps Commander

A.J. HILL

The first Australian officer to command a corps and to become a lieutenant-general was Harry Chauvel (1865-1945).[1] He was a regular soldier with scarcely 21 years' service when he succeeded Lieutenant-General Sir Philip Chetwode in command of the Desert Column on 21 April 1917.[2] Like Chetwode, he had begun as a volunteer when, during the Russian scare of 1885-86, his father raised the Upper Clarence Light Horse. Born at Tabulam on the Clarence River in 1865, Chauvel became an expert horseman and bushman; even as a schoolboy he would ride 150 miles (240 kilometres) from Tabulam to his boarding school in Toowoomba where the boys were encouraged to spend weekends camping and living off the country. Chauvel's ambition to enter the British army could not be realised but he soldiered on in the Darling Downs Mounted Infantry while managing the family's remaining property on the Darling Downs. Opportunity came suddenly in 1896 when the Queensland Defence Force needed an adjutant for the Moreton Regiment and Chauvel's application was accepted.

Service in the Queensland Defence Force offered few chances of professional development but his visit to England with the Queensland Diamond Jubilee contingent in 1897 enabled Chauvel to qualify at the School of Musketry, Hythe, and to serve with two regular infantry battalions for a few months. As he was too senior for selection for the Staff College when places were first made available to Australians in 1905, Chauvel's English interlude provided his only professional training apart from the exercises and staff rides he set or attended after the South African War. He was conscious of his lack of formal training but made up for it as far as he could by the study of military history, especially the operations of mounted troops in the American Civil War.

In the South African War, Chauvel served as a company commander in the Queensland Mounted Infantry and for a time was adjutant. He was at Sunnyside, the relief of Kimberley and the crossing of the Vet River where, with a few soldiers, he captured a machine gun. After the occupation of Pretoria and the battle of Diamond Hill, he commanded Chauvel's Mounted Infantry, a small independent force of his own Queenslanders, New Zealand Mounted Rifles, Canadians and British Mounted Infantry and Yeomanry with four guns. Returning to Brisbane in January 1901, he was eager to go back to South Africa but it was not until May 1902 that he was given command of the 7th Australian

Commonwealth Horse—only to reach Durban three weeks after the end of the war. For his work in South Africa Chauvel was mentioned in despatches, made CMG and given the brevet of lieutenant-colonel in December 1902.

Service in South Africa was for Chauvel a liberal education in man-management and the care of horses, and he sought to pass on his knowledge to the young officers of the new Australian Army.[3] Except for a short period in South Australia reorganising the SA Mounted Rifles as Light Horse, most of Chauvel's inter-war service was in Queensland where he was Acting Chief Staff Officer, 1904-11. He was appointed Adjutant-General towards the end of 1911 and promoted colonel in 1913. Brigadier-General W.T. Bridges, about to become Inspector-General, wanted Chauvel to succeed him as commandant of the newly opened military college at Duntroon, but Chauvel preferred to go to the Dominions Section of the Imperial General Staff in London. With his wife and children he was on his way there when war broke out in July 1914, and on arrival he learnt that he was to command the 1st Light Horse Brigade which would accompany the infantry division offered by Australia.

*Gallipoli*

Chauvel sailed for Egypt in November with Major T.A. Blamey. His brigade was moving into camp at Maadi when he took command. There followed a period of hard training marked by his insistence on formal discipline and attention to detail of which C.E.W. Bean later wrote: 'The punctiliousness of dress and bearing which the infantry acquired in France was required by him of his men from the beginning and the 1st Light Horse Brigade showed throughout the war the results of this early training'.[4] Early in 1915, Chauvel moved his brigade to Heliopolis where it trained with the New Zealand and Australian (NZ & A) Division under Major-General Sir A.J. Godley.

When the Australian and New Zealand Army Corps landed on Gallipoli, the four Light Horse brigades and the New Zealand Mounted Rifles (NZMR) Brigade remained in Egypt. General Birdwood's request for a thousand volunteers from the mounted regiments to reinforce his depleted divisions was strenuously resisted by Chauvel and the other brigadiers; their demand to go as units, although dismounted, was eventually accepted and Chauvel landed at Anzac on 12 May. Godley placed him in command of No. 3 Section of the Anzac position, including Pope's Hill and Courtney's and the notorious Quinn's Posts, a crazy collection of ill-dug trenches in tactically impossible positions. The section was held by Monash's 4th Brigade and a remnant of Marines who were relieved by the Light Horse. When Monash took Chauvel around the posts before handing over, it must have been a sobering experience for a brigadier who had prepared himself and his brigade for mobile operations. Like Monash and the others, he would have to learn on the job, a process which can be very expensive in war. There had been no briefing for Chauvel before he left Egypt and now suddenly he was

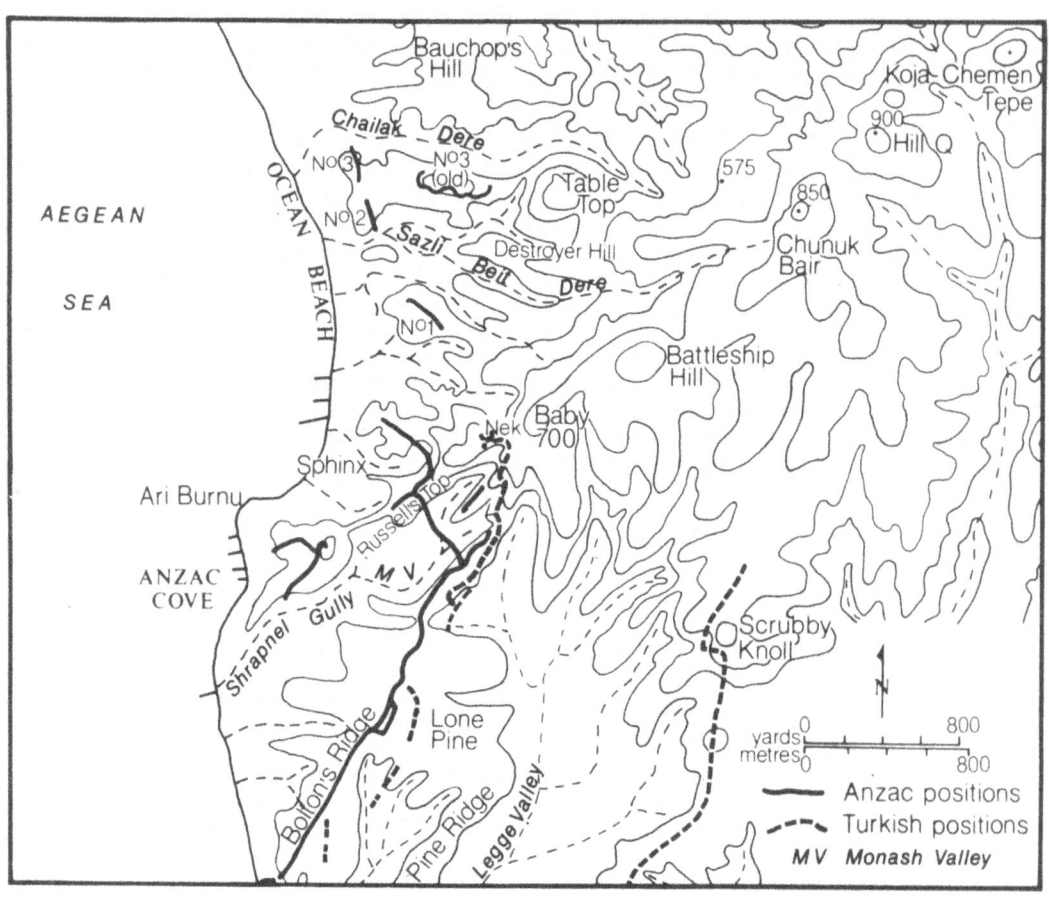

*1* Gallipoli. The Anzac positions, May 1915

commander of the most perilous position at Anzac.

It was also a stiff test for the two brigadiers whose one common ground was their dedication to soldiering. Why Godley preferred Chauvel to Monash is not apparent; it may have been a regular's preference for another regular or perhaps a distaste for Monash's strong objections to his planning for the unsuccessful assault on Baby 700.[5] Be that as it may, the two men managed pretty well until 1 June when the heroic 4th Brigade was relieved. Monash unburdened himself to his diary on 17 May recording Chauvel's 'interference with internal administrative matters...He is always grateful for my help, when he gets into a tangle. [He] is very fidgetty, and frequently calls to consult me during the night.'[6] On 20 May after the defeat of the last general attack by the Turks, Chauvel wrote to his wife: 'I am living cheek by jowl with Monash here and find he is really a very fine soldier. He has been of very great assistance to me the last few days, and very willing indeed considering I have ousted him out of the command of the sector.'[7] The

7 Chain of command: Brigadier-General Chauvel with Major-General Godley (in helmet) and Lieutenant-General Birdwood (hatless), probably in Monash Valley, May 1915 (AWM Negative No. H 15753)

difference in the tone of these two quotations is revealing.

From their positions on the high ground around Baby 700, Dead Man's Ridge, the Chessboard and the Bloody Angle, the Turks dominated much of No. 3 Section, firing down the valley where Chauvel and Monash had their HQ and even into the back of Quinn's Post. So close was the front line of Quinn's to the Turks that it was under constant attack by grenades. As it was only ten minutes from Section HQ to the beach, a Turkish lodgement would endanger the whole of Anzac, and both sides knew it.

Chauvel held this position from 12 May until his brigade was relieved in early September, except for about six weeks in June and July when he was in hospital and convalescing in Egypt. He brought order to the defensive arrangements in No. 3 Section by appointing permanent commanders and staffs for the three posts (it may well have been this reorganisation which led to Monash's diary entry), and when the 4th Brigade was relieved by the NZ Infantry Brigade at the end of May he seized upon an idea for suppressing the Turkish snipers. The Wellington Battalion had formed an anti-sniper group under a subaltern; Chauvel sent for him and 'gave him carte blanche to select as many marksmen as he liked out of the two Brigades... He organised a party of about thirty and planted them... about the area and got the snipers down to such an extent that latterly we could actually get mule convoys up and down the

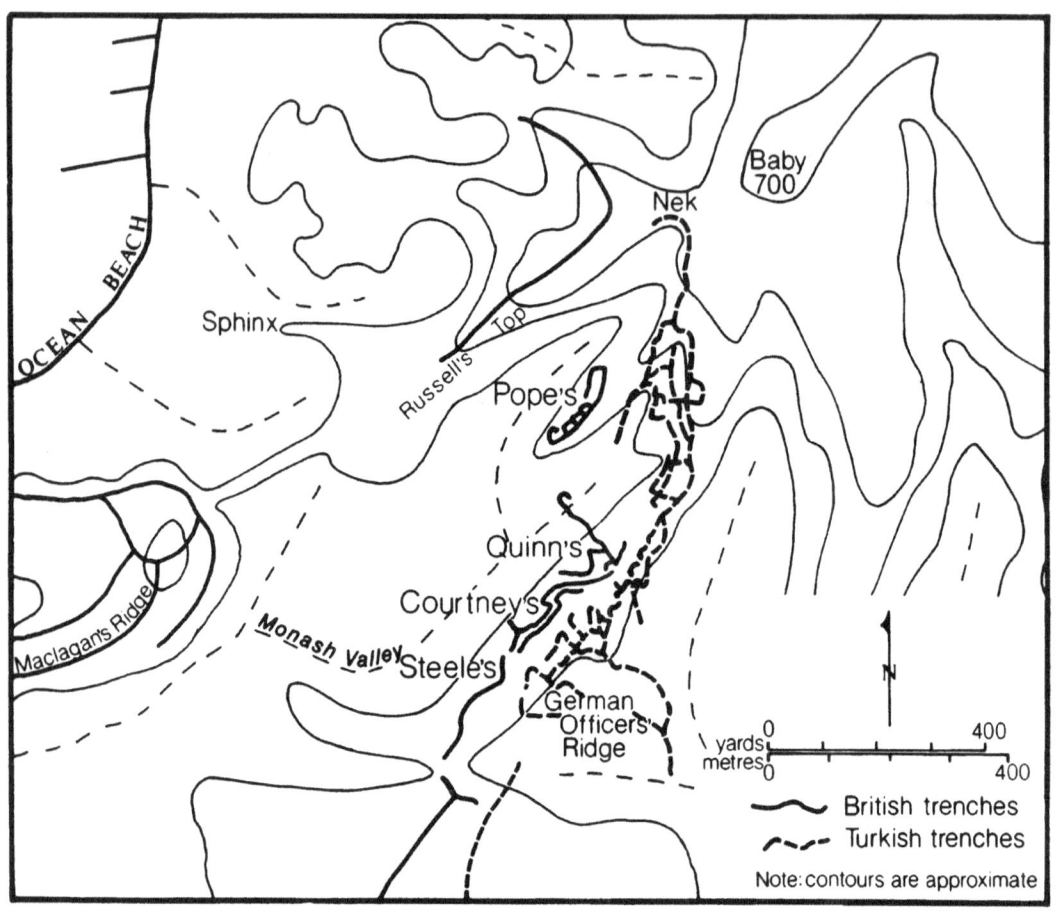

2 Quinn's Post, May–August 1915

valley in broad daylight.'[8]

It seems that Chauvel took time to acclimatise to siege warfare at Anzac. Birdwood at first thought he was slow and lacking in decision but in two major actions at Anzac, in which he influenced the outcome, his performance was exactly the opposite. On 29 May at 3.20 am the Turks fired a mine under Quinn's and poured men into the consequent confusion, penetrating along the trenches each side of the crater. Chauvel ordered the reserve battalion, the 15th, to eject the Turks from Quinn's—which was held by the 13th. As it happened, the commanding officer of the 15th was away recuperating and the commanding officer of the 13th was wounded soon after the leading company of the 15th reached him. Chauvel therefore ordered Pope, commanding officer of the 16th, to take command at Quinn's. He had already asked the artillery to bombard the area in front of Quinn's to prevent the Turks from reinforcing their assault troops and he sent a company to prevent any attempt at penetration between Quinn's and Pope's. Then, leaving

Monash in command of the troops in the valley, he went up to Quinn's to see for himself.

He soon found Pope and Durrant, the acting commanding officer of the 13th. Having questioned them and being unwilling to allow the Turks time to tighten their hold, Chauvel ordered Pope to counterattack. He agreed to delay the assault but insisted that it must be made. When the time came, the Turkish fire suddenly ceased as they, too, were attacking; aided by this lull the 15th seized their objectives. By about 8 am the surviving Turks, having been cut off, were persuaded to surrender. Writing privately to Bean in 1922, Durrant described the action, concluding: 'Amidst all this confusion and excitement there was no soldier whose demeanour was cooler than General Chauvel's. He stayed with us for hours until the situation was adjusted to his satisfaction. I remember being particularly struck with this at the time and so were many others.'[9]

Chauvel's handling of this action showed no lack of decision but rather a strong grasp of the tactical situation and the ability to manoeuvre the available troops so that he could regain the initiative. He has been criticised for persisting in ordering the charge made by companies of the 13th and 15th Battalions.[10] He did not, as P.A. Pedersen claims, ignore the arguments of the local commanders, nor could he or anyone else have been aware that only seventeen wounded Turks were left in the captured part of the post. Chauvel was aware of Godley's orders to regain the position and he must also have been expecting the Turks to try and exploit their initial success as indeed they were about to do. According to Pedersen, 'Chauvel's conduct reflects the uncertainty of a commander in his first major action', which is strange in the light of the facts: he had sent the 15th to the aid of the 13th; he had arranged for the artillery to shell the attackers' communications; he had moved troops to cover his flank; he had sent a senior officer up to take command and had gone forward himself. When he had obtained information from the local commanders he gave his orders. His decision was, in the eyes of some, a wrong one but it was a decision and it was acted on. It was also a lucky one.[11]

The other occasion on which Chauvel's decisiveness was important and in stark contrast to the command of 3rd Light Horse Brigade, was on 7 August when the 3rd was to attack Baby 700 across the Nek simultaneously with attacks by the 1st Light Horse Brigade on the Chessboard and Turkish Quinn's. The tragic and futile assault by the 8th and 10th Regiments has obscured the disaster that befell their sister regiments, the 1st and 2nd, who suffered 80 per cent casualties compared with 90 per cent in the 8th and 10th. That the task of the latter was impossible is unquestionable, whereas part of the 1st Regiment had a covered approach to their first objective. The 1st Regiment actually fought its way into the third Turkish trench but the first wave of the 2nd Regiment's attack from Quinn's was destroyed but for one man. Major Bourne, in charge of the attacking parties, thereupon ordered the second wave to stand fast in their trench and reported the situation to his commanding officer, who decided that they should desist. Chauvel confirmed that decision. Had he insisted on further attempts the slaughter at the Nek must have been repeated at Quinn's.

Chauvel clearly did not impress Birdwood in the early days. Writing in June to Sir Ian Hamilton, the Commander-in-Chief, he complained: 'From the time of his arrival here, he has always struck me as taking the gloomiest view of everything connected with the Expedition and he never seems to put life into things as he might'.[12] Obviously Birdwood had no knowledge of the way Chauvel had handled the crisis at Quinn's on 29 May, and Chauvel seems to have shown a degree of candour unwelcome to the corps commander. By contrast, the adjutant of the 3rd Light Horse, S.F. Rowell, a future Chief of the General Staff, remembered a different brigadier: 'In such a confined area [No. 3 Section] we had the Brigade HQ under pretty close scrutiny. To our eyes, at least, the Brigade Commander was never fussed and appeared the model of what a commander should be.'[13] Birdwood must have learnt more about his man as he sent him to administer command of the NZ and A Division for a short time in September and again in October. Then on 6 November, Chauvel was given command of the 1st Division which he led until March 1916 when he took command of the only mobile division in the theatre, the Australian and New Zealand Mounted Division. (This was his second multi-national command as it included Australian and New Zealand mounted regiments and British artillery.) Even then Birdwood again offered him the chance of taking an infantry division to France but he was determined to remain with his ANZAC horsemen.[14] Chauvel was promoted major-general in December 1915 and in January 1916 was appointed CB.

*Clearing Sinai*

When the 2nd ANZAC Corps left for France in June 1916, Chauvel became GOC AIF in Egypt with a small administrative headquarters in Cairo. This imposed a heavy burden on a divisional commander who, from April 1916 to January 1917, was to maintain contact with the enemy and carry the offensive from the Canal defences to the approaches to Gaza. Chauvel's position was unenviable. Operationally he was responsible to a British commander, first to Major-General H.A. Lawrence for the period at Romani, then to Lieutenant-General Sir Philip Chetwode for the advance to Gaza. As GOC AIF in Egypt he dealt direct with GHQ in local administrative matters but was chiefly responsible to the GOC AIF in France, Lieutenant-General Sir William Birdwood. Promotions and postings had to be confirmed by Birdwood who handled the administration of the five Australian divisions on the Western Front, directed the operations of the 1st ANZAC Corps and administered the AIF in Egypt. There were difficulties but the system worked and despite its frustrations, Chauvel remained firmly loyal to Birdwood. As he wrote to Gullett in 1921: 'I carried on for nearly three years under two masters, which would have been impossible had one of them been anyone else than Birdwood'.[15] When Sir Edmund Allenby

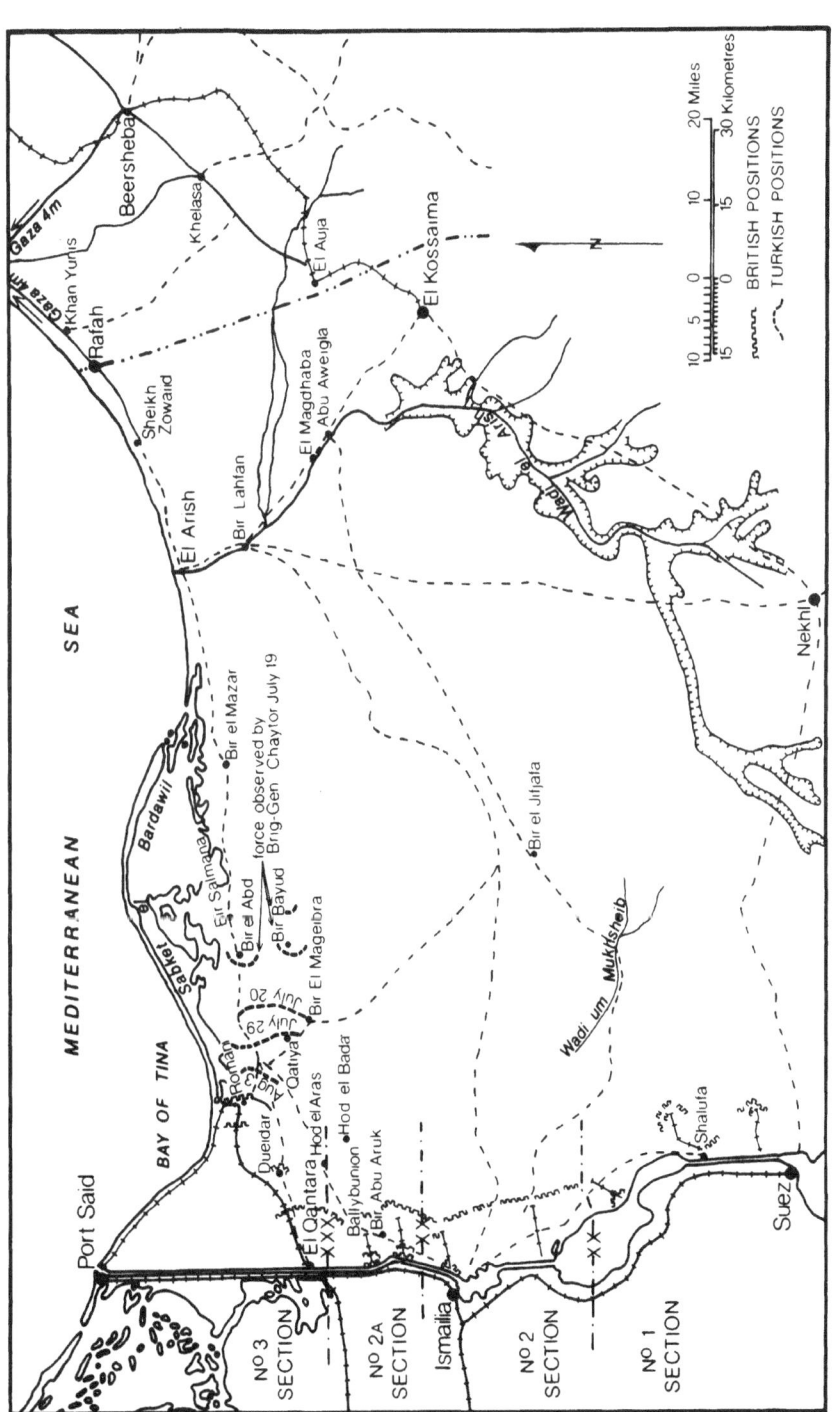

3 The Canal Defences and the Turkish advance July–August 1916

succeeded Sir Archibald Murray as Commander-in-Chief of the British Mediterranean Expeditionary Force in June 1917, he objected to Chauvel's subordination to Birdwood and moved to end it. However, Chauvel refused to have anything to do with Allenby's proposals and warned Birdwood of them. When he found that he had no support from the War Office, Allenby let the matter lapse.

In 1916 Chauvel played a major role in the defeat of the Turkish thrust for the Suez Canal and in clearing Sinai of enemy forces. He stopped and broke the Turkish attack at Romani on 4 August, counterattacked and pursued the Turks to Bir el Abd, 6–11 August; later he destroyed the Turkish garrisons at Magdhaba on 23 December and Rafah on 9 January 1917. Thus he cleared the way for the advance of Eastern Force to Gaza.

If battles were to be judged mainly by the numbers engaged, Romani would be of small account. Fought in blistering heat (with shade temperatures around 50°C) by four mounted brigades with help from the remnants of the 5th Yeomanry Brigade and a British infantry brigade, it ended the Turkish threat to the Canal and wrenched the initiative from a stubborn and well-commanded enemy. Chauvel's part in this affair was important. The concept of the battle appears to have been shaped by Murray himself, although not without serious disagreement with Major-General H.A. Lawrence, the corps commander who would fight the battle. Murray developed a strong position, held by the 52nd Division (Major-General W.E.B. Smith), between the sea and the railhead at Romani, the refused right flank of which was to be held by the ANZAC Mounted Division. The Commander-in-Chief reasoned that the enemy was unlikely to attempt a frontal assault on the prepared position but would rather try to envelop its right flank while putting in a containing or holding attack elsewhere. Murray wanted to delay and wear down the flank attack until the moment came to attack the enemy's flank using mounted troops, some of which were to be detached from Chauvel's division.

The shortage of mounted troops in Egypt was such that GHQ had detached two of Chauvel's four brigades for duty as Section mobile troops leaving him with only two brigades;[16] both were under strength and weary owing to constant and protracted patrolling and frequent clashes with the advancing Turks. Chauvel, after careful reconnaissance, chose a position from Katib Gannit to Hod el Enna, on which to delay the Turkish envelopment, together with a fall-back position. Both positions were reconnoitred by the officers of the 1st and 2nd Light Horse Brigades, gun positions were chosen by the artillery and telephone lines laid, but no defences were dug for fear of observation from the air. On 3 August it was apparent that the Turks were about to attack so Chauvel ordered the 1st Brigade to take up their positions at dusk. As on 29 May 1915 his luck held, for the delaying position he had selected had also been chosen by Kress von Kressenstein, commanding the Turks, as the forming up place for his attack!

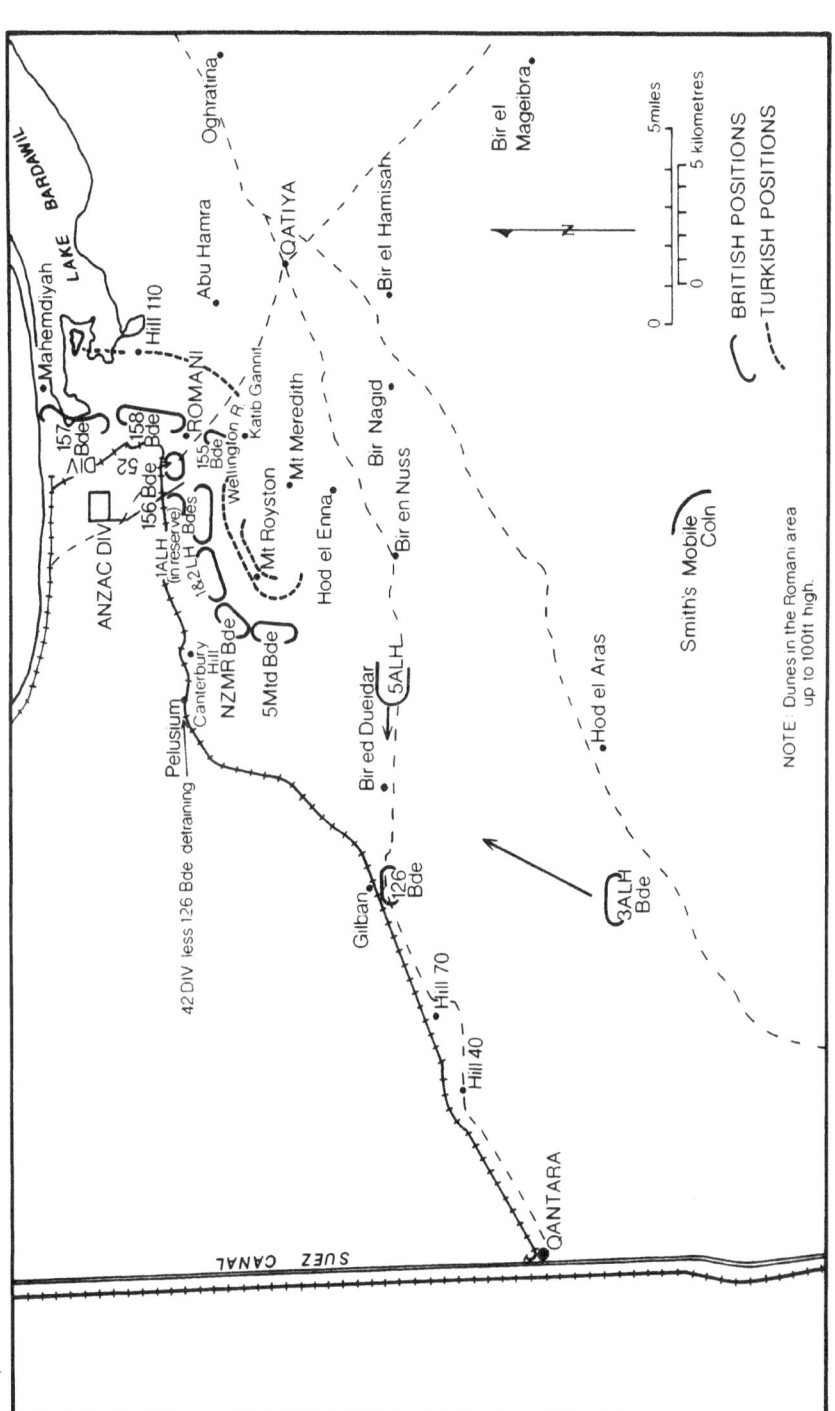

4 The Battle of Romani

After an hour's delay owing to this upset the Turkish attack developed at about 2 am and soon sheer weight of numbers began to tell as the 1st Light Horse Brigade had no more than 800 rifles available. They fell back to their second position, troop covering troop. Chauvel had warned the 2nd Light Horse Brigade to be ready to go to help the 1st, but he would not move them until daylight gave him a clearer view of the situation which he obtained from the headquarters of the 1st Brigade on Wellington Ridge. He then ordered Brigadier-General Royston to extend the 1st Brigade's line westwards with two regiments. Although these troops were quickly in position the threat to the exposed flank increased until deep sand and increasing heat quenched the *élan* of the Turkish infantry. Thirsty and exhausted, they stumbled to a halt at 8 am.

In the course of that long, hot day Chauvel could do little to swing the battle against the Turks, if only because two of his four brigades, the New Zealanders and the 3rd, had been detached by Murray and positioned to attack the enemy's flank. Twelve and twenty miles back respectively (19 and 32 kilometres), and under Lawrence's command, they were not put in motion until after 7.30 am. Chauvel asked the reserve infantry brigade commander of the 52nd Division to take over his position so that he could counterattack, but was refused as that officer had been warned for a possible counterattack by the 52nd. Luckily, a nearby battalion commander on his own initiative occupied ground held by the Light Horse, helping Chauvel to extend on the threatened flank; he later sent two companies across when a further extension by the Light Horse became necessary.

The prime cause of this faltering at a crucial stage of the battle was the location of the corps commander at Qantara, 23 miles (37 kilometres) away. One of his two telephone lines had been cut at 7 am, leaving only a bad line through Port Said. Lawrence did not go forward to coordinate the actions of the two divisions even after the direct line was cut. Meanwhile the Turks were beginning to dig in, and it was not until the early afternoon in fierce heat that the New Zealanders began to come into action with strong support from the 5th (British) Mounted Brigade. Stricken by thirst and casualties, about two battalions of Turks surrendered around 6 pm. Chauvel had been able to obtain General Smith's agreement to attack southwards with his reserve brigade on the left of the ANZAC Mounted Division, but they were so late in moving that Chauvel stopped his own advance in spite of its progress and a good haul of prisoners. In view of the darkness, the rough ground and the weariness of his men, this was a wise decision.

The corps commander did not use the night to relieve the 1st and 2nd Light Horse Brigades with infantry from the 52nd Division, but ordered a frontal attack by Chauvel with support on the flanks from the 52nd Division and the newly arriving 42nd. Lawrence also ordered the 3rd Light Horse Brigade to advance on Bir en Nuss and attack towards Hod el Enna. Chauvel's advance, together with that of the 52nd Division, began at dawn. Very soon the Turks were surrendering in masses for they were in a desperate condition and their main body was already in full retreat; as Chauvel wrote later: 'It was the empty Turkish water bottle that won the battle'.[17] He received orders about 6.30 am to take command of all mounted troops in the area and pursue the enemy, but it was 10.30 am before the scattered brigades could be organised and the advance begun. Chauvel ordered the fresh 3rd Light Horse Brigade to ride for Hamisah and get in on the flank of the Turks. This very nearly came off when the 9th Regiment attacked boldly, taking over 400 prisoners, but shelling by 5.9s caused Brigadier-General Antill to withdraw his brigade. Kress von Kressenstein, the enemy commander, conducted 'a most masterly withdrawal'—Chauvel's phrase[18]—falling back on the entrenched positions dug during his advance and showing marked aggressiveness whenever Chauvel's light forces came within reach. He failed to turn their flank but the Turks continued their withdrawal until their force was clear of Sinai. They left behind about 4000 prisoners, and over 1200 of their men were buried on the battlefield of Romani. The number of their wounded is unknown.

After the pursuit Chauvel visited each brigade and thanked them for the way they had fought; it became his custom throughout the campaign to speak to troops who had distinguished themselves or suffered heavy casualties. It was his own division which had suffered at Romani; of the 1130 casualties over 900 had been incurred in ANZAC Mounted whose 1st and 2nd Brigades had born the brunt throughout. 'That the brunt of the fighting fell on my troops was exactly what I anticipated and was our luck and we would not have had it otherwise.' Writing to H.S. Gullett just after the war he said:

> Romani was the first decisive victory attained by British Land Forces (excluding, of course, the campaigns in West Africa) and changed the whole face of the campaign in that theatre [Egypt and Palestine] wresting as it did from the enemy the initiative which he never again obtained. It also made the clearing of his troops from Egyptian territory a feasible proposition.

H.S. Gullett's claim that 'so that as leadership went, Chauvel won Romani singlehanded' is no exaggeration. He came calmly through the test of controlling a withdrawal under pressure and saved a threatened flank from envelopment, riding from brigade to brigade under fire. That Murray understood his and his division's worth was implicit in his letter to the Chief of the Imperial General Staff after Romani: 'These Anzac troops are the keystone of the Defence of Egypt'.[19]

In the next phase of this campaign, Chauvel was responsible for the destruction of the Turkish outposts in northern Sinai in a series of operations in December 1916 and January 1917 conducted by the Desert Column. Under Lieutenant-General Sir Philip Chetwode, an officer of outstanding ability, Chauvel had a free hand in the advance to El Arish, the reduction of Magdhaba and the capture of Rafah. These operations were remarkable for their long approach marches by night over waterless country, for the accuracy of their navigation and for the skill and determination of the troops.

Fortunately the Turks evacuated El Arish just as Chauvel was preparing to seize it with ANZAC Mounted and the Imperial Camel Corps Brigade.[20] Magdhaba was taken on 23 December after a sharp fight in which, owing to the absence of water nearer than El Arish and the stubborn resistance of the garrison, Chauvel decided to break off and march back to his base. At this critical moment and unknown to Chauvel, the 1st Light Brigade was preparing to assault; the commander, C.F. Cox, ignored the order to withdraw and his assault went in. Its quick success turned the tide and in the dusk the Turks surrendered. At Rafah on 9 January 1917 there was a similar crisis aggravated by the approach of strong enemy reinforcements. This time Chetwode had taken the field although he left Chauvel to plan and direct the operation. Chetwode knew that the whole force was committed and that the artillery and machine guns were seriously short of ammunition owing to his decision to leave ammunition vehicles to come on after the main body. He decided to break off the action and Chauvel was preparing his orders when part of the enemy defences was captured by the New Zealanders. Another position fell to the Camels and again the Turks gave up in the failing light. Rafah was a costly success; there were nearly 500 casualties compared to 146 at Magdhaba, but the way was now clear for an advance on Gaza.

Chauvel's decision to break off the action at Magdhaba was criticised at the time as lacking in resolution; that view was held not by Chetwode or Murray but was held at the brigade level. Chauvel's conduct at Quinn's and Romani had shown no lack of determination; but at Magdhaba Chauvel was influenced by his engineers' report that no water was available and the knowledge that his division was the only desert-worthy mobile formation in Eastern Force. Aware of the mounting difficulties of finding reinforcements for the AIF in France as well as for his own force, he could not incur serious casualties unless for a much greater prize than an isolated Turkish garrison of scarcely 1400. He would have agreed with Wavell's acid comment to Churchill: 'a big butcher's bill was not necessarily evidence of good tactics'.[21]

A few days after Rafah Chauvel was made KCMG. This was in recognition of his services and the gallantry of ANZAC Mounted 'at El Arish, Magdhaba and Rafah'. No mention of Romani where *he* had the star role but the KCB went to Lawrence![22]

*Gaza and Beersheba*

The commander of Eastern Force, Lieutenant-General Sir C. Dobell, gave Chetwode the task of seizing the isolated city of Gaza by a swift *coup de main* towards the end of March. Chetwode planned to use the 53rd Division to assault Ali Muntar, the key to the defences, giving Chauvel the task of preventing the escape of the garrison by getting across their communications. The task of covering his flank against interference from the northeast and east was given to the newly organised Imperial Mounted Division composed of Australian and British brigades.

The first battle of Gaza opened on 26 March 1917. ANZAC Mounted moved off at 2.30 am, crossed the Wadi Ghazze in dense fog and advanced northwards on Beit Durdis over open, firm ground. By 7.30 am the fog was clearing and by 11 am Chauvel's troops had cut the main road out of Gaza and had reached the sea. The Turks were surrounded and the amazed commander and staff of the Turkish 53rd Division were captured in their carriages as they drove towards Gaza. By contrast the fog had delayed the infantry advance; their attack did not start until almost noon when all surprise was lost. Chetwode on learning from a prisoner that the Gaza garrison was six battalions, not the two he had expected, placed the Imperial Mounted Division under Chauvel's command and ordered him to attack Gaza from the north with his own division.

It took Chauvel more than an hour to reorganise for this attack, but he was able to launch three brigades against Gaza by 4 pm and by dark all had entered the outskirts of the town. But even as this was happening, Chauvel could hear the sound of rifle and machine gun fire from the east where the Imperial Mounted Division was holding off Turks advancing to the relief of Gaza. Chetwode, deciding that the conditions for a successful *coup de main* no longer existed, ordered all troops to withdraw. Chauvel's protests were overruled and he turned to the business of concentrating his division in the darkness and organising the withdrawal of the entire mounted force. This was achieved in style; all guns including those captured, the wounded, all fit prisoners and even the few dead were brought away. Chauvel himself saw to it that some badly wounded Turkish prisoners were given medical attention and left close to the road, each man with a full water bottle.

It was a frustrating and exhausting battle for all divisions, more especially the 53rd which had captured Ali Muntar with heavy loss. ANZAC Mounted was furious at being withdrawn when they believed they were in a position to take a surrender in the morning. Chauvel, who was deeply disappointed, explained the failure in a letter to Birdwood but concurred in the wisdom of withdrawing in the face of the advancing reinforcements.[23] His own performance had been exemplary and was immediately and warmly acknowledged by Chetwode.

In the second thrust for Gaza Chauvel's role was minor but the aftermath of that disastrous battle brought him great opportunities. Dobell, the commander of Eastern Force, was replaced by Chetwode and Chauvel took command of the Desert Column. The return of two Yeomanry brigades to Egypt enabled Murray to reorganise the Desert Column into three divisions, each of three brigades, an enviable appointment for the first officer of any dominion army to command a corps. Then Murray himself was replaced by General Sir Edmund Allenby at the end of June. In August Allenby reorganised the Egyptian Expeditionary Force into three corps directly under his own command. Chauvel was promoted to lieutenant-general and it was he who persuaded Allenby to accept the title Desert Mounted Corps in place of 2nd Cavalry Corps, an inappropriate title for a force composed largely of mounted rifle brigades and a camel brigade.

It should be understood that the Desert Mounted Corps was a British formation including an important Australian and New Zealand component. Of its nine horsed brigades, four were Australian, four British and one New Zealand; two of the three camel battalions were ANZAC.[24] Except for one mountain battery, all the artillery was British and the corps HQ was a British unit. Chauvel saw to it that some of the key administrative appointments were held by proven Australians; he brought in Colonel R.M. Downes as Deputy Director of Medical Services and Lieutenant-Colonel W. Stansfield as Assistant Director of Supply and Transport. However, Allenby appointed R.G. Howard-Vyse as Brigadier-General, General Staff as, 'rightly or wrongly' according to a British historian, 'it had been considered that Chauvel might need some coaching in the early days of his big command'.[25] As Howard-Vyse came fresh from over two years of trench warfare in France and Chauvel had been successfully directing mounted operations in Sinai and Palestine for more than a year, it is conceivable that the coaching may have been done by the latter. Fortunately they worked well together and formed a friendship which was to last until Chauvel's death in 1945.

Allenby planned to destroy the Turkish Eighth Army by first seizing Beersheba and its wells then making an enveloping attack on the main positions between Beersheba and Gaza. Beersheba was to be attacked from the south-west by Chetwode and from the south-east and east by Chauvel. The bombardment of Gaza, starting four days before D Day, and a brilliantly concocted deception plan were designed to fix the enemy's attention on his seaward flank. Even so, Chauvel was obliged to stage a series of night marches across the enemy's front to position his corps so that on D Day he could move quickly against Beersheba and seize the wells on which would depend the further development of the battle. The operation had to be based on deception, surprise and speed; if the wells were not taken on the first day the Desert Mounted Corps would have to fall back for water as there was none in the desert around Beersheba.

While many are aware of the dramatic charge of the 4th Light Horse Brigade at sunset on 31 October, during which the wells were seized, few know about the sheer professional skill which enabled Chauvel's two mounted divisions to reach their deployment area undetected and on time (the Yeomanry Division and the Camel Brigade had been detached for other duties in this phase of the battle). Chauvel had ascertained the location and extent of the ancient wells of Khelasa and Asluj as far back as May and this alone made possible his attack from the east. While Chetwode was assaulting the main defences of Beersheba, Chauvel's first move was dictated by the Turkish positions on Tell es Sabe about three miles (5 kilometres) to the east, which dominated the bare slopes leading down to the Wadi es Sabe. He gave ANZAC Mounted the task of subduing Tell es Sabe and cutting the road to Hebron, while keeping the Australian Mounted Division (formerly Imperial) in hand for seizing the wells.

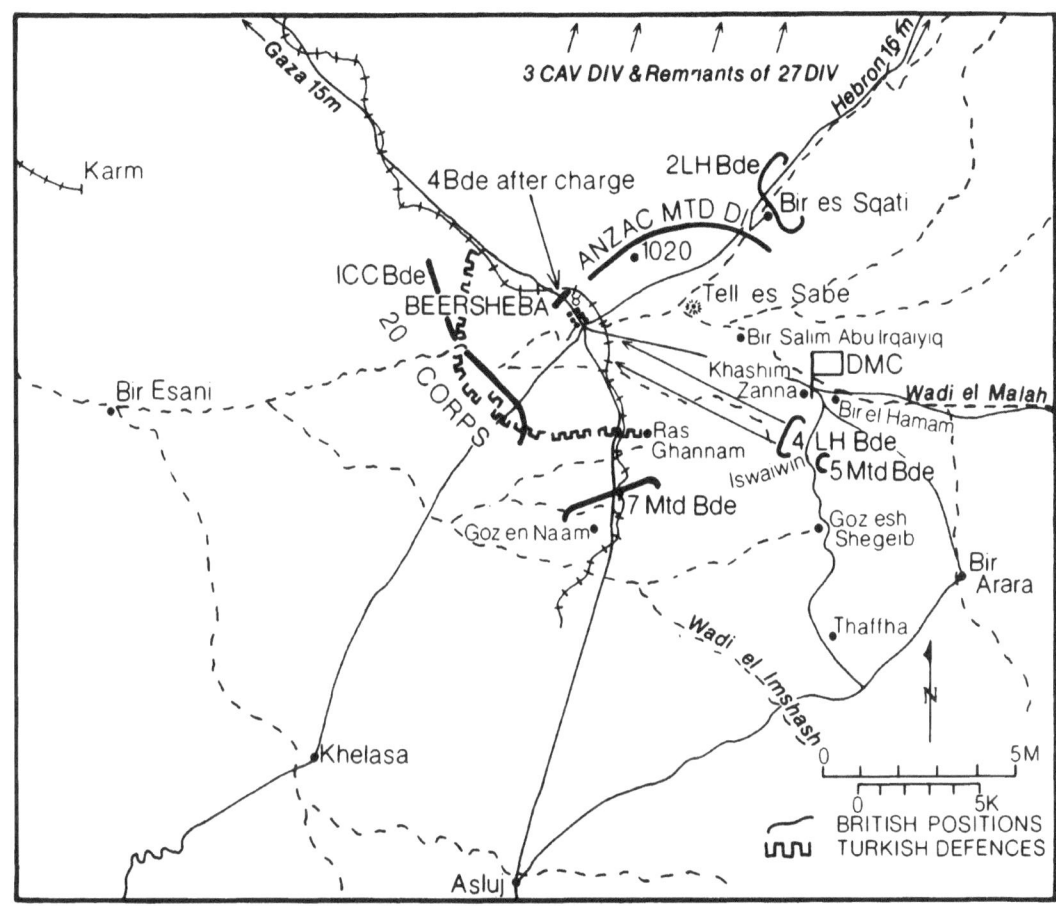

5 The Capture of Beersheba, 31 October 1917

Although they were in position by dawn, the ANZAC attack developed slowly and it was not until 3 pm that the final assault on Tell es Sabe was made. With sunset at 4.30 pm this left little time for the completion of Chauvel's mission. His experience of earlier dismounted attacks must have prepared him for this situation, but he had captured the only real obstacle to an advance into Beersheba—Tell es Sabe—and he had a division in reserve while ANZAC Mounted was preparing to advance to its final objective. He knew that there was neither wire nor horsetraps around the trenches to his front so his order, 'Put Grant straight at it', to General Hodgson, commanding Australian Mounted, was pre-eminently the quickest and simplest way to rush the wells. This was triumphantly achieved by the 4th Light Brigade under William Grant at the cost of only 63 casualties. At the same time the whole corps, except the brigade watching the Hebron road, was moving on Beersheba—anything but a '"forlorn hope" charge' as claimed by H.S. Gullett.[26] Chauvel's casualties for the day were 197 against 1500 prisoners, nine guns and all but two of the wells.

Allenby's plans for the development of the Third Battle of Gaza now suffered a measure of distortion owing to Turkish reactions and the water problem. He was drawn into a struggle on his right for Tell Khuweilfe and its wells; this inevitably involved Chauvel's corps. By the time Chetwode was ready to attack the Turkish centre and Chauvel had been freed from the fighting at Khuweilfe he had only four tired brigades for the pursuit. Chetwode's attack on 6 November was successful but Chauvel's jaded men and horses achieved little against the stubborn Turkish rearguards in that waterless wilderness. Too late the Yeomanry Division, the NZMR Brigade and others were ordered back to Chauvel; but long, hard marches were needed to bring them into action against the fleeing Turks; on 9 November, having sent a division back to water its horses, Chauvel had only two brigades available for pursuit.

That the 'cavalry' failed to cut off the Turkish Eighth Army aroused criticism at the time but Allenby himself had an adverse passage removed from a report to the Chief of the Imperial General Staff. He must have been as keenly aware as Chauvel that he—and the enemy—had brought about the dispersion and exhaustion of the Desert Mounted Corps by the time it was needed for pursuit. Allenby was careful to avoid a repetition of this situation when he launched his next major offensive. As for Chauvel, the award of the KCB on 1 January 1918 recognised his contribution to Allenby's victory.

In the great reorganisation consequent upon the German spring offensive of 1918, six of the nine British regiments in the Yeomanry Division were replaced by Indian regiments and a fourth division, part British part Indian, was added to the Desert Mounted Corps; but these two formations, called 4th and 5th Cavalry Divisions, would not be ready for operations before the summer. In the meantime Allenby began to operate beyond the River Jordan. He carried out a major raid to Amman in March and a demonstration across the river on 18 April. Then he relieved the infantry on that flank by the Desert Mounted Corps and ordered a major attack beyond the Jordan.

*Across the Jordan*

What the official histories call 'the Second Trans-Jordan Raid' or 'the Es Salt Raid' was nothing of the kind. It is significant that the enemy commander-in-chief refers to it as 'the Second Battle of the Jordan'.[27] Chauvel was given a demanding mission. He was to surround and capture the Turkish forces at Shunet Nimrin and Es Salt and, after clearing the country from the Jordan to Jisr ed Damiye–Es Salt–Madeba, he was to secure the railway near Amman. Finally, as soon as he had gained Amman and Es Salt he was to prepare for operations northward with a view to advancing rapidly on Deraa, a vital rail junction.[28] For this operation he would have only his two veteran divisions and the Camel Brigade. GHQ also indicated that considerable help could be expected from the Beni Sakr tribe near the objective area.

Chauvel believed neither in the usefulness of the Arabs nor in the logistical feasibility of such an operation; he was also aware that recent

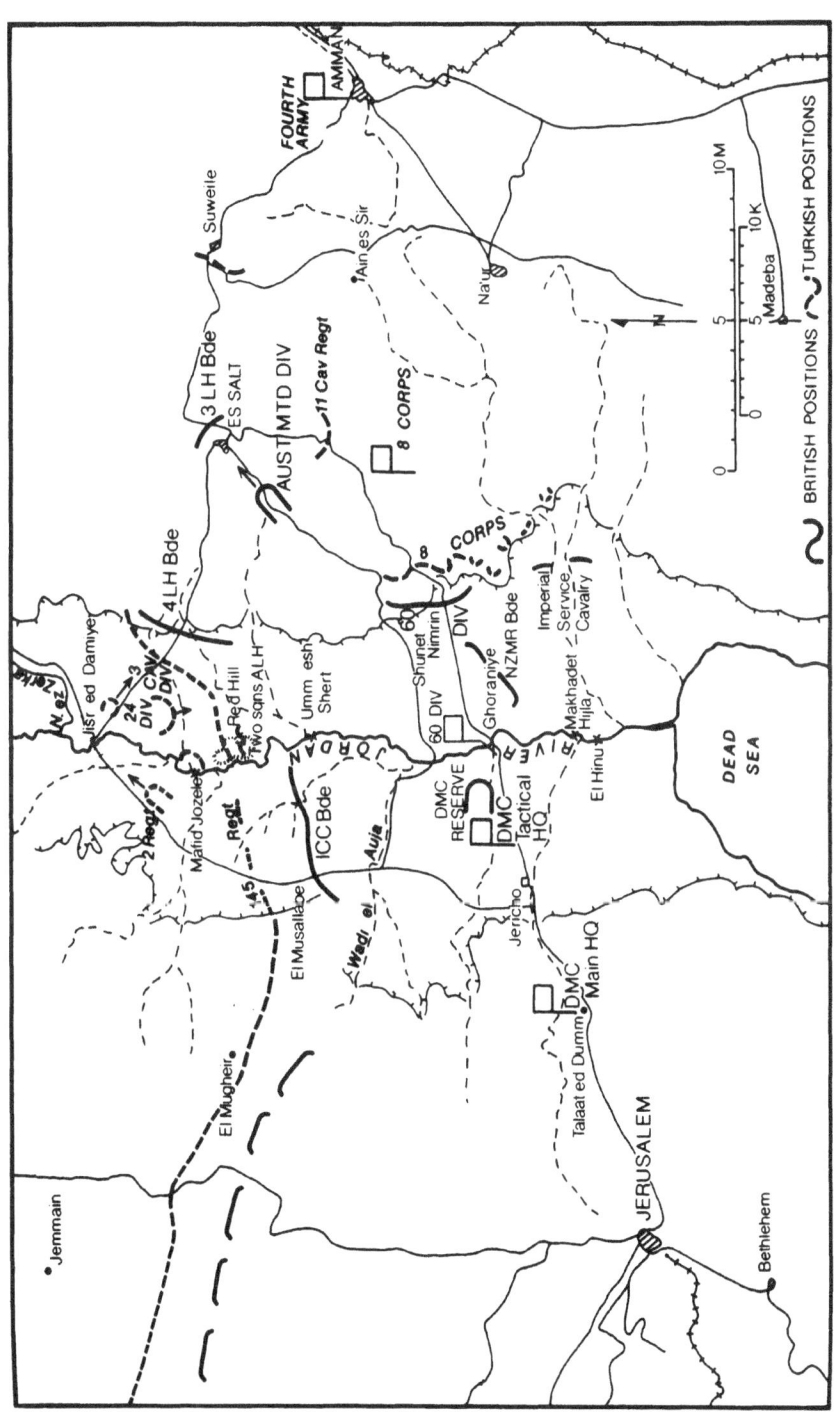

6  The Battle for Es Salt, 30 April–4 May 1918

attacks had alerted the enemy. To give weight to his attack he obtained the 60th Division (but less a brigade) and the 20th Indian Brigade, and also secured from GHQ a modification of his objectives. Allenby fixed D Day as 30 April to suit the Arabs, who were to cooperate by cutting the alternative Turkish line of communications from Shunet Nimrin.

There were some notable gaps in the intelligence available to Chauvel. Neither air nor ground reconnaissance had spotted the preparation of a pontoon bridge five miles (eight kilometres) south of Jisr ed Damiye nor improvements in an old track into Shunet Nimrin. GHQ had not discovered that the Turks were preparing to attack Allenby's Jordan flank and that large numbers of fresh troops were concentrating close to Jisr ed Damiye. Finally, the GHQ estimate of the troops on Chauvel's front across the Jordan was at least 2000 below their actual strength.

The first day went well enough. Shunet Nimrin was attacked by the 60th Division supported by ANZAC Mounted but their successes were only local. The Australian Mounted Division moved rapidly eastward on Es Salt and northward against Jisr ed Damiye, surprising the Turks by the vigour of their attack. Chauvel's shortage of troops immediately began to tell; the two brigades of 60th Division were not strong enough to capture Shunet Nimrin and the 4th Light Horse Brigade was unable to reach the bridge at Jisr ed Damiye. When Chauvel visited the brigade that afternoon he was concerned at the width of its front and ordered it back to a more secure position astride the track from Jisr to Es Salt, but he would not give Grant the extra regiment he wanted. Es Salt was entered before dark about the same time that a new Turkish pontoon bridge at Mafid Jozele was reported.

Next day the battle swung against Chauvel. Not only had the Beni Sakr disappeared but a powerful attack fell upon the 4th Light Horse Brigade which saved itself only by a precipitate withdrawal, losing nine of the twelve guns supporting it. At the first news of this attack Chauvel sent Chaytor, the commander of ANZAC Mounted, to take control of the threatened flank with such troops as he could scrape together. Chaytor was able to hold off further attacks while covering the last remaining track up to Es Salt. Australian Mounted prepared to attack Shunet Nimrin from Es Salt and Chauvel called for another effort by the 60th Division, but the Turks beat off all attacks. Air reports showed that fresh Turkish forces were closing in from all sides and the release of 60th Division's third brigade on 2 May came too late to affect the battle. On 3 May the pressure on Australian Mounted around Es Salt became so serious that Chauvel decided to withdraw. There was no chance of his regaining the initiative, ammunition was low and around Es Salt the troops were living off the country. Allenby approved Chauvel's decision and the orders went out at 4 pm. The withdrawal was well handled and by midnight on 4 May the last of the Mounted troops re-crossed the Jordan.

The second battle of the Jordan cost Chauvel's force 1649 casualties, nine guns and other equipment. They brought back almost 1000 prisoners and inflicted an estimated 2000 casualties. Once again

Chauvel's luck had held, as a whole Turkish regiment which had crossed the Jordan at Mafid Jozele in readiness to counterattack on 1 May had been ordered to march back and cross at Jisr, thus being out of the fight throughout 1 May. Allenby pointed out Grant's faulty defensive layout and blamed Chauvel for not warning Grant more emphatically of the dangers of his position. The real causes of this failure, so carefully avoided in the British Official History, were the alerting of the Turks by the repeated operations ordered by Allenby, the inadequacies of the intelligence before the operation, the failure of GHQ to allot adequate forces and the failure of the Arabs to play their part in the battle. Chauvel showed great determination and hung on as long as he could; he moved quickly and effectively when his left flank was threatened; he recognised the futility of going on with the battle and he withdrew successfully. The fighting qualities of the three divisions had provided the only foundation for the enterprise, the chief result of which was that the Turks were convinced that the next attack would be in the same area and by the same troops. It may have been an early and unspoken appreciation of this that caused Allenby to answer Chauvel's expression of regret at his failure with: 'Failure be damned!'[29]

*The final offensive*

Allenby's final offensive destroyed three Turkish armies and freed Palestine and Syria from Turkish rule. This time Allenby got it right. He used his infantry to smash a way through the Turkish defences on the coastal flank while Turkish eyes were fixed on the Jordan flank. When the way was clear Chauvel's corps, concentrated, rested and ready, poured through the gap created by the infantry and rode hard, without having to fight until they had almost reached their objectives, Afule and Beisan, about 40 miles (65 kilometres) behind the crumbling Turkish front. Having seized those centres they went on to block the few Jordan crossings, and rounded up the fleeing Turks. Liman von Sanders, the enemy C-in-C, narrowly escaped capture in Nazareth—a great disappointment for Chauvel as this was his own scheme.

The battles of Megiddo, to use the official title, opened on 19 September 1918. This time Chauvel had three cavalry divisions (Australian Mounted having been armed with swords had thus become 'cavalry'). ANZAC Mounted had been detached for an independent role in the Jordan Valley. The movement of the corps by night from the valley to the orange groves north of Jaffa and its secret concentration close behind the 21st Corps, itself heavily reinforced, was a triumph of staff work and discipline covered by GHQ's superb deception plan. Chauvel launched his corps with two divisions forward on parallel axes despite Allenby's stated preference for having both on the one axis. So complete was the surprise and so rapid the penetration that on 20 September Chauvel was able to send Australian Mounted southward towards Jenin to meet the masses of fugitives trudging northward.

On 22 September when Allenby visited Chauvel at Lajjun, he raised

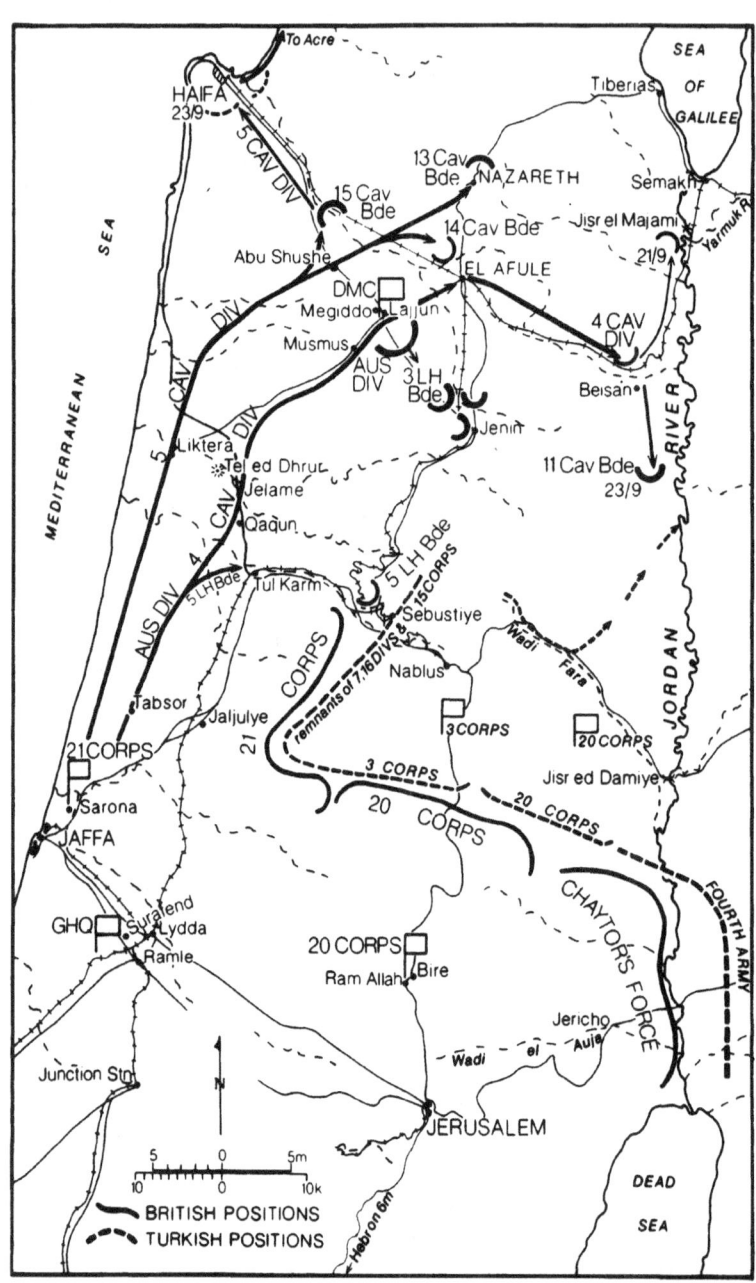

7 The Destruction of Army Group F, 19–25 September 1918

the question of advancing to Damascus but gave no orders beyond requiring Chauvel to secure the line Tiberias-Nazareth-Haifa. Chauvel quickly prepared plans for the capture of Damascus and issued preliminary orders to the 4th Cavalry Division so that when Allenby sent a staff officer to explain his intentions there was little more to be done. Chauvel also directed the Australian Mounted Division to the Sea of Galilee and the bridges over the Yarmuk in time to prevent the enemy from organising a solid defence of the area. These operations on 25 September concluded the battles of Megiddo. Allenby cabled Melbourne acknowledging that 'the completeness of our victory is due to the action of the Desert Mounted Corps under General Chauvel'. In his message to his army he said: 'Such a complete victory has seldom been known in all the history of war'.

Chauvel's plan for the capture of Damascus was for the 4th Cavalry Division to advance eastwards to Deraa then to Damascus by the Pilgrims' Road; if he could not destroy the remnants of the Turkish Fourth Army they were to be driven into the arms of the other two divisions around Damascus. This was to be done in cooperation with the Arab Northern Army which would come under Chauvel's orders. The rest of the corps with Australian Mounted leading, would advance over the Golan Heights and make for Damascus on the Quneitra Road.

Despite a series of rearguards, Chauvel was at the gates of Damascus by the evening of 30 September and had cut the road through the Barada Gorge to Beirut; the guns of 4th Cavalry Division could be heard to the south of the city. He entered Damascus next morning while the 3rd Light Horse Brigade continued the pursuit northwards. This was the beginning of a period unique in the experience of Australian commanders, for Chauvel was now responsible for a city of 300 000 people over whose control a political struggle had begun involving not only Arab and Syrian leaders but his own liaison officer, Lieutenant-Colonel T.E. Lawrence. Allenby had not sent a political adviser with Chauvel and his political briefing was meagre; moreover, neither Allenby, nor Chauvel, nor Feisal, commander of the Arab forces, was aware of Lawrence's personal aims in Damascus and Chauvel had no contact with him until they met by chance in the city. Fortunately for Chauvel, Captain Hubert Young, one of Lawrence's assistants who well understood what was going on, presented himself and gave Chauvel something of the political insight which ought to have been provided by GHQ before Damascus was entered. Acting on Young's advice, Chauvel ordered a show of force through the city. This promptly restored the calm that Lawrence and the Arabs had failed to achieve; Lawrence was later to lampoon it as a 'triumphal entry'. Chauvel also established himself in the Turkish military commander's house on Young's advice (Lawrence had suggested he go to the British consulate), and from there proceeded to bring order to Damascus, which meant not only securing food and policing, but also caring for hordes of sick and wounded Turks and about 20 000 prisoners.[30] Simultaneously his troops were going down with malaria and pneumonic influenza at an alarming rate; there were over 1200 admissions to hospital in the first week of October.

As Allenby was determined to continue his drive, Chauvel also had to direct the pursuit from Damascus. This had become a follow-up with dwindling forces; it ended 200 miles (322 kilometres) further north beyond Aleppo which had been entered by the Arabs on 25 October. On 30 October an armistice was concluded with the Turks.

Since 19 September, Chauvel's three division had marched and fought their way over 300 miles (480 kilometres), captured well over 70 000 prisoners[31] and great numbers of guns and vehicles. All this at a cost of only 533 battle casualties! Now he faced the problems not only of the repatriation of the AIF but also of dealing with large elements of the Turkish army under uncooperative commanders. Allenby insisted on his remaining in command of the Desert Mounted Corps, but as he could maintain only one depleted cavalry division around Aleppo his position was highly insecure. The Turkish commanders responsible for implementing the armistice were evasive and dilatory; unrest around Aleppo became so serious that Chauvel was ordered to occupy the main towns across the border. The removal of Mustapha Kemal and another troublesome army commander improved the situation so that Chauvel was able to go to Cairo for Christmas and return to England in April 1919. There he was made GCMG; he had also received the French Croix de Guerre, the Egyptian Order of the Nile twice and had been mentioned in despatches on ten occasions. As the senior officer of the Australian Army, he returned home in September 1919 to become Inspector-General. He succeeded Brudenell White as Chief of the General Staff, 1923-30.

Harry Chauvel was not an 'educated soldier' such as Haig or Brudenell White but he had learnt from his experience in South Africa and he applied himself to his profession between the wars. In the First World War he rose more rapidly than any other Australian commander. He was fortunate that the opportunity of leading a mounted division came when it did because it was for such a role that he had prepared himself. However, he suffered the same limitations as all Australian commanders in war, in that he was always subordinate to the commanders of a major power. Within those limitations his career in Sinai and Palestine was an outstanding success from Romani to Damascus. As Birdwood put it to the Minister for Defence on 4 October 1918: 'His record is second to none not only with the Australian troops, but I may say among the more senior officers of the British Army serving anywhere in this war'.[32] That success was based on the timeless military virtues—courage, coolness under fire, knowledge, the capacity for decision.

Chauvel was very careful of his prime resource, his soldiers; it is significant that there has never been any criticism of him related to throwing away lives for questionable ends. A humane, even humble man, living by a strong Christian faith, his gentleness went out even to the enemy; few commanders burdened with the problems of a sudden withdrawal by night, as Chauvel was at First Gaza, would have spared a thought for a handful of sorely wounded Turks. He brought Downes and

8 Lieutenant-General Sir Harry Chauvel, Commander, Desert Mounted Corps, Damascus, October 1918. (AWM Negative No. B 326)

Stansfield to his staff in the Desert Mounted Corps where they wrought miracles for the fighting troops in the fast-moving, far-ranging operations under Allenby. Greater recognition of courage and skill in battle also received his attention and he pressed Allenby to restore the honours lists relating to ANZAC Mounted after the Surafend incident.[33]

In his relations with the British, Chauvel worked on the basis of cooperation rather than confrontation. This was not good enough for some of his more aggressively nationalistic subordinates, but Chauvel had assessed his situation and never forgot that his first line of responsibility was to Birdwood in France. Birdwood had agreed to or accepted the detachment of Light Horse brigades on special operations in December 1915 so that Chauvel had to work persistently to reassemble ANZAC Mounted and the scattered regiments of the disbanded 4th Light Horse Brigade. In the end he succeeded, but it seems to have been unnecessarily protracted except that there was no charter in 1916, not even for Birdwood, such as Blamey took to the Middle East in 1940. Yet one cannot escape the impression that in this area a more independent and forthright stance may have served Chauvel better when dealing with GHQ. But who among the Australian generals of 1914–18 would have spoken to Murray or Allenby as Blamey spoke to Auchinleck about the relief of the 9th Division in 1941?[34] By that time the Australian Army and, to some degree, the country had come of age.

# General Sir John Monash: Corps Commander on the Western Front

## P.A. PEDERSEN

'If only the Generals had not been content to fight machine gun bullets with the breasts of gallant men, and think that that was waging war.' Churchill's lament captured the bitterness which is still provoked by debate on the performance of British commanders on the Western Front. Etched on the central headstones in that graveyard of military reputations were the names of the Commander-in-Chief, Field Marshal Sir Douglas Haig and his army commanders. Conversely, historians have lauded General Sir John Monash (1865–1931) as a commander of exceptional vision and ability. Yet he served in the comparatively junior appointment of commander of the Australian Corps and then only in the last six months of the war. Unlike Haig, Monash was a citizen soldier and in civil life, an engineer whose intellect, determination and capacity for hard work ensured his rise to the very top of his profession. These personal qualities made telling contributions to his success as a general but the good fortune with which he was liberally blessed was a factor no less potent.

Until the publication in 1982 of A.G. Serle's biography, little of substance had been written on Monash the commander other than his own account, *The Australian Victories in France in 1918* and C.E.W. Bean's penetrating, but tendentious commentary in the *Official History of Australia in the War of 1914–18*. Predictably, a Monash legend grew, claiming for its subject achievements which were exaggerated and credit that rightly belonged to others. The most serious casualty of this myth has been the attempt accurately to assess Monash's place among his contemporaries and among the Great Captains of history.[1]

*Civilian engineer*

John Monash was born in Melbourne on 27 June 1865, the son of Jewish parents recently emigrated from Prussia. In 1874 the family moved to Jerilderie, a remote country town in New South Wales, exposing the boy to the bush and its earthy inhabitants. When Jerilderie's only schoolteacher recognised he could no longer challenge Monash's obvious intellect, he urged his parents to send him back to Mebourne. He enrolled at Scotch College in October 1877 and left three years later as equal dux of the school, dux in mathematics and modern languages and winner of the matriculation exhibition in mathematics. He had also won the English prize almost every year.

In March 1882 Monash began at Melbourne University, intending to become a civil engineer. He was soon disenchanted by the soporific and repetitious lectures of professors whom he censured in later years for their failure to keep abreast of modern thought and research. Abandoning formal instruction, he spent hours at the University and Melbourne Public Libraries, reading literature and history. Not surprisingly, Monash failed his first year. A more deliberate approach to study produced nothing better than third class honours for 1883 and equally mediocre results in 1884. Deliverance from despair came through his deep involvement in student politics. He was instrumental in the foundation of the Students' Union while his fluent pen championed student grievances in the University paper. Troubled by the fatal illness of his mother, he ignored study altogether in 1885 and, short of money, sought full-time employment in engineering.

Joining the firm responsible for the Prince's Bridge over the Yarra, Monash displayed almost a natural aptitude for field engineering work. For the next two years he was engaged in design and construction planning, opened a quarry ten miles (sixteen kilometres) from the bridge site and ran a supply yard. His approach was one of constant inquiry: 'He was most enthusiastic and embarrassingly curious. He always wanted to know the reason of things.'[2] In 1888 he switched to the company constructing the Melbourne Outer Circle suburban railway and was soon in charge of the entire works. Monash's reasonable hopes that this might lead to more profitable employment were dashed by the onset of an economic depression before the project was completed. His marriage to Hannah Victoria Moss in April 1891 and the birth of their daughter Bertha two years later made this financial hardship even more difficult to bear. Emotionally, the union was disastrous, an unending round of quarrels which flowed from the couple's basic incompatibility. Ironically, Monash now scored the academic success which had eluded him during the years of comparative serenity. Just before his wedding he finished his engineering degree, winning the *Argus* scholarship as top student. By the end of 1893, he had gained his master's degree in engineering, finally completed his Bachelor of Arts and equipped himself with a Bachelor of Laws as well.[3]

Until April 1895 Monash was employed by the Melbourne Harbour Trust, designing the swing bridge over the Maribyrnong River, the first of its kind in Victoria, transit sheds for the Yarra wharves and roads and drainage schemes. He detested the monotony, inertia and absence of challenge typical of government service, while the security it offered was temporary, for the Depression of the 1890s eventually forced his retrenchment. Almost immediately, he joined an old friend, J.T. Noble Anderson, in a partnership as patent agents and civil, mining and mechanical engineers. Desperate measures were often taken as the pair sought to establish their company. Their fees were the lowest allowed by the London Institute of Civil Engineers; they approached shires to engage them as consultants rather than employing a shire engineer and, hoping to win future lucrative contracts, they accepted many jobs for expenses only.

While the company struggled, Monash's work in legal engineering as an advocate and expert witness increased, rewarding his earlier foresight in studying law. His penetratng lucidity was always apparent, Sir Robert Menzies regarding him as the greatest advocate he had heard.[4] Meanwhile Anderson had secured an agreement making Monash and Anderson Victorian agents of Carter and Gummow, a firm holding the Australian patent rights for Monier reinforced concrete construction. Monier work seemed a profitable alternative when a slump overtook conventional engineering activity. But impecunious shires defaulted on contracts, forcing legal action which was often costly and protracted. Monash's health suffered and he occasionally surrendered to fits of depression. A.G. Serle writes:

> It was agonizing to have to implore his major creditors to stay their hands. For years he was to be humiliated by his indebtedness; suffering scores of demands for payment and solicitors' letters. He was often months behind with his rent and life assurance payments...[5]

Salvation came from the Monier Pipe Company, formed in 1901 to capitalise on the suitability of reinforced concrete for pipe manufacture. Dame Nellie Melba's father, David Mitchell, provided the working capital and held a 40 per cent share, the same as the partnership of Monash and Anderson. By 1905 the latter had been dissolved and a new company established with Monash as superintending engineer and Mitchell its chairman of directors. The scale of projects increased. Erected in 1911, the Janevale bridge boasted the largest spans in Australia for a structure of its class. Extensions were added to the Melbourne Town Hall and Hospital. The South Australian branch formed by Monash in June 1906 was equally successful. At the end of 1913, he estimated his worth at £30,000, the present day equivalent of over $1 million. Monash was at the zenith of his pre-war fortunes. In 1912, he was elected to the University Council and made President of the Victorian Institute of Engineers. Two years earlier he had taken his family on a world tour. Monash was astonished by the advanced technology of Britain, Europe and America which made Australia seem 'a little provincial place', while its engineers, he told the Institute, could not 'think big' and saw no reason to remedy their ignorance of developments elsewhere.[6]

Monash's civil experience was the first and in some ways, the most important single factor underlying his later success as a commander. He was as relaxed with railway fettler as with university professor, and could bend both to his will without the reassuring presence of the hierarchical system of discipline to which he later had recourse as a soldier. His unrestrained criticism of his Institute colleagues required moral courage, an indispensable quality in a commander but one frequently exorcised by the military system's obsessive discouragement of contrary views.

Monash insisted on punctuality, tidiness and fine attention to detail. He was also a man of patience. As Chairman of the Inventions Board in 1909-10, he encouraged inventors even if their ideas were bizarre.

It is as an engineer that Monash the civilian is best remembered. Railway, bridge and building construction had given him experience of large-scale projects with their requirement for the concentration of materials and the careful organisation and direction of labour. As Major-General H. Essame observed, Monash was guided by principles equally applicable to high command: the need for foresight, flexibility, cooperation, economy, delegation of authority and an awareness of time. While Essame's contention is irrefutable, its usefulness is limited by the generality of his criteria. Any competent commander or businessman would apply them irrespective of whether he had been trained as an engineer or not. The distinction lies in the definition of engineering as 'literally a mode of looking at things', and its practitioner, 'the man with the genius first to recognize the real conditions of the technical problem before him and then by skilful effort to discover an adequate solution'. Monash's pioneering of reinforced concrete showed that he possessed these attributes in high degree.[7]

Monash, then, embodied what Morris Janowitz later perceived as the converging needs of the civilian and military establishments. Both faced the same problems of administration, research and development and the maintenance of initiative and morale. Hence it was vital that the military leader 'develop more and more of the skills and orientations common to civil administration and civil leaders'.[8] Thirty years before Janowitz, Monash had reached the same conclusions after reflecting on the performances of the five divisional commanders who served under him in the Australian Corps. Only one, Major-General E.G. Sinclair-MacLagan had been a regular officer in 1914. Monash felt that the civil experience of the others conferred a positive advantage upon them as generals: 'This advantage rested upon a wide civilian training as engineers, architects or as captains of industry—a training far more useful for general applications to new problems than the comparatively narrow training of the professional soldier'.[9]

Monash omitted the considerable advantage which, in his own case, accrued from civilian pressures rather than civilian training. For over 30 years he was confronted by adversity, whether his own penury, academic failures and the completion of his education by part-time study, a turbulent marriage or the fight to support a young family with limited means. Most of his business career was a constant struggle for survival. Monash's triumph over such an environment required ruthlessness, cunning, persistence, confidence in his own judgment and the capacity to translate thought into action. In other words, Monash in peace had developed 'robustness', that ability to withstand the shocks of war which Field Marshal Wavell considered was the most important quality of a general. The argument about the hardening effect of civil experience is reinforced by the difficult career of Monash's great contemporary, Sir Arthur Currie, the commander of the Canadian Corps.[10]

In comparison, Haig and most other senior regular officers led a leisurely and sheltered life in the pre-war period. While commanding the two divisions at Aldershot in 1913-14, Haig's daily routine consisted of supervising training and office work in the morning, sport in the afternoon and two hours professional reading after dinner. Unlike Monash or Currie, he was not exposed to the exacting test of competition, a test offering 'constant practice at conflict and increasing experience of its psychological conditions'.[11] Monash did not equivocate on the regular army as a career:

> I do not regard and have never regarded permanent soldiering as an attractive proposition for any man who has some other profession at his command...if a man could command an income no larger in private practice than he could in military employment, I would recommend him to stick to private practice every time. There is something about permanent military occupation which seems to confine a man's scope and limit his opportunities, and after he has had a few years under the circumscribed conditions of official routine, he generally finds himself wholly out of touch with civil occupation.[12]

## Militia officer

Monash's opinion was hardly baseless. In July 1884 he had joined the University Company of the 4th Battalion, Victorian Rifles, rising to sergeant within a year. In April 1887, he was appointed a probationary lieutenant in the North Melbourne Battery of the Victorian Garrison Artillery which defended Port Phillip Bay and several smaller harbours from attacks by hostile warships. It was a precision arm, demanding exact calculation to score direct hits on moving targets. Monash's technical training was particularly suited and soon applied to this environment. He evinced a thorough grasp of gunnery, comprising drills, gun construction and maintenance and the working of prediction equipment and sighting scales. He designed a full-scale wooden working model of a five-inch (127 millimetre) breech loading gun which proved a useful training aid. The intimate relationship between technology and the development of modern weapons fascinated him while his belief that 'Fighting Machinery' had replaced physical force and brute courage indicated an awareness of the change wrought by technology on the nature of warfare.

By 1897, Monash was a major and the commander of his battery. His friend and fellow militia officer, George Farlow, recalled:

> His orders were models of conciseness and at the same time completeness. Nothing was overlooked...He never buzzed about the tents of his men to see if they were properly provided for but what he did do was to think out all things and detail officers to work out the details and report to him as to their satisfactory development.[13]

Unfortunately, Farlow did not mention the extent to which Monash's thoroughness trespassed on the responsibilities of his subordinates. This would be one of the most frequently misunderstood aspects of his performance throughout the First World War. Nor did Farlow allude to the critical relationship between commander and soldier. Monash's 21 years in the Garrison Artillery provided his only experience of command at a level in which he was directly involved with the soldiers under him. His moral courage was evident in his vigorous defence of seven men recommended for dismissal without a proper trial for alleged fraudulent use of railway coupons in July 1905. Testifying to his leadership ability were the touching words of his Staff Company Sergeant Major whose military career had begun in the British Army 25 years previously: 'during the whole of that time I had never had a better officer and a gentleman to deal with. Nor have I had to deal with a better Artillery Officer, either Permanent or Militia.'[14]

Monash rarely saw soldiers in his next appointment. Promoted lieutenant-colonel in March 1908, he assumed command of the Victorian Section of the newly raised Australian Intelligence Corps (AIC). It was responsible for the collection of information on the military resources of the Commonwealth and Pacific, the compilation of a library and the preparation of military maps. Except for staff clerks and draughtsmen, the AIC was composed exclusively of officers, most of whom possessed civil skills of unique relevance to the corps. Clearly this was no command in the conventional sense. Rather, Monash's role was more that of the business manager, directing diverse talents towards the task at hand—that is, the relationship between a commander and his staff. The fundamental principle of his command of the Victorian Section was uniformity of thought. 'It is essential that every tendency for the work to be carried on in "watertight compartments" shall be discouraged...the utmost co-operation between officers must be practised'.[15]

Government embarrassment at the parlous state of topographical information elevated map-making to the most important of all AIC tasks. Monash's surveying and engineering experience were ideal qualifications for this work and he assumed direct control. His understanding of the technical aspects was crucial to Monash's style of command for it enabled him to visualise ground instantly by studying the map depicting it. He was not concerned with maps as a means of accurate position finding; rather, map accuracy was only that which gave a true conception of the *general* form of ground and the *general* relationship of levels. The planning of manoeuvre schemes pitting brigades against each other during annual camps was another responsibility. Monash compiled the tactical portion of the Victorian Commandant's Annual Report after the camps in 1909, 1910 and 1911, work which he described as involving 'the observation, analysis and draft criticism of the tactical handling of four composite brigades'. His observations as an umpire were astute. At Seymour in 1910, he noted that the form of operation orders was not understood while their promulgation was 'not only slow but ineffective'.[16]

Monash's intimate grasp of the importance of staff work in both its operational and administrative forms was reflected in the lectures and exercises which he used to train his section. Logistics problems featured prominently in the intelligence staff tours he directed in 1908 and 1912: railway movements, the transport required to move a force and its food, forage and communication needs. Monash regarded these as operations of war in an age when that term was applied exclusively to combat:

> I believe that the task of bringing the force to the fighting point, properly equipped and well-formed in all that it needs is at least as important as the capable leading of the force in the fight itself.
> In fact it is indispensable and the combat between hostile forces is more in the preparation than the fight.[17]

Stimulating Monash's thoughts further was his attendance at two war courses conducted by the Director of Military Science at Sydney University, Colonel H.J. Foster, in October 1909 and 1911. Tactical schemes at brigade level were practised without troops with students filling the various positions on the brigade staff. Monash concluded: 'Staff work in war is exactly on the same lines as here in this room. Troops are just as non-existent as they are at exercise.'[18] He was developing the art of imagining the position of his own troops, the location of the enemy and how both were affected by orders he issued. Liddell Hart contended that this was 'a vital faculty of generalship...It is that flair which makes the great executant'.[19] Imagination was also crucial to Monash's intense study of military history. He would visualise the situation and ground described in an account of a campaign and apply to them the principles of command, staff work and logistics. Then he would read the account in full, noting violations of the principles. In 1912, he won the army's inaugural Gold Medal Essay competition on the topic 'The Lessons of the Wilderness Campaign—1864', from which tactical, administrative and organisational lessons applicable to the defence of Australia were drawn.

Monash's final pre-war appointment began in June 1913 when he assumed command of the 13th Infantry Brigade, one of the formations raised after the introduction of the Universal Training Scheme. His task, the training of a large number of raw soldiers and officers, was a difficult one, but by now Monash's style of command was fully developed. At their first meeting he enjoined his four battalion commanders to achieve 'an harmonious whole—a healthy rivalry, yet mutual sympathy—and to apply the good ideas of one regiment for the benefit of others because the ultimate aim is the efficiency of the whole'. 'The Development of a Soldierly Spirit', a note he made in January 1914, listed those qualities Monash considered essential in a soldier: obedience, respect for authority, unselfishness, self-sacrifice, mutual help and cooperation, self-respect, personal tidiness and cleanliness, courage, determination and optimism. The 13th Brigade camp at Lilydale in February 1914 proved the efficacy of Monash's training policies. It culminated in two days of brigade manoeuvres before which he emphasised to his officers and NCOs the need 'to let all ranks know what is going on'. A battalion position was assaulted by two battalions while two more delivered a flank attack after a concealed march of four miles (about six kilometres). Machine gun sections provided covering fire on the approaches. The Inspector-General, Overseas Forces, General Sir Ian Hamilton was 'well pleased with the technical parts of the work...and with the go and keenness of the officers and men'.[20]

Monash's pre-war military career was the second factor influencing his development as a commander. Most Australian generals of the First World War had spent their entire careers in one corps before 1914, but Monash served with distinction in three, gaining a thorough knowledge of weapons from heavy artillery to the machine gun. His appeals for a community spirit in the 13th Brigade were repeated in every formation he led and formed the basis for discipline in them. Monash's inexperience in the practical handling of infantry was shared by all of his Australian contemporaries but his brief command of the 13th and his umpiring of brigade commanders in the AIC gave him a better start than most. Studies of Monash as a general which conclude that he gained his military education after 1914 are guilty of a gross distortion. Most of his attitudes and methods were formed before the war and did not change during its course. After a pre-war career spanning 30 years, this was not surprising. It remained only to build on such an immovable foundation with the third factor, his wartime command of an infantry brigade and division.

*Gallipoli*

On the outbreak of war, Monash was appointed Deputy Chief Censor in the Department of the Chief of the General Staff but within a month he took charge of the 4th Infantry Brigade. He reissued '100 Hints for Company Commanders', a pamphlet containing his thoughts on officers' responsibilities which he had prepared for the 13th Brigade. The recurrent theme was the officer's duty to his men. He must put their needs for rest, food and hygiene before his own and his constant thought must be 'the preservation of the soldiers' fighting spirit and physical condition'. When the 4th Brigade reached Egypt on 31 January 1915, some weeks after the first contingent of the AIF, Monash appealed to all ranks to strive harder to compensate for their late arrival, thereby furthering 'the efficiency and prestige of the brigade'. According to Monash, it invariably shone in the various brigade and divisional manoeuvres conducted by its parent formation, Major-General A. Godley's New Zealand and Australian Division which, together with Major-General W.T. Bridges' force comprised the Australian and New Zealand Army Corps under Lieutenant-General Sir William Birdwood. This was not empty boasting. By mid-February, both Birdwood and Godley had 'quite made up [their] minds that it is the best Australian brigade in Egypt'. Training differed little from what Monash had witnessed in Australia: long advances in column over the open desert usually concluding with a triumphant attack. Even Monash's plea for more sub-unit work would not have rectified the appalling inadequacy of this program for operations on the tortuous, scrub-covered slopes of the Gallipoli Peninsula.[21]

The Cape Helles and Anzac landings were both contained. At Anzac, where Monash came ashore early on 26 April, the plan had gone completely awry. The intended advance inland to sever the Turkish lines

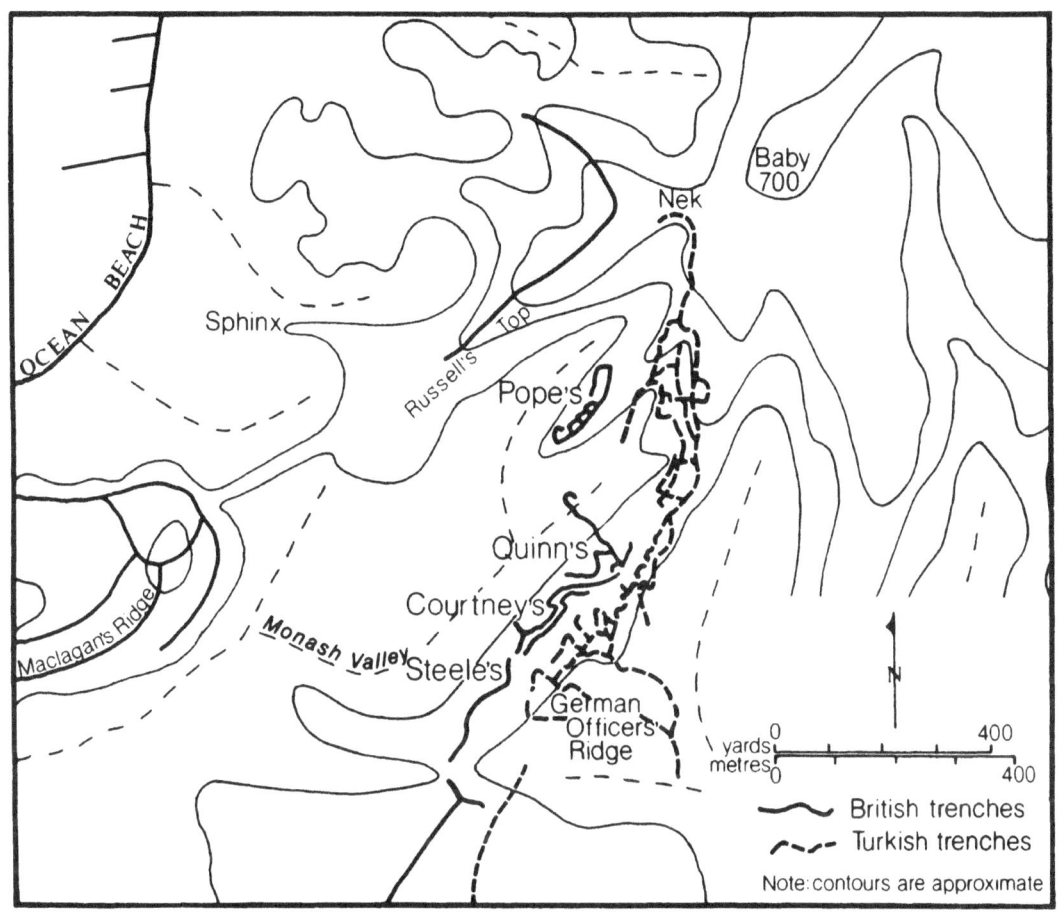

8 Monash Valley, May–August 1915

of communication to Helles had been stopped 1000 yards (925 metres) from shore, leaving Birdwood's corps clinging desperately to a triangular-shaped line one-and-a-half miles (2.5 kilometres) long. The intermixture of units was chronic, particularly at the head of Monash Valley where Pope's Hill and Quinn's and Courtney's Posts formed the apex of the triangle. This sector was allotted to Monash and he wrote, two days after the landing, that it was 'held by parts of the 3rd, 10th, 11th, 12th, 13th, 14th and 16th Battalions...' Pressed by Turkish counterattacks, he did not attempt any immediate internal redistribution but divided his sector between his commanding officers in ratios according to the strengths of the units then in their charge. Only after the 1st and 3rd Brigades withdrew for rest to the beach did Monash begin the reconstitution of his companies under their original commanders, but only as opportunity permitted. When the task was completed on 30 April, Bean praised Monash's organisation of the defences as 'an object lesson in covering fire'. Machine guns on Pope's and Courtney's interlocked before Quinn's while its guns could enfilade those two positions.[22]

Two days later, the New Zealand and Australian Division launched an ill-conceived attack on Baby 700, the dominant feature at Anzac. Godley and his staff prescribed neither boundaries nor routes while the New Zealand Infantry Brigade did not receive its orders until three hours before the operation was due to commence. Its attack, intended to be concurrent with Monash's, began one and a half hours afterward and dissolved before terrible fire. The 4th Brigade seized much of its objective but, despite reinforcement by two battalions of Royal Marines, the position degenerated and a distraught Monash ordered a withdrawal. He had argued against the scheme ever since its inception and now he complained to Bean: 'They've tried to put the work of an Army Corps onto me'.[23] His brigade was now only 1800 strong, weaker than any of the brigades of the 1st Division when they left the line. Monash sought relief but Godley refused because the needs of the Helles offensive had consumed all the available fresh troops.

Forced to devise a system which minimised the strain, Monash allotted Pope's and Courtney's permanently to the 13th and 14th Battalions, one company from the supports in each relieving daily one in the firing line. Every 48 hours the 15th and 16th alternated on Quinn's, the most dangerous post, an arrangement which ensured that half the men in both lines were always fresh. Even after the more senior Brigadier-General Trotman of the Royal Marines took charge of the sector, Monash remained its dominant figure. Trotman freely admitted that his marines were 'quite useless' and they were replaced by Colonel Harry Chauvel's 1st Light Horse Brigade on 13 May. Monash was also junior to Chauvel but their relations were always harmonious although he observed that Chauvel was 'very fidgetty' and 'grateful for my help when he gets into a tangle'.[24]

The 4th Brigade was still the core of the defence. Rightly judging the newly arrived Light Horse to be too inexperienced to garrison Quinn's, Birdwood and Godley bolstered its depleted defences with the 13th Battalion, leaving Pope's as the mounted troops' only responsibility. But their contribution was timely, for the 15th Battalion had lost heavily in a raid ordered by Godley on 12 May to ascertain Turkish strength at Quinn's. On 19 May it was the Turks' turn to suffer as they incurred 10 000 casualties in their unsuccessful general offensive. The 4th Brigade was finally relieved after repulsing a Turkish attempt to capture Quinn's on 29 May. Bean described its defence of Monash Valley as one of the four finest feats of the AIF during the war. The conclusion was sound for the continued existence of the tenuous foothold at Anzac depended on the retention of Monash's sector, by far the most difficult in the line. At Quinn's the opposing trenches were separated at one stage only by a bombstop and had the Turks broken through, an unimpeded passage to the beach would have been theirs.[25]

One of the great offensives of the war was launched in August as the substantially reinforced Australian and New Zealand Army Corps attacked the three crests of the Sari Bair range: Chunuk Bair, Hill Q and

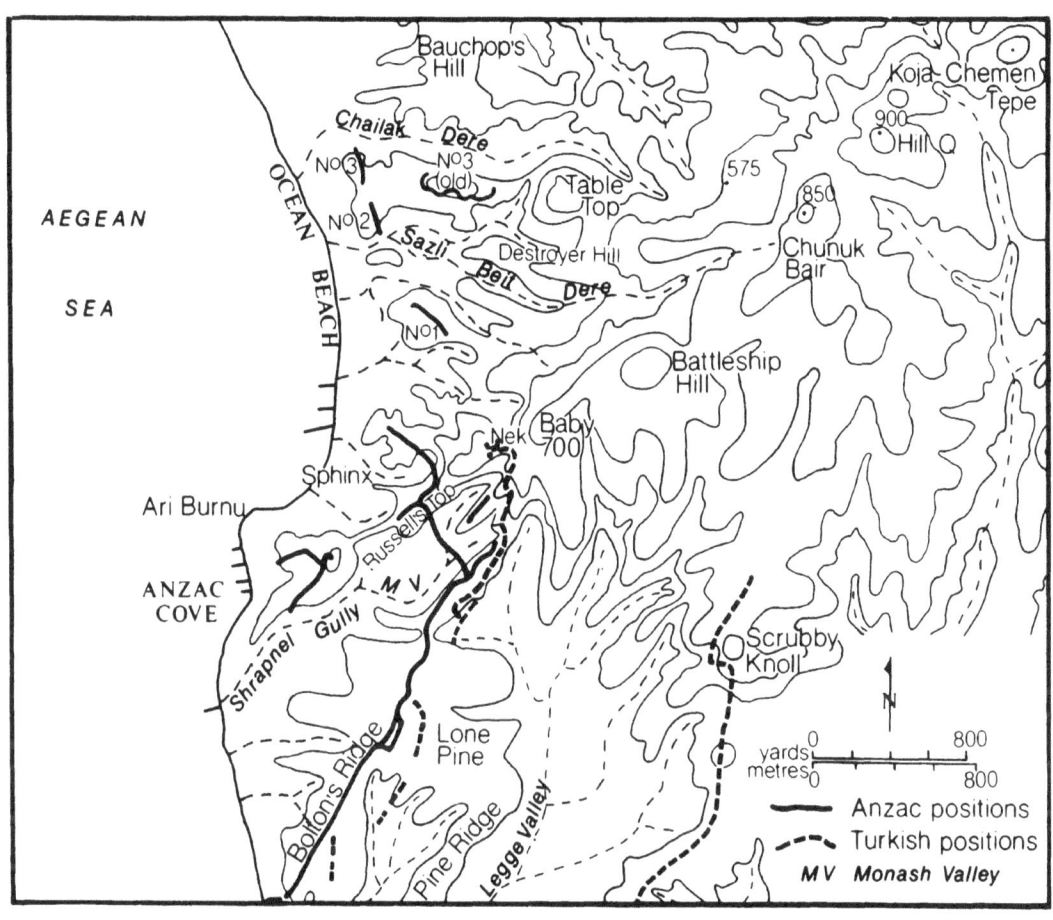

9 Gallipoli. The Anzac positions, May 1915

Hill 971. Each of these undefended peaks looked down on the rear of the Turkish position at Anzac and further east, the Narrows and the road to Helles. The 4th Brigade was to capture the northernmost crest, Hill 971, and the 29th Indian Brigade, in column behind it, would take Hill Q. The march began at 9.35 pm on 6 August and the column commander, Brigadier-General H.V. Cox, intended that Hill 971 would be secure six hours later. Instead, daybreak found Monash well short of the feature and uncertain of his exact position. Although Cox reinforced him with Major C.J.L. Allanson's 6th Gurkhas, Monash argued strongly against continuing the advance. A furious Cox finally relented. Subsequently, Allanson's criticism was scathing: 'Monash seemed temporarily to have lost his head, he was running about saying "I thought I could command men, I thought I could command men..." ...he said to me "what a hopeless mess has been made of this, you are no use to me at all".'[26]

But the remarks of others suggest that Monash had not lost his composure and that he performed creditably. He revived the flagging advance twice, told his commanders to leave behind sub-units to deal with parties of Turks fleeing from the New Zealanders' attack on Chunuk Bair and, as that resistance stiffened, deployed his battalions to meet it. Steady progress was impossible in the tortuous terrain. Monash can be criticised, however, for his inactivity when the attack was renewed on 8 August. Even though three of his battalions were involved, he placed one of his commanding officers in charge and then did not give him the communications necessary for the task. Meeting torrential machine gun fire, the battalion commanders squabbled among themselves about whether to withdraw. Because the supporting field ambulance had not been informed of the attack, wounded who would normally have been picked up were abandoned.

The fighting continued even after the loss of Chunuk Bair on 10 August removed the last chance of success at Gallipoli. On 21 and 27 August, Hill 60 was attacked in an attempt to join with the right flank of the 9th Corps at Suvla. With the strength of his brigade reduced to 890, these were Monash's last major offensive operations of the campaign. He had good reason to remember the assault on 21 August, 'Everybody got rattled' because Godley altered the artillery arrangements shortly before it was due to begin. Thereafter, Monash adopted Napoleon's aphorism: 'Order, Counterorder, Disorder', and insisted that once issued, orders should not be modified unless 'absolutely necessary for the safe conduct of the operation and the men'.[27]

The August battles surpassed even those of Monash Valley as an illustration of what men could endure. Beginning the offensive with bodies wracked by the dysentery and fever which had become epidemic at Anzac, they responded to call after call in the broiling summer heat. If a man could hold a rifle, he was deemed fit. Though spared any illness, Monash, at 50, was too old as a brigade commander to weather operations of the nature and prolonged intensity of Sari Bair. His inaction on 8 August might plausibly be explained by his tiredness after the night march. Birdwood forgot that he had been robust enough to withstand the sedentary battle at Anzac. Admitting that Monash looked after his brigade 'thoroughly well', he nonetheless could not 'look upon him as a leader in war'. Command of the two new divisions formed after the doubling of the AIF in Egypt in early 1916 went to others and Monash sailed with his 4th Brigade for France on 2 June 1916. Cox, now his divisional commander, had been highly impressed with its retraining and wrote in May that any 'little weaknesses' Monash had 'would not count so much in a Divisional command'. Godley concurred and Birdwood agreed to give him the 3rd Division which was raised in Australia and began to arrive in England in July.[28]

*Commander, 3rd Division*

The training of this formation ranked among Monash's finest achievements. It was based on familiar principles. He emphasised to the three brigade commanders the need for loyalty, *esprit*, 'obedience of orders—no dodging, cohesion of units—helpful spirit...Making the

9 Major-General John Monash, Commander 3rd Australian Division, 25 May 1918, at Glissy, just before receiving command of the Australian Corps. (AWM Negative No. E 2350)

best of all situations'.[29] Most of the 72 points listed by Monash after the first divisional march dwelt on the reluctance of junior officers to exercise their authority. '100 Hints' was revised and reissued for the third time. All units down to platoon were rotated through the same exercises to ensure tactical uniformity. An 'engineer platoon', composed mainly of miners, was formed in each battalion for the rapid digging of trenches. Monash obtained from Birdwood the latest details of the Somme fighting and incorporated them in the training. His own brief experience on the Western Front with the 4th Brigade was also useful. A brigade trench system was dug 'replete...in every detail with bomb stores, observer stations [and] complete wiring'.[30] Each brigade occupied it for five days at a time, conducting raids and patrols. On 6 November, four artillery batteries, signallers, engineers and four aircraft supported the 'assault' of five battalions on a trench system under which a mine had been exploded. The 3rd Division began to leave for the Western Front three weeks later.

The fourth and final factor in Monash's military development was the unbroken run of good luck which began with his appointment to divisional command. His staff was excellent, thanks to the care taken by Birdwood in its selection. The training facilities in England were far superior to those used by the 4th and 5th Divisions in Egypt. On arrival in France he joined Godley's 2nd ANZAC Corps as part of General Sir Hubert Plumer's Second Army. Plumer's exceptionally close partnership with Charles Harington, his chief of staff, symbolised the cooperative spirit which marked the relations of the Second Army's staff with the fighting troops. Monash agreed wholeheartedly: 'Harington's doctrine that all staffs exist to *help* units and not to make difficulties for them is the only one that can possibly lead to success and I am constantly preaching that doctrine myself'.[31] This was certainly true of his division's participation in the attack on the Messines–Wytschaete Ridge, the capture of which secured the southern flank of the Ypres salient and enabled Haig to launch the offensive forever associated with his name.

Even by the high standards prevailing in the Second Army, Monash's preparation was meticulous. He himself remarked: 'Everything is being done with the perfection of a civil engineering construction so far as regards planning and execution'.[32] Every aspect of the attack was covered in 36 instructions, the first of which appeared on 15 April 1917, thus avoiding the issue of a lengthy operation order before the attack. Monash's preliminary bombardment program listed 446 targets. But his requests for tanks to clear potentially troublesome strongpoints and additional Stokes mortars for gas bombardment were refused. In his advocacy of the use of artillery and other arms, Monash was already inclining towards the theory that he elaborated after the war:

> the true role of the Infantry was not to expend itself upon heroic physical effort...but, on the contrary, to advance under the maximum possible protection of the maximum possible array of mechanical resources...guns, machine guns, tanks, mortars and aeroplanes...to be relieved as far as possible of the obligation to *fight* their way forward....[33]

This theory was not original for Pétain had been preaching it since 1884. Nor was Monash's use of conferences unique. Careful briefings were a byword in Plumers' Army while many of the best British generals such as Maxse, Cavan and Jacob employed the method as well. But Monash was its most renowned exponent. In attendance were the infantry and artillery brigadiers and their staffs, the heads of the operational and administrative staffs and of all arms and services and frequently, the battalion commanders. All would be asked for their opinions after Monash had described the operation from start to finish. Inevitably they were lengthy affairs because, said Monash: 'I want to leave nothing to chance...we are going to talk these matters out to a finish and will not separate until we have a perfect mutual understanding among all concerned'. The conference on 29 April 1917 lasted over four hours and Monash's lucidity and breadth of mind were the outstanding features.[34]

Haig, who had earlier congratulated him on the obvious quality of the 3rd Division, came under his spell on 24 May: '[Monash] is in my opinion a clear-headed, determined commander. Every detail had been thought of.' Their contributions at these conferences also served to balance Monash's frequent intrusion into his commanders' responsibilities. Instead of allowing them to prepare their brigade plans within the framework of his divisional scheme, he told them virtually how they were to employ their battalions. As Brigadier-General Jobson wrote after receiving Monash's plan for the assembly: '[It] is the best possible and should work satisfactorily'.[35] Yet Monash's concern was justified. His division occupied the extreme right flank of an offensive front ten miles (sixteen kilometres) long, a responsible role for a formation about to participate in its first major battle. He had anticipated gas shelling during the approach march. Four separate routes were used and when the distressed infantry reached the assembly trenches they found water especially stored there.

Nineteen mines were exploded along the Army front as the infantry moved forward behind a dense creeping barrage to win the most complete victory of the war thus far. As some of the mines were begun two years earlier, assessments of this classic set-piece invariably reduce its significance by alluding to the time spent in its preparation.[36] But the mines were not among Plumer's reasons for success, which included thorough staff work and training and efficient counter-battery fire.[37] The tactical key was his setting of limited objectives which the infantry could reach in sufficient strength to consolidate. Unlike the Somme or Arras, the advance was deliberately stopped before resistance hardened. The quick counterattacks at which the Germans were so adept invariably broke against the heavy standing barrage in front of the newly captured line. Called 'Pétain tactics' after the commander who first employed it, this method of attack was the only viable one in the static warfare of 1914–17. It had been grasped by Plumer, Maxse and now Monash:

> I am the greatest believer in the theory of the limited objective. So long as we hold and retain the initiative we can in this way inflict the maximum of losses when and where we like. It restores to the offensive the advantages which are natural to the defensive in an unlimited objective.[38]

He adhered strictly to the principle within his own division for the two attacks in which it participated during the Third Battle of Ypres. The

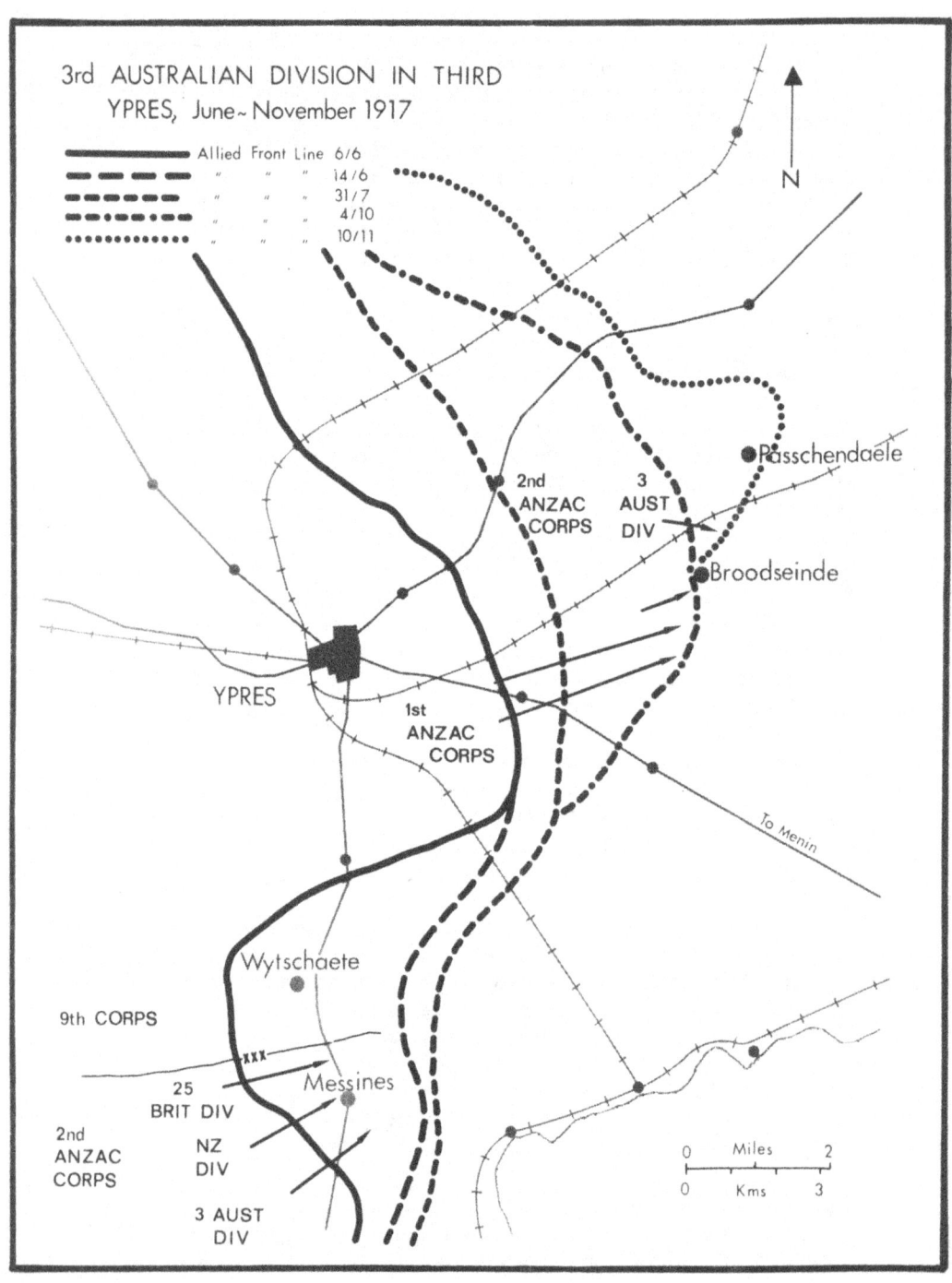

10 3rd Australian Division in Third Ypres, June–November 1917

Army scheme for the assault on Broodseinde on 4 October 1917 contained only one halt before the final objective but Monash's plan for the 3rd Division included two more. His attack was divided into a series of progressively shorter advances, each launched by fresh battalions. The issue of frequent staff circulars continued the smooth battle procedure established before Messines and permitted the accommodation of a hurried change in plan after Plumer decided to include the still-fresh 1st ANZAC Corps in the attack. Broodseinde was a smashing success but Haig's decision to continue the offensive despite torrential rain, turned the next blow, the assault on Passchendaele on 12 October into a defeat. There was little time to prepare for this attack on positions which were infinitely stronger than those at Broodseinde. Any slim chance of success vanished with the refusal of Monash's request for a postponement after the 66th Division failed to secure his start line. Floundering before Passchendaele in the waist-deep mud of the Ravebeek, the attacks were repulsed by machine guns untouched by the feeble barrage. Each yard gained cost the 3rd Division 35 men.

Messines, Broodseinde and Passchendaele showed Monash as a forceful, even bullying battlefield commander. In a war in which the absence of radio meant that generals seldom had 'the faintest idea of what had happened or what was happening' after the battle began, Monash would complain if the information he received from his brigadiers was an hour old. They had to provide enough runners to gain a comprehensive picture of the situation which was passed on to Monash by frequent telephone messages, usually from the brigadiers themselves. Hence Monash's unwavering belief that he should remain at his headquarters during battle was usually correct. It was the only place 'where I can have before me, all the time, a complete picture of what is going on, and...can at all times reach every possible subordinate...with the minimum of delay'.[39]

But sometimes the authority imposed by a commander's presence among his subordinates is essential. One such occasion arose on 8 June when a breakdown in communications and conflicting reports left both Monash and Brigadier-General W.R. McNicoll unaware of the progress of an attack by the 44th Battalion to recapture ground lost in a misdirected barrage. Monash did not appear at McNicoll's Headquarters. After Messines, he made no effort to reconnoitre any of his battlefields either. He prepared his plans after building up a mental picture of the terrain and its defences from maps and aerial photographs and discussions with his staff and his commanders as he had done before the war. This was the method of a commander with a powerful creative imagination and it usually worked for him. Again there were exceptions. Had Monash seen the state of the ground before the attack on Passchendaele, his objections to it might have been much stronger. His failure to do so was the likely cause of Birdwood's criticism of Monash's performance in this operation. But Birdwood was the absolute antithesis of Monash as a commander: 'He was always "buzzing about",' observed Monash, 'waking people up, perambulating all over the place, barely

10  Entrance to the General Staff office, headquarters 3rd Australian Division, in the old ramparts of Ypres, 21 October 1917. From here Monash directed the operations at Broodseinde and Passchendaele Ridge. Note the signal wires leading to brigades and other formations. (AWM Negative No. E1184)

ever at Headquarters and not *really* exercising any command at all.'[40]

Unlike Birdwood, Monash never went near the trenches. His chief staff officer in the 3rd Division, Colonel G.H.N. Jackson, said that he 'never got him nearer than the third line and he did not seem to enjoy even that'. Creative imagination, or what Jackson called 'seeing through other men's eyes', rendered such visits unnecessary from Monash's standpoint. As Harington recalled: 'He would tell you which duck-board needed repairing'. The effect on the morale of his men who never saw their commander under the conditions they were experiencing was another matter. No other Australian general shared Monash's view on this question.[41]

Monash's final battle as a divisional commander began shortly after the great German offensive on the Western Front opened on 21 March 1918. Arriving at Franvillers at 1 am on 27 March, he was directed to fill the ten-mile (sixteen kilometre) gap which had yawned the previous afternoon between the Ancre and the Somme to expose Amiens. Monash's ability to visualise a plan unfolding was never more evident than in his dictation of his orders. Bean concluded:

*General Sir John Monash*

> [It] shows Monash's great powers of grasp and of lucid expression at their best—the officers to whom they were read at the time recognized, with a flash of pride, 'the old man's' masterly touch. The situation that called for each phase of action was clearly explained and the action then crisply ordered.[42]

Dispatch riders rushed his instructions to the 10th and 11th Brigades then moving to join him. The staff representatives he had brought with him began the preparation of their own orders. The 11th Brigade began to arrive at 8 am and by the following day, Monash felt 'quite secure against all but an attack on a grand scale'. The German drive on Amiens was blunted before it reached his line, allowing its occupation undisturbed. Nevertheless, the 3rd Division's deployment was an outstanding example of efficient battle procedure and, not least, as Bean noted, Monash's 'quick, cool, successful planning'. But his uncharacteristically feeble coordination of an abortive sally towards Sailly-Laurette on 28 March suggests that he had been exhausted by the effort. If tiredness was the cause of that failure, then the criticism of his neglect of 8 August 1915 also applies.[43]

*Commander, Australian Corps*

The Sailly-Laurette setback was overshadowed by the fulsome praise Monash received for his performance on 26–27 March. It impressed Birdwood at a time when he was considering his successor as commander of the Australian Corps. Comprising the five Australian divisions, the corps had been formed on 1 November 1917, one month after Birdwood's promotion to general made inevitable his eventual appointment to Army command. Though reluctant to relinquish it, Birdwood deferred to Haig's opinion that he was denying the opportunity to an Australian. He selected Monash as his replacement not only because he had proven himself as a divisional general, but because he was senior to the only other real contender, the chief of staff of the Australian Corps, Brudenell White. Birdwood assumed command of the Fifth Army and took White with him as its chief of staff. White's principal admirers, the correspondents C.E.W. Bean and Keith Murdoch, were dismayed, feeling that the abler soldier had been displaced by a man with a pronounced capacity for self-advertisement. Thus began a sordid intrigue to replace Monash which persisted several weeks after he assumed command and formed an unnecessary distraction at a vital stage in the war.

For Monash, command of the corps was 'something to have lived for...'[44] With some 166 000 men, it was by far the largest in the British Expeditionary Force (BEF). As he admitted, the timing of his appointment was another example of his good fortune. The long winter rest out of the line had raised morale in the Australian Corps to a peak, while on the battlefield it was gaining a tactical ascendancy over the Germans which it never lost. This domination was exemplified by the consistent

*11* Monash with the senior staff officers of the Australian Corps. Bertangles Chateau, 22 July 1918. From left: Brigadier-General C.H. Foot, Chief Engineer, Brigadier-General R.A. Carruthers, DA & QMG, Brigadier-General T.A. Blamey, BGGS, Brigadier-General L.D. Fraser, BGHA and Brigadier-General W.A. Coxen BGRA. (AWM Negative No. E. 2750)

success of Peaceful Penetration, that process, begun in April, whereby patrols seized prisoners and cut off posts in the German line. Australians, or men who had lived in Australia for many years, now led the divisions. Major-Generals T.W. Glasgow and C. Rosenthal were appointed to command the 1st and 2nd Divisions, Major-General J. Gellibrand replaced Monash in the 3rd and Major-Generals Sinclair-MacLagan and J.J.T. Hobbs remained in command of the 4th and 5th. By mid-1918 they had been fighting for over three years and the appellation of 'amateur', a derisory reference to their pre-war militia experience, was hardly fair. Monash's relationship with his chief of staff, Brigadier-General T.A. Blamey, resembled the Plumer–Harington combination. He praised Blamey as having 'an extraordinary facility of self-effacement, posing always and conscientiously as the instrument to give effect to my policies and decisions'.[45]

Monash's philosphy was unchanged. The training of staff officers had to be broadened because its present form encouraged them to work in 'watertight compartments'. Fortnightly conferences with the divisional commanders helped the creation of 'a unity of thought and policy and a unity of tactical methods throughout the whole corps'. To introduce himself as corps commander, Monash addressed men from every arm and service, usually beginning with an appeal to their pride and their unit's prestige and concluding with 'Standards and soldierly behaviour;

public spirit; every man pulling his weight'. Junior officers were the target of a separate appeal; they were 'to create and foster...a sense of responsibility to themselves, their commanders, their comrades, their men, their country and their cause'.[46]

## The Australian Corps at Hamel

Monash's first major operation as corps commander was the attack on Hamel spur on 4 July 1918. Its defences blocked any eastward movement of the line between Villers-Bretonneux and the Somme. North of the river, Rosenthal's recent success at Morlancourt and Sailly-Laurette had advanced his line so far beyond the spur that its guns were engaging his right flank in enfilade and in rear. Fearing heavy casualties at a time when the Australian Corps could ill-afford them, Birdwood and White had rejected the attack when General Sir Henry Rawlinson, to whose Fourth Army the corps was allotted, proposed it as a feint in April. In mid-June Monash conceded that it was the most useful operation that could be mounted on his front but, possibly still uncertain of its cost, he recommended deferral. Meanwhile, Brigadier-General A. Courage's 5th Tank Brigade, supporting the Fourth Army, completed its re-equipment with the Mark V tank. Monash and Blamey attended demonstrations of the improved machine and were impressed by the view of the Tank Corps commanders that it 'would so increase the capacity of infantry and artillery that decisive defeat might be inflicted on the Germans before winter'.[47]

It is unclear whether Monash or Rawlinson first concluded that an attack on Hamel was possible immediately if tanks were used. But on 19 June Courage discussed the operation with Monash, and next day sent him his plan for an attack on a frontage which gradually increased to 7500 yards (7000 metres), by four companies of tanks in three echelons. No preliminary bombardment was necessary as the German wire would be crushed by the first echelon, moving as quickly as possible to the rear of the position to prevent withdrawal or reinforcement. Leading the infantry onto its objectives was the second and largest echelon while the third echelon mopped up any remaining opposition. Heavily influenced by the attack at Cambrai in November 1917, Courage's scheme embodied the doctrine of Lieutenant-Colonel J.F.C. Fuller, later to become the most famous of Tank Corps officers. Essentially a tank plan, it sacrificed some of the accepted tactical requirements of other arms to ensure that the tanks' mobility was unimpeded. Thus, there would be no creeping barrage because its linear shape and slow pace imposed unacceptable restrictions on tank movement.

Monash did not alter the concept of the operation but Sinclair-MacLagan, whose 4th Division had the major role, was unconvinced. Attacking without a barrage at Bullecourt in April 1917, it suffered grievous losses when the supporting Mark IV tanks failed to arrive on one night and could not reach the German wire on the next. Sinclair-MacLagan and his brigadiers urged the retention of a creeping barrage with the infantry advancing close behind it. As the leading echelon could not move between the infantry and the barrage, its tanks would augment those of the main body. It was a dramatic revision of the concept: it 'ceases to be primarily a tank operation', wrote Monash. 'It becomes an infantry operation in which the slight infantry power receives a considerable accretion by the addition of a large body of tanks.'[48]

Monash could adduce strong arguments in favour of Courage's scheme and against Sinclair-MacLagan's. The Mark V tank was much more reliable than its predecessor and would be advancing over ground which was no obstacle. The intelligence available suggested that strong resistance was unlikely. But if Monash had overruled them, Sinclair-MacLagan and his commanders would have been forced to execute a plan in which neither they nor those they led had any confidence. The optimism always emphasised by Monash would be extinguished by the spectre of Bullecourt before the attack began. Monash's deference to Sinclair-MacLagan's objections was a potent example of his use of psychology before the battle but far from the only one. He insisted that the tanks and infantry advance level with each other, despite the anxiety of the tank commanders who feared that shells falling short would hit their machines. The latter were made subordinate to their infantry counterparts. Each battalion rehearsed with the tanks alongside which it would attack, the tank crewmen discussing every aspect of the operation with the infantry and taking them for joy-rides. This was the final reassurance.

Monash added important details. Aircraft were to drop ammunition to the most forward troops, imitating the technique used by the Germans on the Lys and the Aisne. After desultory barrages by heavy guns, shellholes were plotted for use as cover once the infantry reached their objectives. Barrages and diversionary operations on the flanks would confuse the Germans as to the sector under attack. Harassing missions and flavoured smoke were fired daily so that the Germans would not regard them as unusual at zero hour.[49] Courage's suggestion that aircraft fly over the German lines and the tanks to drown the noise of their assembly was adopted by Monash. He increased Courage's request for smoke on the flanks to a screen along the entire front. To preserve secrecy, ammunition, stores and guns were moved forward by night with 'police' aircraft reporting next day on camouflage and signs of unusual activity. Conferences assumed a new importance because as little as possible was committed to paper. Monash's final meeting on 30 June was attended by 250 officers and lasted four and a half hours as he worked through an agenda listing 133 separate items. Subsequently, 'no fiddling with the plan was permitted'.[50]

A major change threatened as the attack was about to be launched on 4 July 1918. Four of the eight American companies participating had been withdrawn the previous morning. The American C-in-C, General J.J. Pershing, was not informed of their involvement until 2 July and considered that the use of his partly trained troops contravened the agreement with Haig whereby they were attached to British troops not for operational use but to gain experience. At 4 pm on 3 July Monash

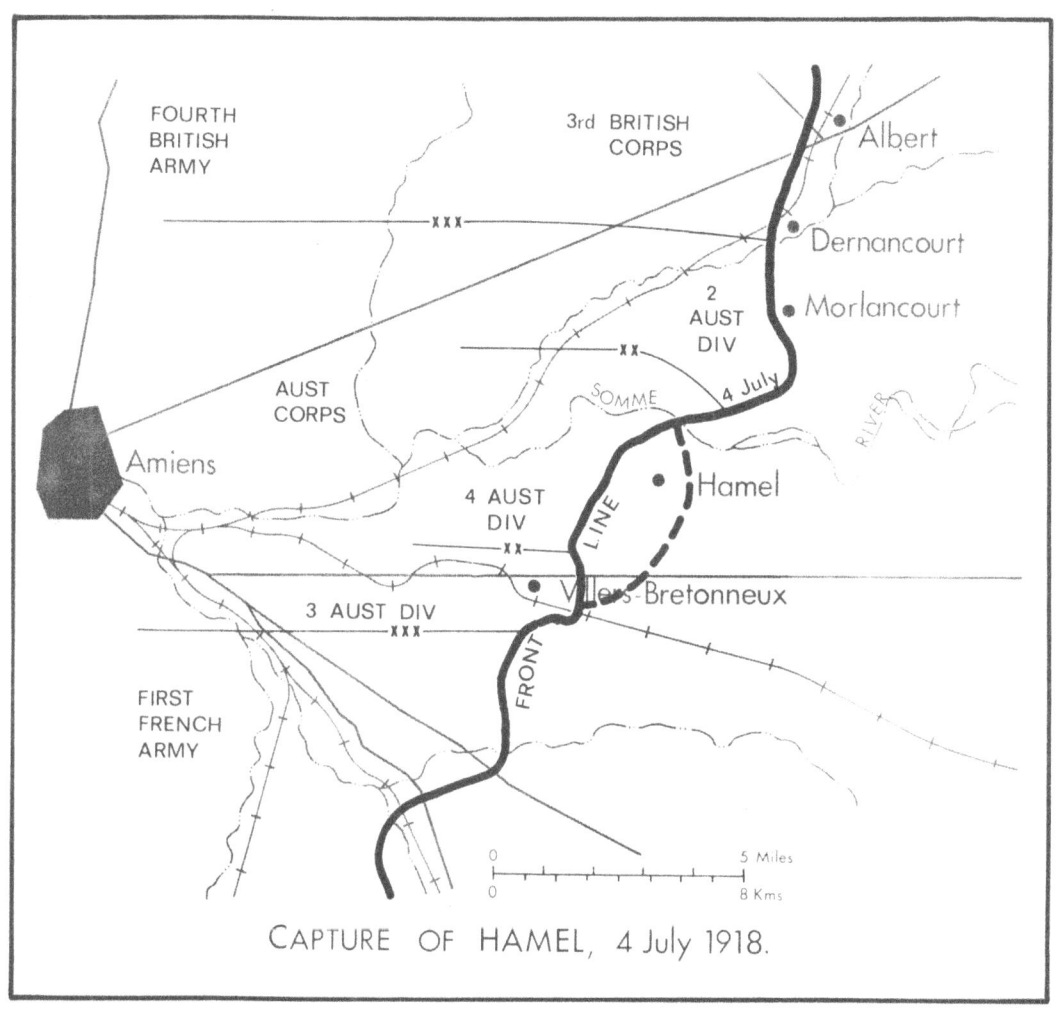

11 Capture of Hamel, 4 July 1918

learned that the remaining four companies were to be withdrawn as well. He told Rawlinson that it was too late to comply and that the dislocation if the Americans did not take part would force the abandonment of the attack. In the meantime he intended to proceed unless ordered to the contrary. Rawlinson had until 7 pm, when the infantry commenced its final move to the start line, to do so. Haig, whom Rawlinson contacted with a few minutes to spare, directed the attack to continue. It began at 3.10 am and was all over 93 minutes later. At a cost of 1400 casualties, over 1600 prisoners and 177 machine guns were captured. The closeness between execution and intention made Hamel the ideal illustration of Monash's famous analogy: 'A perfected modern battle plan is like nothing so much as a score for an orchestral composition, where the various arms and units are the instruments and the tasks they perform are their respective musical phrases'.[51]

Although four independent arms, infantry, artillery, armour and aircraft had been coordinated on the battlefield, the cooperation between infantry and tanks attracted the most attention. The use of tanks enabled each of the eight attacking battalions to assault on a frontage which Ludendorff had allotted to a division on 21 March. One of two staff sheets published by GHQ on the battle asserted: 'The value of tanks in assisting infantry to advance was conclusively proved'. They crushed German machine guns as soon as the infantry located them. But when the liaison broke down, as at Pear Trench which was missed by the tanks in the darkness and heavy fog, the infantry losses were severe. That incident supported Courage's view that without them, the infantry would have suffered severely or been unable to advance at all. Each tank had carried ammunition and water for the infantry while four carrier tanks dropped the ammunition and defence stores normally carried by 1200 men. But the true texture of the operation must not be forgotten: the tanks, like the aircraft and the artillery, formed part of the maximum array of mechanical resources *supporting* the infantry. There was absolutely no resemblance to Guderian's armoured thrusts in France in 1940 or Russia in 1941. Contrary to the views of some historians, Monash cannot be considered the first exponent of blitzkrieg.[52]

Monash's own claims for the battle were exaggerated. It was launched neither from a desire to assuage the 'anxiety and nervousness of the public,' nor to rekindle the offensive spirit in his superiors, but because of local tactical reasons.[53] General Mangin's Tenth French Army had attacked three times before 4 July and then two weeks later, smashed Ludendorff's last great offensive. Supported by over 2000 guns, 1000 aircraft and nearly 500 tanks, 24 French divisions advanced six miles (10 kilometres) to take 15 000 prisoners and 400 guns in what the Germans acknowledged was the turning point of the war. It was this battle, the counterstroke on the Marne, and not Hamel, which influenced Allied thinking in the manner described by Monash. His role in the origins of the famous offensive before Amiens on 8 August was further convincing proof that as a corps commander, he was far removed from the strategic direction of the war.

*'The black day of the German Army'*

In 1919, Monash told his close friend, Major-General J.H. Bruche, that while credit for this offensive was not wholly his, he was 'the prime mover in the events from July 1st to 21st'. He contended that in several meetings with them after 4 July, he impressed upon Rawlinson and his chief of staff, Major-General Sir A.A. Montgomery, the desirability of action on a decisive scale which his corps could undertake if its front were shortened from three to two divisions and if the Canadian Corps advanced on its right. On 15 July, they told him the Candians would be available before asking Monash about his flank protection needs in tanks and guns. Monash claimed that this conversation 'was really the genesis of the whole plan,' although he could only 'with very great difficulty' persuade Rawlinson to advocate it to Haig.[54]

This version of events belies reality. The Amiens stroke originated as the northern pincer of a double offensive proposed by Marshal Foch in April to eliminate the salient created by the German attack the previous month. Ludendorff's thrust in Flanders forced a temporary postponement but in May Foch expanded the scheme considerably. Rawlinson would launch a surprise assault, using tanks, with the Morcourt–Harbonnières ravine, five miles (eight kilometres) away, as it objective. Haig intimated to Rawlinson that he would receive the Canadian Corps for the operation and that leapfrogging would be necessary. The German attack on the Aisne necessitated another delay but Rawlinson resurrected the scheme after the successes of Hamel and Peaceful Penetration demonstrated the weakness of the German defences and the ascendancy gained by the Australians.

So much for Rawlinson's disinterest! On 17 July he sent Haig extremely detailed proposals. Yet this was a mere two days after his formative conversation with Monash! There is no doubting Monash's keenness to exploit the Hamel success, but the reduction of the salient it had created on its right flank was the plan he had under review. Monash was naturally preoccupied with it as it seemed likely to be the corps' next major operation. Even if he was thinking in terms of a wider offensive, he had no idea that Foch, Haig and Rawlinson had been doing so since April. As a corps commander, he was not privy to planning at their level. Bean's conclusion was appropriate: 'Monash did not devise the August offensive, though of course he was responsible for many of the details in the plan for his own Corps'.[55]

Monash was indeed responsible for most of the details. The Australian Corps was to attack between the Somme and the Amiens–Nesle railway, penetrating 9000 yards (8.3 kilometres) on a frontage which gradually expanded from 7200 yards to 9000 yards (6.6 kilometres to 8.3 kilometres) while advancing through three objectives. Currie's Canadians would attack alongside with Lieutenant-General R.H.K. Butler's 3rd Corps clearing the northern bank of the river. Monash protested that the first objective (about 2000 yards away) fell short of the German gun line, allowing them to withdraw their artillery during the pause before the attack on the second objective.[56] Rawlinson acceded, lengthening the initial advance to 3500 yards (3200 metres). A second objection was overruled. Rawlinson refused to extend Monash's line north of the Somme and left Butler in charge of the sector. Monash had always disliked the use of a prominent natural or artificial feature as a boundary: 'It creates a divided responsibility and necessitates between two independent commanders...a degree of effective co-operation which can rarely be hoped for'.[57] Like Birdwood and White when Rawlinson adumbrated the plan to them in May, Monash was concerned by the long, finger-shaped Chipilly Spur, the guns on which could enfilade the length of his advance. Rawlinson compromised by moving Butler's first objective east of Chipilly.

The most novel features of Monash's plan were the arrangements for the assembly of the attacking formations and the extent of his leapfrogs. The 2nd and 3rd Divisions were to move through the 4th and 5th to attack the first objective. Once taken, the 4th and 5th would leapfrog them to begin the advance on the second objective, 5000 yards (4600 metres) away. The reserve battalions of the brigades which had captured it would then carry out a third leapfrog to the final objective. By placing the 4th and 5th Divisions closest to the start line, Monash had shortened the approach marches of the formations with the longest advances by almost three miles (4.8 kilometres). He claimed that this gave 'a proper distribution of troops to objectives and ensures a minimum of fatigue under ordinary conditions'. Monash acknowledged that in such complexity lay grave risks but he trusted the intelligence of his troops and 'the sympathetic', loyal and efficient cooperation 'of staff and commanders at all levels'.[58] He assisted them by ordering the preparation of roads and light railways for motor traffic, mules and men. All engineers and pioneers were brigaded for the task and most of the 13th Light Horse was allotted to traffic control. It was a repetition on a much larger scale of the care Monash had lavished on the 3rd Division's approach march before Messines. Bean called his plan for the assembly of the corps 'John Monash's masterpiece', adding, 'the elaborate placing of the brigades and the timing of their starts so that each punctually took up its post in the intricate task, affords what will probably be the classical example for the launching of such operations'.[59] The scale of the leapfrogs was also astonishing. They had frequently been executed *within* divisions, as Monash had done at Broodseinde, but rarely *by* divisions. Now he proposed two leapfrogs and by two divisions side by side to ensure that fresh troops attacked each objective.

The infantry received maximum assistance. Each division was supported by 24 tanks and during the second phase, each brigade by an artillery brigade, an engineer and a machine gun company and a field ambulance. Such a grouping gave brigade commanders the freedom needed to undertake a deep penetration without a creeping barrage. Monash described this phase as resembling an advance in open warfare: 'The place of the protective barrage will be taken by tanks and the advance will be supported by artillery'. Conversely, the attack behind a creeping barrage during the set-piece first phase reproduced 'the conditions of Hamel on a much larger scale'.[60] But easing the task of the infantry did not preclude the acceptance of heavy casualties if Monash thought the objective valuable enough. The repair of the St Quentin Road as the infantry advanced along it was worth the sacrifice of an entire pioneer battalion because of the 'enormous effect' produced by roadbound armoured cars as they passed through the infantry to raid German rear areas.

Other aspects of the operation reflected Monash's experience in previous commands. Orders were conveyed in 21 'Battle Instructions'. No alteration to the plan was permitted after the final conference on 4 August which lasted over four hours as Monash ran through the various divisional plans. Smoke was included in the barrage. So that the loss of even one tank 'would impair no part of the battle plan other than the fraction to which the Tank had been allotted', the principle 'one Tank, one Task', was rigidly enforced. The measures introduced by Monash before Hamel to conceal the concentration of tanks and infantry were applied throughout the Fourth Army. But on 6 August, the Germans raided the 3rd Corps and took 250 prisoners, prejudicing secrecy and reawakening Monash's fear about that formation's ability to take Chipilly Spur. At his instigation, Rawlinson told Butler to regain the lost ground immediately. Monash warned Sinclair-MacLagan to form a defensive flank in case Chipilly should remain in German hands. The failure of Butler's attacks on 7 August was the first indication of how closely Monash had anticipated likely events. On that day, the men were listening to his appeal to every one of them to carry on to 'the utmost of his power, for the sake of AUSTRALIA, the Empire and our cause' in this battle 'which will be one of the most memorable of the whole war'.[61]

Many of them were thrilled by this message and it was to prove highly prophetic. The attack began in dense fog at 4.20 am. By midday the Australian flag had been hoisted over Harbonnières. The armoured cars ranged as far ahead as Vauvillers, Framerville, Proyart and Chuignolles, firing on headquarters, transport columns and billets. As Monash expected, Chipilly had not fallen and Sinclair-MacLagan (4th Division) had to protect his flank by deploying his reserve along the river. At the end of the day even the most advanced troops received a hot meal and a drink. For the loss of 2000 men, the Australian Corps had captured 7925 prisoners and 173 guns. Sir Joseph Cook's son, a major in the 4th Division, wrote to his father: 'The organization of the show was wonderful. Monash seems to be making good. I have seen nothing to equal it. It puts fresh heart into one to see evidence of the master hand.'[62]

The Canadians were equally successful and by nightfall General Debeney's 9th and 31st Corps which had attacked later, joined them on the final objective. Ludendorff's famous comment described the strategic effect of the battle: 'August 8th was the black day of the German Army...[It] put the decline of its fighting power beyond all doubt...The war must be ended.'[63]

Nevertheless the German reaction was exceptionally swift. Next day Crown Prince Rupprecht's Army Group Headquarters believed it had deployed sufficient reserves to halt the advance. Its confidence was justified in view of the confusion among Allied commanders. On 26 July Foch had ordered the continuation of the attack after the first day to the line Roye-Chaulnes, another five miles (eight kilometres). One week later, Haig told the Generalismo that he had instructed Rawlinson to drive a further fifteen miles (24 kilometres) to Ham, across the Somme. These modifications transformed the attack into an unlimited offensive but Rawlinson continued to think of it in terms of the limited operation he had always favoured. Not until 6 August did he direct Currie to exploit towards Roye-Chaulnes on the 9th while Monash, pivoting on the Somme between Méricourt and Etinehem, advanced his right flank in

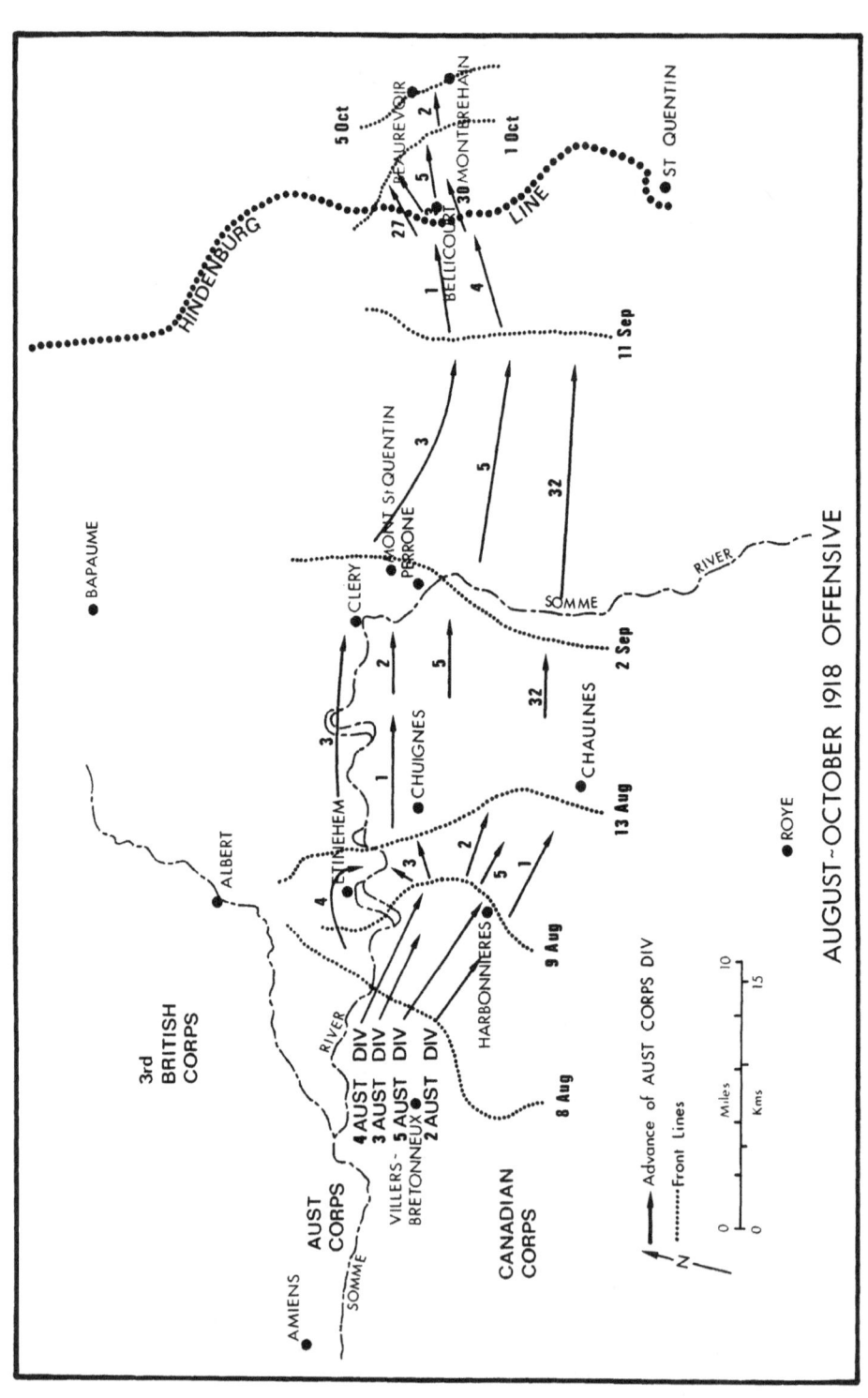

12 August–October 1918 Offensive

conformity. Although disturbed by the conflicting views of his superiors, Monash, too, thought of the objective as 'Strictly Limited', explaining to Bean on 5 August that he intended 'to go only for guns and leave this junction [Chaulnes] to be dealt with by our guns'.[64] What he had learnt at Messines applied to this as it did to every other attack:

> in every offensive operation, large or small, a definite limit was set to the task to be performed...under no circumstances whatever, no matter how tempting, were these limits to be exceeded...To allow troops a free hand to exploit a local victory by continuing their advance indefinitely, had often led to complete disorganization and an inability to resist the shock of the enemy's inevitable reaction.[65]

*Exploitation*

The conduct of the exploitation accounts for the naming of the battle after its opening day. The Australian 1st Division, just arrived from Flanders would advance on the left of the Canadians towards Lihons, with the 5th Division initially, and then the 2nd on its own left flank. Because Rawlinson did not issue his preliminary orders until 6 August, Monash was unable to brief Glasgow of the 1st Division until 7 August, and the other divisional commanders on the morning of 8 August. Formal orders did not arrive until the evening of 8 August as Rawlinson had waited for the results of Debeney's attack. They were based on a false premise because heavy tank losses to direct artillery fire had reduced the number available to the Australian Corps to 21, far from the 'strong body of tanks' envisaged for the operation.[66]

The haste surrounding it militated against success. At midnight Monash was still waiting for Currie to fix zero hour. Then Montgomery countermanded the replacement of a tired Canadian division even though its relief was underway. New orders had to be drafted but there was insufficient time for their dissemination. Consequently, the 1st Division had no chance of reaching the start line and so Elliott's 15th Brigade attacked in its place, without a barrage and with a single tank lent by the Canadians. Glasgow finally leapfrogged at 1.30 pm while Rosenthal (2nd Division), supposed to attack alongside him, was unable to begin until 4.30 pm. All formations lost heavily as the attack reached the foot of Lihons Hill in the midst of the old Somme battlefield. Lihons was not captured until 11 August. By now there were only 38 tanks left in the Fourth Army, so that these were predominantly infantry attacks, reminiscent in most respects, including casualties, of those in 1917. The lessons of the earlier operations went unheeded. Glasgow was informed at midnight on 9 August of Currie's decision to attack at 8 am the next morning, while the barrage plan arrived too late for issue to the assaulting companies.

The cause of the trouble was Rawlinson's consistent failure to coordinate the attacks. He had twice left the setting of zero hour to Currie, an incredible lapse. As formations other than the Canadian Corps were involved, the start time was clearly Rawlinson's responsibility. Like everyone else in the Australian Corps, Monash had to await

Currie's decision. He could only order his divisional commanders to confer closely with Currie's to save time. This is not to say Monash was blameless. He failed to ensure that Rosenthal understood his intention for the attack on 9 August. The latter's methodical orders were devoid of any sense of urgency and suggest that Rosenthal was thinking of a deliberate attack launched only when all preparations were complete.

Monash redeemed himself with a brilliant display of generalship on the northern bank of the Somme for which Rawlinson had finally given him responsibility. The 13th Brigade (4th Division) was to seize the Etinehem Spur while McNicoll's 10th Brigade (3rd Division) swung the line forward on the southern side to Méricourt, straightening the bulge caused by the attack of Glasgow and Rosenthal. Monash regarded the operation as 'a species of investment' for these U-shaped spurs did not have to be taken by frontal assault but could be sealed off at the base to prevent the escape of their defenders. Tanks were to accompany both columns, mainly for the psychological effect upon the Germans who would hear but not see them.

Launched at 9.30 pm on 10 August, the 13th Brigade's attack succeeded brilliantly but McNicoll's was bombed by aircraft soon after it commenced. Fully alerted, the Germans repulsed the column with a hail of machine gun and anti-tank fire. Feeling ran high in the 10th Brigade but there was no denying the boldness of the scheme. Bean called it 'ingenious'. Méricourt and Proyart fell on 11 August. The day before, Haig, warned by Currie of increasing resistance, ordered Rawlinson to remain on his present line while Byng's Third Army prepared an assault against Bapaume. Visiting Monash, the C-in-C told his divisional commanders: 'I have the most complete confidence in your corps commander—I want you to know that I trust him completely'. On 12 August the King knighted Monash on the steps of the Australian Corps Headquarters at Bertangles Chateau.[67]

A brief but much needed rest followed. After complaints from Monash about the impossibility of relief while the Corps had to hold a line some nine miles (14.7 kilometres) long, Rawlinson reinforced it with the 17th British Division. Between 12 and 23 August, every Australian division enjoyed a respite which was essential for the maintenance of the continuous offensive in which it was engaged and during which Monash 'regarded it as a fundamental principle to employ whenever possible absolutely fresh and rested troops for an operation of any magnitude or importance'.[68] It had been a 'fundamental principle' in the 4th Brigade and the 3rd Division and his employment of each division in the attacks on and after 8 August demonstrated its application in the Corps. On 22 August 1918, the 1st and 4th Canadian Divisions come briefly under his command, expanding it to eight divisions and virtually making Monash an army commander. Liddell Hart used his example to refute the post-war contention 'which has become a dogma, that a commander cannot possibly handle three or more sub-units'.[69] He concluded quite rightly that the size of the corps was an important reason for Monash's success for it endowed him with a flexibility denied his contemporaries who commanded a conventional corps of three divisions.

Byng's Third Army offensive began on 21 August and received valuable assistance on its right flank from the 3rd Division which cleared Bray three days later. On 23 August Rawlinson increased his cooperation with Byng as both armies attacked on a frontage of 33 miles (53 kilometres), the right extremity of which was the Australian line south of the Somme. Monash selected the objectives in that sector, ordering the capture of all but the northern portion of the Chuignes Valley. The attack was a replica of Hamel but Monash made some uncharacteristic blunders. On 21 August he had expanded the objectives and then, less than 24 hours before the battle, added the Froissy Beacon to them, necessitating an advance across the remainder of the valley. The importance of Froissy was obvious from any map. It dominated the surrounding valleys and had to be taken if the gains further south were to remain tenable. The operation represented a departure from two of Monash's strengths: an ability to recognise the tactical significance of ground and fixity of plan. But according to Bean there was 'never...a more completely successful fight made by one single division' than that of the 1st Division on 23 August. At a cost of 1000 casualties it took one-quarter of the 8000 prisoners captured on this day by the Third and Fourth Armies. The inconvenience forced upon him had tested Glasgow's generalship and Monash was generous in his praise, attributing much of the success to him.[70]

In accordance with directives issued by Haig in the last week of August, Byng continued the main thrust on Bapaume with Rawlinson advancing astride the Somme to protect his right. On 24 August, the latter ordered Monash to 'keep touch' with the 3rd Corps north of the river while on the southern bank, 'no opportunity will be missed of making ground towards Péronne'. When Haig denied him reinforcements because his was the subsidiary offensive, Rawlinson revised this policy on 25 August. The Fourth Army would 'mark time and await events elsewhere'. Believing that the fruits of Chuignes would be lost, Monash disagreed and fell back on an order which had been superseded to circumvent his Army commander. On 26 August he told his divisional commanders that aggressive patrolling was the means by which the offensive would be continued on his front: 'Close touch will be kept with the enemy in this way and advantage will be taken of any opportunity to seize the enemy's positions and to advance our line'.[71]

This was the only time Monash rejected the limitations imposed by his superiors and it demanded great moral courage. He was adamant that the advantages already won must not be frittered away by allowing the enemy even a brief respite, for then even greater demands would have to be made on his tired troops. Monash allotted each division a lane along which it moved on a single brigade frontage. The leading brigade would be relieved only when it reached the limits of its endurance, calculated by Monash to be at least two days. By employing these tactics, only one-sixth of his infantry was committed at once and no time-consuming divisional reliefs were necessary so that the momentum was maintained.

Two field guns were added to each battalion to destroy pockets left behind to delay the advance, and at Monash's direction, one-fifth of the ammunition carried by each battery was to consist of smoke to blind the enemy. By 28 August the 3rd Division had reached Cléry after capturing Suzanne, Vaux and Curlu while on the other side, the 2nd, 5th and 32nd British were approaching the Somme bend.

*Mont St Quentin*

At Cléry the Somme abruptly abandons its northerly direction to follow a westerly course to the sea after making a left-angled turn past Péronne. This section of the river was canalised but, extending 2000 yards (1840 metres) to the east of the canal, were marshes too deep to be waded. Overlooking the river and the main bridge which carried the road and railway into Péronne was Mont St Quentin, 300 feet (96 metres) high, devoid of cover and defended by an excellent division, the 2nd Prussian Guards. This feature also dominated Cléry, lying at the foot of the Bouchavesnes Spur and connected by a footbridge to Ommiécourt on the opposite bank. Péronne, too, was a formidable obstacle with Vauban's ramparts towering 60 feet (20 metres) above the Somme. From the line reached on 28 August, Monash knew that the Germans would have to withdraw over the river next day. On the afternoon of 28 August, he ordered Rosenthal (2nd Division) to seize the Péronne bridge and then attack the Mont while the 5th and 32nd Divisions were to capture crossings further south. In summary, Monash was proposing a frontal assault against the river line on 29 August, hoping to swamp the Germans before their retirement was complete.

The gamble failed. The 2nd Division could not cross the marshes, while all but the Cléry crossing were either destroyed or drenched by torrential fire. Monash did not mention the reverse in *Australian Victories*. In one of several glaring distortions in that work, he claimed to have dismissed the frontal assault immediately as 'a costly enterprise, and fraught with every prospect of failure...'[72] Only now did he adopt the plan of turning the line of the Somme from the north which he claimed had been forming in his mind for some time.

Monash outlined the scheme at 5 pm on 29 August. Each division would sidestep to the left, taking over most of the frontage held by its neighbour, which brought Rosenthal's 2nd Division opposite Ommiécourt. Once Gellibrand's 3rd Division had taken Clery and Bouchavesnes, Rosenthal was to cross behind Gellibrand's front and attack Mont St Quentin from the northwest. Hobbs's 5th Division would follow through the 2nd to take Péronne. But if Hobbs secured the main crossing at Péronne, Rosenthal would follow him. During the night engineers were to build as many crossing points as possible for the infantry on the 30th. Bridge repair generally was an important factor in the plan. Ever since the advance had commenced, Monash gave as much attention to the restoration of captured territory as to driving the Germans eastward. Engineers began the repair of ruined crossings as soon as the infantry reached them, permitting rapid switches of men and material from one bank to the other. Work was almost completed on the Feuillères crossing, 3000 yards (2770 metres) west of Ommiécourt, even as Monash planned the assault on the Mont.

All did not go according to plan. The need for haste left commanders little time to brief their subordinates. Finding the Cléry crossing unapproachable at 4 am, Rosenthal's 5th Brigade made for Feuillères. As this would take two hours, hopes of launching the attack at 5 am had vanished. It was postponed for 24 hours. In the meantime, the exhausted troops of the 9th and 10th Brigades (3rd Division) cleared Cléry, enabling the 5th to move up behind the town after completing the crossing. At 5 am on 31 August, its three assaulting battalions, each a pathetic average strength of 330 men, attacked under a heavy bombardment, making as much noise as possible to convince the Germans their numbers were greater. The defenders were overwhelmed but prompt action was needed on the flanks if the gains were to be held. At 8.35 am Monash told Gellibrand: 'Casualties no longer matter', as he ordered him to seize the rest of the Bouchavesnes Spur.[73] Because Hobbs had been unable to find a crossing, Monash directed him to send his reserve brigade along Rosenthal's route through Cléry thence to clear Péronne. The move was completed ten hours later at 8 pm. The 5th Brigade was then clinging to a line just below the summit to which it had been forced by heavy counterattacks. Attacking through the 5th at 6 am on 1 September, the 6th Brigade (2nd Division), 1334 strong, retook the summit in heavy fighting which raged throughout the day. Equally severe was the battle for Péronne with the 14th (5th Division) stalled at the town centre and Elliott's brigade at the southern ramparts. Monash ordered Hobbs to renew the attack and the rest of the town fell next day. Rawlinson, who had been sceptical beforehand, referred to the victory as the finest feat of the war. Bean's appraisal was incisive:

> Among the operations planned by Monash it stands out as one of movement rather than a set-piece; indeed within Australian experience of the Western Front it was the only important fight in which quick, free, manoeuvre played a decisive part. It furnishes a complete answer to the comment that Monash was merely a composer of set-pieces.[74]

Monash compared the battle to Stonewall Jackson's 'swift turning movements at night', leading A.J. Smithers to assert that the Civil War commander was the inspiration for the plan. This claim is hardly credible for it omits the frontal assault originally proposed. Then again, had Hobbs secured the Péronne crossing, there would have been no enveloping sweep. Yet Monash's 'turning movement' was predictable. The proposed encirclement in the final stage of the Passchendaele attack and the recent Etinehem–Méricourt operation were only two examples. Throughout the battle Monash never lost sight of his aim and he did not hesitate to adopt a new plan to attain it. His insistence on the capture of Bouchavesnes was not the only example of his ruthlessness. Faced by Hobbs's earnest pleas that his division was approaching the limits of

*12* The Australian Prime Minister, William Hughes with General Monash on the Western Front, 14 September 1918. (AWM Negative No. E. 3851)

endurance, Monash was 'compelled to harden my heart and to insist that it was imperative to recognize a great opportunity and to seize it unflinchingly'. The demands on commanders and staffs were great for the nature of the battle precluded close control by Monash. Although reserving special praise for Rosenthal and Hobbs, he admitted candidly that the victory was due 'first and chiefly to the wonderful gallantry of the soldiers'. More than any other battles, Mont St Quentin and Péronne showed the quality of the instrument at Monash's disposal and how fortunate he was to command it.[75]

*Breaking the Hindenburg Line*

Disappointed by this reverse and wracked by the Canadian breaching of the Drocourt-Quéant Switch near Bullecourt on 2 September, Ludendorff ordered a retirement to the Hindenburg Line which faced the entire length of Rawlinson's front. Advancing in lanes on brigade frontages, the 3rd, 5th and 32nd Division halted on the old British reserve line, overrun by the Germans six month earlier. Monash foreshadowed that the front line would have to be taken by a set-piece attack. Rawlinson agreed, suggesting to Haig that it could be captured cheaply if assaulted on a wide enough front with adequate artillery support. Included in the objectives was the Hindenburg Outpost Line which gave excellent observation over the Hindenburg Main and Reserve system beyond. After the Outpost Line had fallen, Rawlinson and Byng advocated attacking them before the defenders' morale had recovered sufficiently to make the cost prohibitive. On 13 September, Haig told Rawlinson to launch the preliminary operation as soon as possible. It was fixed for 18 September.

Monash's command now comprised only the five Australian divisions on a shortened front. The 1st Division would attack towards Hargicourt and the 4th towards Le Vergieur. By this stage of the war the methods of the corps were so well understood that the battle 'presented only a few novel features'. As only eight tanks were available, Monash ordered the construction of dummy tanks to be placed in positions highly visible to the Germans before zero hour. Also compensating for the tank shortage was the doubling of the machine gun barrage by adding the guns of the 3rd and 5th Divisions to those of the 1st and 4th, making 250 in all. It was extremely successful, one captured German battalion commander describing the barrage as 'absolutely too terrible for words'. Despite some heavy fighting, notably on Sinclair-MacLagan's front, 'as a general rule, the mere sight of Australian troops decided the enemy to surrender immediately'. Fighting in their last battle of the war, both divisions took all their objectives. Bean, who had doubted the capacity of these exhausted formations to execute so ambitious a plan, conceded Monash's intimate grasp of what his command, despite its state, was capable of achieving: 'he was clearly right in his estimate that the Germans in front of us were so broken that it did not matter what trenches his infantry was in—that infantry would not stand and face our men.'[76]

An incident occurred which showed just how tired the men were. Attacking on either flank, the 3rd and 9th Corps failed to reach the Outpost Line. When the 1st Brigade (1st Division) was ordered, on leaving the trenches, to assist the renewal of the 3rd Corps' assault on 21 September, 119 men mutinied, walking to the rear in protest.

Monash knew the meaning of exhaustion and what men could achieve in spite of it. Compared to the 4th Brigade in August 1915, the Australian Corps in 1918 was a fresh and healthy formation. Monash himself was tired. Blamey noted that he became very thin and the skin hung loosely on his face. He would ride in his car for long periods in silence. Monash's attitude was moulded by the obvious decline in the British Army which, as he pointed out to brigade after brigade during his visits, was no less overworked than the Australians, and by the demoralisation of the Germans. He wanted to hit them as hard and as often as he could, a policy with which his divisional commanders concurred. Only the Australians and Canadians were able to win constantly and hence it was inevitable that 'they should be called upon to yield up the last particle of effort of which they were capable'. Bean concluded: 'In this decisive fighting...Monash was right to work his troops to the extreme limit of their endurance'. He appealed to the men on the grounds of prestige and pressed higher commanders to give greater publicity to

Australian efforts. The Australian was a sportsman and he wanted to see his score on the board. To release more divisions for rest, he consistently urged a reduction of the corps' front. His policy of utilising the maximum of mechanical resources allowed him to attack with inferior numbers: 'I welcome any pretext to take the fewest possible numbers into action. So long as [battalions] have 30 Lewis guns it doesn't very much matter what else they have.'[77]

The attack by the Fourth Army on the Hindenburg Line was the main British contribution to a series of hammer blows by the BEF, the French and Americans to conquer territory held by the Germans since 1914 as distinct from recapturing ground lost in the offensives between May and July. Rawlinson allotted the principal role to the Australian Corps and called on its commander to shape the plan on which his army fought. Monash described the task as 'the most arduous, the most responsible and the most difficult' of any he undertook throughout the war.[78] On the Australian front, the St Quentin Canal, the formidable obstacle on which the defence rested, ran through a tunnel which began at Bellicourt and emerged near Gouy, 6000 yards (5500 metres) further north. Troops were comfortably accommodated in the tunnel, immune from bombardment and able to launch prompt counterattacks through many concealed passages and airshafts. Above them the fortifications protruded some 1200 yards (1100 metres) westward and included the village of Bony, while 1500 yards (1380 metres) further west, Gillemont and Quennemont Farms formed an uncaptured section of the Hindenburg Outpost Line.

Monash's plan was derived largely from the complex scheme for 8 August. Supported by tanks and a creeping barrage, the infantry would attack across the tunnel, regarded by Monash as a 'bridge', seizing both the main Hindenburg Line and its supporting Le Catelet Line one-mile (1.6 kilometres) to the west. An open warfare attack by infantry, tanks and mobile artillery on the reserve or Beaurevoir Line would follow. Monash thought all objectives could be captured in one day and as there was no hope of surprise, recommended a bombardment lasting at least four days to damage the defences and demoralise the occupants. Rawlinson made two alterations to this scheme. Monash readily concurred with the first change, that the assault on the Beaurevoir Line would be dependent on the results of the initial advance and hence it might not be taken on the first day. When Rawlinson added that the 9th Corps would attack directly across the canal near Bellenglise, he dissented vehemently.

Monash had recognised that unless the salient formed across the tunnel was widened quickly, the Germans could concentrate all their reserves against what would be a very small breach in the line. Consequently, the 3rd and 9th Corps could pass across the tunnel and extend the breach either side of the Australians. Even though the needs of his own corps would preclude the use of the tunnel by others for at least two days, Monash thought that his scheme would incur far less casualties than Rawlinson's. The trouble was that the outcome depended entirely on the establishment of a bridgehead across the tunnel. Rawlinson realised that success was more likely if the attack was launched on a wider front, forcing the dispersion of the Germans along its length, and insisted on the 9th Corps crossing the canal.

The thoroughness and precision of Monash's planning were exceptional. As he had done in the 3rd Division, he handled personally many matters for which the corps staff should have been responsible. Calculation of time and space was a typical example as Monash reckoned that two artillery brigades must withdraw from the barrage at 9 am, thus giving them 'four hours to bring up teams and waggons, limber up and march to join their [infantry] brigades for the next advance'.[79] He ordered the preparation of four approach routes which engineers would extend during the attack in the manner of 8 August. Anticipating the congestion as the attack converged on the tunnel, he drafted the intricate instructions for road repair and traffic control himself. Gellibrand was struck by Monash's concluding remark that 'this operation is more a matter of engineering and organization than of fighting'.[80] It seemed to justify further Monash's extreme confidence in the outcome of a set-piece battle:

> given a resolute infantry and that the enemy's guns are kept successfully silenced...nothing happens, nothing can happen, except the regular progress of the advance according to the plan arranged. The whole battle sweeps relentlessly and methodically across the ground until it reaches the line laid down as the final objective.[81]

The opposite was the case although it might be argued that the circumstances were extenuating. Monash had welcomed Rawlinson's offer of the 27th and 30th American Divisions to replace the 1st and 4th Australian. They would carry out the first phase of the operation and its preliminary phase, the capture on 27 September of that portion of the Hindenburg Line which the 3rd Corps had failed to secure. It was from there that the main assault would be launched two days later. The American task was difficult but it was nonetheless simpler than the open warfare advance to the Beaurevoir Line which Monash allotted to the infinitely more experienced Australian divisions. He lectured to the American commanders and formed an Australian Mission, the members of which were attached to every American formation down to battalion, to advise on tactics, equipment and supply. Soon 'it became possible to talk to the whole American Corps in our own technical language'.[82]

But the 27th Division's preliminary attack failed badly. Unconfirmed reports of Americans reaching the Hindenburg Outpost Line ended discussion on whether to start the creeping barrage for the main attack from the 27th's present position. It would begin on the Outpost Line 1000 yards (about 1000 metres) ahead to avoid any Americans who might be lying in between. Monash's request for a day's postponement to clarify the situation was rejected by Rawlinson because of the effect on other formations. He promised additional tanks instead. When Haig visited him on 28 September, Monash described himself as being in 'a state of despair'. Haig reassured him that 'it was not a serious matter and that he should attack tomorrow as arranged'.[83]

Monash's anxiety was unrelieved and with good reason. Next day, the 30th Division reached Bellicourt and took the southern entrance to the tunnel, assisted by the 9th Corps' brilliant crossing of the canal which vindicated Rawlinson's appreciation. But the 27th was halted largely on the Outpost Line although some Americans were reported near Bony. The situation seemed so dangerous that Gellibrand's 3rd Division, following the 27th, defied orders and launched local assaults on the American objectives, capturing Quennemont Farm. Alarmed by the lack of information, Gellibrand left his headquarters and went forward to see for himself. On his return he was unable to dispel the erroneous impression held at Corps Headquarters that the 27th Division had taken most of its objectives and that all his division had to do was mop up the intervening machine guns to join them. Despite his protests, Gellibrand was ordered to assault frontally against the Hindenburg Line at 3 pm. Intense fire stopped the advance after a few hundred yards.

Only then did Monash issue the obvious order. Next day the 3rd Division would attack from its right flank where its junction with the gains made by the 5th and 30th Divisions created a salient behind the Hindenburg Line. Cooperating with the 5th, it was to advance northward astride the Le Catelet Line, both formations securing the northern end of the tunnel and Bony. In bitter fighting which continued on 1 October, the remainder of the Hindenburg system was cleared in the Australian sector. Tiredness seemed the most plausible explanation of Monash's erratic handling of the battle. Supporting the contention was his acquiescence to Rawlinson's request for one more attack, seemingly to fill the day remaining before relief. The 2nd Division's costly and pointless capture of Montbrehain on 5 October was the Australian Corps' last fight of the war.

*Monash, the corps commander*

Monash's performance before and during the attack on the Hindenburg Line showed that he was not infallible as a commander. He had no 'feeling' for the battle because he did not visit his subordinate commanders fighting it. The mistake at Chuignes would not have occurred had Monash made a personal reconnaissance. He often overreached himself by performing the functions properly discharged by his staff, a habit which was inappropriate in the 3rd Division and throughout his command of the Corps. Perhaps his style was inevitable after a pre-war career which was a testimony to the results gained from personal effort with no assistance from others. His vanity and ambition were deprecated by his senior commanders even though they displayed the same failing. But it mattered little, for as Bazley observed, Monash 'had the restraint and intelligence to keep [them] under control; [they] never trapped him into mistakes in the field'.[84]

Few disliked him. Monash was the recipient of 'perhaps the greatest volume of unmitigated approval of any one man in the whole of [Haig's] diary'. He won 'undying respect' from nearly all the AIF generals while

13 Breaking the Hindenburg Line. Troops of the 11th Brigade (3rd Division) and tanks moving into battle near Bellicourt, 29 September 1918. (Australian War Memorial, Negative No. K114)

those at Corps Headquarters 'swore by him'. Gellibrand, perhaps his sternest critic, 'could admire and follow him with comfort and pleasure despite the fact that I was well aware of his failings'. Albert Jacka's high opinion of him typified the soldiers' views. Under Monash, the men 'went into action feeling with justification, that, whatever might be ahead, at least everything was right behind them'. Their achievements were great. At a cost of 21 243 casualties, just over a quarter of whom were killed, the Australian Corps, led by Monash, took 29 144 prisoners and 338 guns as well as liberating 116 towns and villages. The Australians formed some 8 per cent of the British Army in the line, yet these figures represented about 22 per cent of the captures of the entire British Army in the last phase of the war on the Western Front.[85]

Irrespective of the quality of his troops, this success could not have been won without Monash's skill as a commander. His technical mastery of all arms and tactics, particularly surprise and deception, was unsurpassed among his contemporaries. Cooperation and coordination between arms and between units in the Australian Corps was unexampled in the BEF. He had evolved a philosophy of warfare which gave the infantry every conceivable assistance and did not commit them to assaults on distant objectives which they would reach too weak to hold. He gave his commanders every chance to have their say before the battle but he did not hesitate to bully them ruthlessly once it began. His concern for the welfare of the men enabled Monash to make demands on them at which other commanders would have baulked. He was always aware of the precarious state of the Germans opposite him and sought to exploit it at every turn. He applied psychology to create and maintain *esprit de corps* in every command he held. It was accompanied by unflinching moral courage. He was equally successful as a trainer, numbering among the few divisional commanders who did not staff their training battalions in England with undesirables from their formations in France. The battles in the latter half of 1918 each reflected different facets of his ability, from his concessions to Sinclair-MacLagan over the use of the tanks before Hamel, to his assembly arrangements on 8 August, to the rapid switching of brigades and steadfast maintenance of the aim at Mont St Quentin. Finally he had good fortune, that quality which Napoleon prized.

It is doubtful if any of the other Australian generals could have achieved the same result as Monash. They were, as Serle noted, 'simply not comparable in intellect, articulateness or personal magnetism'.[86] In any case the comparison is fruitless. How can it be concluded, for example, that Hobbs would have driven the corps as relentlessly or that Rosenthal would have ensured the repair of the Somme bridges as the advance passed them? Yet these questions must be answered if an accurate judgment is to be formed. The same applies to White. As a divisional commander, Monash sought his advice on the frontage his battalions should occupy between the Somme and the Ancre in March 1918. However, their discussion on the use of machine guns and artillery before the German offensive proved conclusively that Monash's tactical thinking was ahead of White's. Would Monash have been a better chief of staff than White? After all, he performed the functions of that post often enough. Nothing more can be said except that both men were outstanding in their respective roles. Speculation on whether Monash could have replaced Haig, a legend fostered largely by Lloyd George's attempts to discredit the C-in-C, must be treated in similar vein. In any case, when Haig's dismissal was mooted, Monash was an unknown divisional commander, one of about 60 in the BEF. When Monash gained prominence as a corps commander, Haig did not need replacing for the counter-offensive had begun and was continuing largely unchecked.

It is as a corps commander that Monash must be judged. His formation always operated as part of the Fourth Army and apart from a single instance at the end of August 1918, he was strictly bound by Rawlinson's orders. Hence his decision-making was limited and it is safe to say that he did not influence the outcome of the war. The great opportunities which fell to higher commanders such as Allenby at Megiddo, were simply beyond his reach. For this reason, he is disqualified from a place alongside Marlborough, Wellington or Napoleon in the hall of Great Captains. Unpalatable though it may be to his admirers, the strategic claims made for him must be dismissed. But Monash's reputation as an exceptionally capable and successful corps commander stands. The Australian Corps was fortunate to have him as its leader in its last and greatest battles. Monash's contribution to his country did not end with the war. Appointed Director-General of Repatriation and Demobilization in November 1918, he supervised the return to Australia of some 180 000 men. In August 1920 he accepted the chairmanship of the infant State Electricity Commission of Victoria, planning the development of the State's power scheme. Aged 66, Sir John Monash died on 8 October 1931.

**Part Two**
The Second World War

# Lieutenant-General Sir John Lavarack: From Chief of the General Staff to Corps Commander

A.B. LODGE

With a military tradition strongly founded on the qualities of the citizen soldier, perhaps it is not surprising that in some quarters in Australia after the First World War the opinion was held with conviction that permanent, or Staff Corps, officers were not suited to the task of leading men in battle, and that their role should be strictly limited to staff duties. As the most senior Staff Corps officer of the inter-war period to be appointed to an operational command during the Second World War, Lieutenant-General Sir John Lavarack (1885-1957) provides an interesting opportunity to test this theory at the highest level. He was Chief of the General Staff from April 1935 to October 1939, and commanded the 1st Australian Corps during the bitter Syrian campaign of 1941.

*Chief of the General Staff*

In July 1932 Admiral Sir Herbert Richmond published an article in the British *Army Quarterly* on imperial defence. This article prompted Colonel J.D. Lavarack, then commandant of the Royal Military College, Duntroon, to submit his own views to the same journal in early 1933. The opinions offered by Lavarack were those which were to guide his actions as Chief of the General Staff in dealing with the *bête noire* of the Australian army during the 1920s and 1930s, the Singapore strategy.[1]

Since the 1923 Imperial Conference Australia's defence had been dependent upon the concept of imperial defence, which was based on British seapower. For Australia to be protected against its most likely enemy, Japan, there needed to be a base in the Far East, Singapore, from which the British fleet could operate, and an assurance that the fleet would come when needed. In his article Lavarack argued that the Royal Navy no longer had ships in sufficient numbers to protect both the Atlantic and the Far East, and as a war in the Far East would most likely coincide with hostilities in Europe, 'the possibility of detachment of adequate forces [to the Far East] would be remote' He continued: 'The conditions of the problem have fundamentally changed with the disappearance of unchallengeable British supremacy at sea'.[2] To offset these changed circumstances, Lavarack suggested that, as well as cooperating in naval defence, Australia should build up its land-based forces to prevent an enemy obtaining a base on mainland Australia, which, he argued, the enemy would be forced to procure if he were to maintain an effective naval blockade of the entire Australian coastline. Therefore, the larger and more effective the Army and Air Force, the less chance there was of a crippling blockade. And it made sense to build up these Services as they, unlike the Navy, were unfettered by agreements such as that made at Washington in 1922 which limited the construction of capital ships. In conclusion Lavarack wrote:

> security based on the control of sea communications alone is a counsel of perfection which, for practical reasons, is not, and cannot be, of absolute application. Control sufficient to ensure the maintenance of essential trade routes will, however, still be possible, and this, supplemented by the deterrent effect, and actual opposition, of Dominion land and air forces, must furnish the basis on which the Imperial Defence of today must be founded.[3]

Lavarack was soon to be given the opportunity of expressing these views officially, first during the visit of Sir Maurice Hankey, the Secretary of the Committee of Imperial Defence (CID), who arrived in Australia the next year to advise on defence matters, and then from 1935 as Chief of the General Staff (CGS).

At the time of Hankey's visit in late 1934 the Lyons government was trying to restrict defence spending, especially on the Army. These financial constraints had led to a bitter dispute between that Service and the Navy on the degree of trust to be placed in British promises of despatching a fleet to be based on Singapore in the event of war in the Far East. It was hoped that Hankey's tour would resolve the question. Lavarack's article had expressed views held by other senior officers in the Army who also had been arguing that the Singapore base was not equipped to handle the large numbers of ships which would, in theory, arrive to forestall an enemy attack on Australia; and that, in any case, there was no guarantee that the fleet would be despatched. The army leaders urged the government to commit itself no further to the Singapore strategy until work on the base was completed and until there were guarantees of a fleet being based there or sent in time of need. Such advice was timely, for Britain's naval strength in the Pacific was then so weak that the British Chiefs of Staff were of the opinion that if Japan attacked, the Royal Navy would be fortunate to be able 'to disappear into the blue', and the base at Singapore was, according to the First Lord of the Admiralty, 'little more than a hole in the earth'.[4]

The Army had grasped well the deficiencies in the Singapore strategy, but Hankey, who greatly favoured the concept of imperial defence, could not admit to the deficiencies without undermining his own position and discrediting pledges made in London. Consequently, in his report to the Australian government, Hankey stated that: 'Even in the very extreme case of simultaneous trouble in Europe and the Far East without our having allies...the ratio of naval strength in capital ships...is sufficient to enable a numerically superior battle fleet to be sent to the Far East and yet leave a small margin of strength in both theatres...' It is difficult to imagine that as Secretary to the CID (and also to the British

Cabinet) he was unaware that British warships did not exist in numbers large enough to permit the splitting of the fleet between Europe and the Far East without compromising one or both theatres of operations, and that it was increasingly unlikely that a naval force of any effective size would be sent.[5]

By this time Lavarack was Chief of the General Staff designate and as such was required to comment on Hankey's report. As might be expected, he stressed the value of land and air forces which, in his opinion, had been greatly underestimated and for that reason recommended that the defence vote, which he felt unduly favoured the navy, be distributed more evenly. He recommended that no further commitment to the Singapore strategy be taken until the British government had guaranteed the date of completion of the base at Singapore, the strength of the naval forces to be despatched there in emergency, and the time which would elapse before the arrival of the fleet. Concerning Hankey's advice, Lavarack offered the opinion: 'That the whole matter is one for the Australian Government to decide on the representations of its own advisers, and that the advice of Sir Maurice Hankey, whilst receiving the respect and attention due to his high authority, should not be allowed to obscure consideration of the detail of what is a purely Australian problem'. Lavarack's views were very similar to those of the Chief of the General Staff, Bruche, who also presented a paper critical of Hankey's report. But Lavarack's words were not empty carriages running on the official line. He argued that his views had 'been formed during a fairly long and intimate experience of the question at issue', an assertion which is borne out by his article of more than two years before, and by his criticism in 1930 of a paper by F.G. Shedden, Secretary of the Defence Committee, which advocated the primacy of the Royal Navy in imperial defence.[6]

In April 1935 Lavarack replaced Bruche as CGS, and in July he was presented with another opportunity of challenging the Singapore strategy when the question of coast defences as outlined in CID Paper 249C was considered. The Australian Council of Defence had already endorsed the scheme and had passed it to the Defence Committee so that technical aspects, such as the priority of ports to be defended, might be considered. Lavarack went further than this, however, and presented a memorandum in which he questioned the basis of Australia's defence policy—the likelihood of a fleet being despatched to Singapore. CID Paper 249C had stated that the fleet would arrive at Singapore 'with a minimum of delay after the outbreak of war in the Far East', which had been reckoned to be within 42 days after hostilities commenced. Lavarack suggested that this be altered to read: 'Whilst it is intended that the Main British Fleet should arrive at Singapore with a minimum of delay after the outbreak of war in the Far East, possible complications in Europe concurrent with a war in the Far East may cause the despatch of the Fleet to be delayed for a prolonged period'. He continued: 'This, I feel, is more in accord with the true situation as with the growth of the Japanese Navy concurrent with that of Germany...there can be no certainty that under all circumstances the British Main Fleet or any appreciable proportion of it can be despatched to Singapore in the event of a War in the Far East'.[7]

Lavarack requested that his views accompany the report on the scheme when it was placed before the Committee of Imperial Defence in London and stressed that as the foundation of the Australian system of defence had been questioned, the opportunity of giving the matter full consideration should not be allowed to pass.[8]

When Lavarack's suggestions were referred to the next meeting of the Council of Defence, which took place more than a year after he had produced his memorandum for the Defence Committee, he found himself to have incurred official disfavour. The Minister for Defence, Sir Archdale Parkhill (who had been influenced by the views of Hankey and Shedden) put forward the proposal that:

> The special observations of the Chief of the General Staff involve implications of a highly political nature and the subject is one solely for the United Kingdom and Commonwealth Governments. The Services have a definite basis on which to proceed with their plans, and...remarks are to be requested on the purely technical aspect of the scheme.

Lavarack replied that he had based his observations only on strategic considerations and that he had not intended to trespass upon political ground. Nevertheless, Lavarack had been officially censured.[9]

He was also criticised by the British Chiefs of Staff after the 1937 Imperial Conference for suggesting that development of the navy be maintained at the present level to allow for improvements in the army and the air force. 'The adoption of these principles', the Chiefs of Staff argued, 'would therefore seem inconsistent with the basis of Australian policy', which, it must be added, they had endorsed.[10]

Despite the apparent and growing disfavour into which he was falling, Lavarack was not content to remain silent on the subject of defence policy and raised it again in the Council of Defence early in 1938, by which time Parkhill had been replaced as minister by H.V.C. Thorby. At a meeting held on 24 February Thorby criticised the Army's use of the money allocated to it in the defence vote, arguing that too much money had been spent on buildings and not enough on armaments. Lavarack replied that the buildings were necessary to house the troops who were to man the fixed coastal defences, to which the government had given priority. In any case, he went on, there were sufficient small arms and tanks (although there were only thirteen) to fulfil the government's policy of preparing only for defence against raids. It seems that Lavarack, like Bruche before him, was putting the Army's slice of defence money into fixed assets which, unlike weapons, could not be sent overseas with an expeditionary force.[11]

The Treasurer, Casey, was not satisfied and demanded from Lavarack and the Chief of the Air Staff, Williams, explanations as to why their services were so unprepared for war in terms of munitions and other equipment. Lavarack replied that he had not in any year since becoming

CGS received the amount he had requested and referred Casey to a submission for money he had made at the previous Council meeting which had been considerably reduced. As the meeting progressed defence policy came under closer scrutiny and the Chief of Naval Staff, Sir Ragnar Colvin, leapt to the defence of the Singapore strategy as the best means of defending Australia, a view which, not surprisingly, was challenged by Lavarack.[12]

The year 1938 was to be a watershed in Lavarack's career. It appears that his criticism of defence policy plus a feeling within the government that he was withholding information, and was perhaps disloyal to his political superiors, contributed to the appointment of a senior British officer, Lieutenant-General E.K. Squires, as Inspector-General of the Australian Army. When the Munich crisis arose later that year Lavarack could see his misgivings concerning the unpreparedness of the Army about to be realised, and expressed his disquiet in the Defence Committee. According to Menzies: 'He made a bad impression on Ministers at the time of Munich when he seemed to be excited and jittery'. Neither Menzies nor Casey, both friends of Major-General Sir Thomas Blamey, could be considered admirers of Lavarack and so from 1938, as Blamey's star ascended with his appointment as Chairman of the Manpower Committee, Lavarack's declined.[13]

Although Lavarack had been a trenchant critic of Australia's reliance on British seapower up to the time of the Munich crisis, he was not opposed to the principle of imperial defence, as he explained in a letter to Admiral Richmond in 1936:

> All that I do is to query the freedom of action, in all circumstances, of the Royal Navy. I consider that those who believe that the 'period before relief' in the Far East will be only a few weeks are dangerous optimists, and that Australia may have to be prepared to hold her own territories and protect her own local interests for many months, or even a few years...I am a firm believer in Naval defence for the Empire, but I believe that the standard of Naval power is too low for safety...
>
> I have never opposed Australia's Naval contribution, though I have opposed its extension towards the point at which it would entail the annihilation of the other Services.[14]

Lavarack's goal was a balanced policy relying on the three services, but he had been unable to realise it in the face of overwhelming British advice to the contrary and political expediency in Australia. Lavarack might have resigned, but clearly he saw that to do so would do nothing to advance defence preparedness. In 1936 he wrote that to get the defences on their feet 'a war scare is our chief hope'.[15] Two years later such an incentive materialised and had the desired effect, but by that time Lavarack had done serious damage to his career as CGS in attempting to achieve the same end.

Although at times he overstated his case, Lavarack had shown that he possessed a keen appreciation of strategic and political problems, and that despite almost two years spent in the United Kingdom at staff college, two years attached to the British Army during the First World War, a year at the Imperial Defence College in London, and the close military ties between Australia and the United Kingdom before 1941 he could maintain a distinctly Australian outlook. Had Lavarack been the Chief of Naval Staff he may have been less prepared to expose the weaknesses of the Singapore strategy, but it seems likely that at least he would have argued against almost total reliance upon it, in spite of the attractiveness of such a panacea.

*Tobruk*

At the outbreak of war Lavarack was appointed GOC, Southern Command and then stepped down a rank to command the 7th Australian Division as a major-general. He was given few opportunities to exercise command during the Second World War: as GOC, Cyrenaica Command in Tobruk in April 1941; as commander of the 7th Australian Division, and later of the 1st Australian Corps, during the Syrian campaign of June–July 1941; and as GOC, 1st Australian Corps in the Netherlands East Indies in February 1942. The first of these appointments Lavarack held for only one week, and in Java his tasks did not extend past reconnaissance and planning before the proposed operations were abandoned.

On 3 April 1941 Lavarack, who was in Palestine with his division, received orders to prepare to embark in three days to take part in Operation 'Lustre', the plan to forestall a German invasion of Greece. However, the advance into Cyrenaica begun by General Rommel late in March was gathering momentum and Egypt was threatened. On 6 April it was decided at a meeting of the Army, Navy and Air Force Commanders-in-Chief in the Middle East that a stand against Rommel would be made at Tobruk, which might have to hold out for two months until sufficient forces could be built up to launch a counter-offensive. The Army Commander-in-Chief, General Sir Archibald Wavell, decided that the 7th Division would not go to Greece, that one of its brigades would be sent to Tobruk, and that Lavarack would succeed General Neame, who had been captured, as GOC Cyrenaica Command.[16]

Lavarack was informed of the arrangement the next day, and on 8 April he and Wavell flew to Tobruk, where Cyrenaica Command headquarters would be established. Sir Anthony Eden, who was then in Cairo, wrote: 'I was much impressed by [Lavarack's] calm and grip of situation and readiness to take command at Tobruk'. Lavarack relished the challenge which lay ahead. His tasks as the new commander in Cyrenaica were to familiarise himself with the existing defences; to organise the forces at his disposal and establish a command system; and to advise Wavell whether Tobruk could be held for two months.[17]

By the end of his first day in Tobruk Lavarack had grouped his forces, mainly the 9th Australian Division under General Morshead and assorted British units, and was reorganising them, but he had still to decide whether to stay there or to recommend to Wavell that a withdrawal to the Egyptian frontier begin. Lavarack did not think a withdrawal would be necessary. There were sufficient supplies for his force, morale was high among the British and Australians, and he estimated that Rommel's force was not much larger than the combined strength of

Tobruk and Egypt. Therefore he felt that Tobruk should be defended in order to provide a thorn in the side of Rommel's advance to Egypt and to draw off enemy forces which otherwise might have been used at the frontier. Also he was reluctant to sacrifice the large amounts of supplies and equipment in Tobruk, and was unwilling to abandon those men still struggling into Tobruk ahead of Rommel's troops.[18]

Within three days Tobruk was surrounded by enemy forces, and late on 13 April Rommel launched his first attempt to take the fortress. Soon after dawn the next day the attack had been repulsed with heavy losses to the enemy. On 14 April Lavarack left Tobruk, not to become the commander of Western Desert Force (which was formed from Cyrenaica Command) as he had expected, but to return to the command of the 7th Division in Egypt, complete except for the 18th Brigade which remained in Tobruk.[19]

This change in Lavarack's fortunes was not brought about by any inadequacy in his conduct of operations during the week he had been in command in Cyrenaica. From the beginning Lavarack had recognised the necessity for an active defence and, unlike the enemy who were surprised at their failure, that breaching the perimeter did not mean victory for the attacker. While in Tobruk Lavarack made several important decisions and exhibited a determination to adhere to them. The first was to stay and organise an aggressive defence with the forces at his disposal. Second, he based the defence on the outer of the two defensive perimeters at Tobruk, despite urgings from Wavell and Blamey that he use the inner perimeter which he did not consider viable. Third, he organised a defence in depth and ordered work to begin on a second line of defence with switch lines to localise any enemy breakthrough. Finally, Lavarack maintained his aim of defending Tobruk and refused to be enticed into actions which would dissipate his strength, such as complying with requests from GHQ in Cairo to open the road to Bardia and to despatch to the frontier his cruiser tanks, which he considered would have been destroyed in the attempt.[20]

At Tobruk Lavarack had shown that he was a capable organiser and a determined, confident commander. He was returned to the command of the 7th Division not because he had failed in any way in Cyrenaica, but because of the intervention of General Blamey, GOC AIF in the Middle East, who argued that Lavarack's temperament did not make him suitable for high command. There had been a long-standing disagreement between the two senior Australian officers.[21]

## Syria

Several weeks after leaving Tobruk Lavarack was informed that the 7th Division was to take part in an operation against the Vichy French in Syria. The command structure of the campaign reflected the hurried nature of the operation and the lack of balanced forces. The commander, General Sir H.M. Wilson, was located at Jerusalem. His subordinate formations were the 7th Australian Division, an Indian brigade and a Free French force of about brigade strength. It was

*14* Early in the Syrian campaign. From left, Lavarack's ADC, Major-General J.D. Lavarack, Brigadier A.S. Allen and Brigadier J.E.S. Stevens, Commander of the 21st Brigade. Lavarack was shortly to take over the 1st Australian Corps, to be replaced by Allen as Commander of the 7th Australian Division. (Imperial War Museum, Negative No. E 3163)

planned that after the advance reached a certain stage the 1st Australian Corps would take over responsibility for the operations from Wilson's headquarters and that Lavarack would then assume command of the corps. Although his division was to provide the largest proportion of the troops, there was less scope for planning by Lavarack than he had enjoyed during his few days in Tobruk. Wilson had already decided the routes which Lavarack's two brigades would take: the 21st Brigade would advance up the coast road to Beirut and the 25th inland from Metulla towards Rayak.[22]

Nevertheless, Lavarack was able to make adjustments to the plan, and altered the route taken by the 21st Brigade on the first day of the invasion in the event that the coast road, which was susceptible to demolition, had been blocked. Although Lavarack had repeatedly warned Wilson of this danger, the latter for some time seemed to consider it unlikely as he doubted that the Vichy French would offer much resistance. Lavarack, and his staff, had formed a more realistic appreciation of the likely Fench reaction to an advance into Syria and Lebanon. Presciently, Lavarack allotted his divisional reserve, the 2/25th Battalion, to the 25th Brigade which, he concluded three days before the campaign began, 'will be opposed at Merdjayoun, perhaps seriously'.[23]

Once the invasion began Lavarack faced many difficulties: the control of two widely separated brigades advancing through rugged country poorly served by lateral communications; serious shortages of equipment, such as mortar ammunition and tanks; and, the greatest handicap,

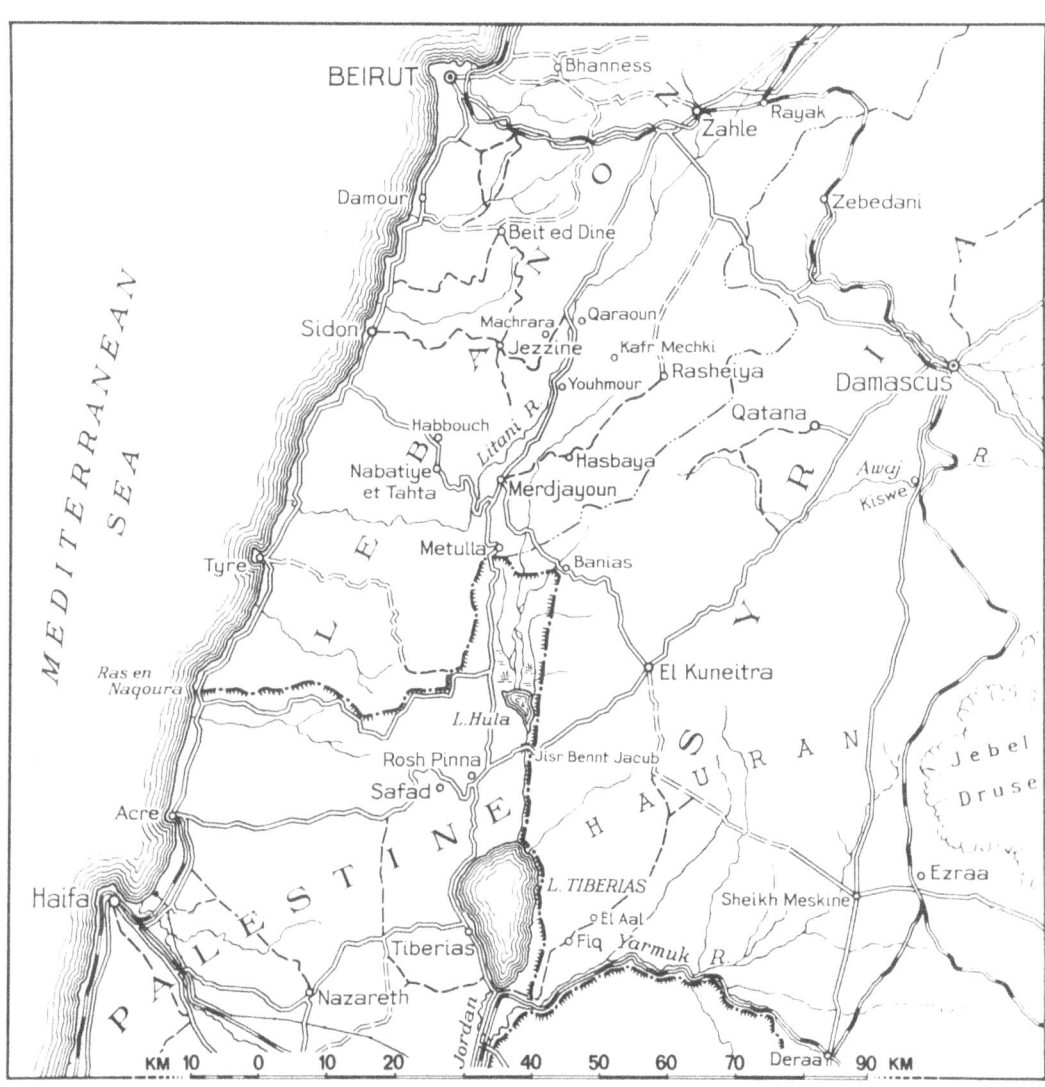

13  Area of 7th Australian Division operations in Syria, June–July 1941

the lack of a reserve because his 18th Brigade was still in Tobruk. The shortage of units prevented the maintenance of momentum in the advance as troops were switched from one line of advance to another to compensate for the lack of a reserve formation. It was during one operation of this kind that Lavarack committed an error which contributed to the success of the Vichy counterattack at Merdjayoun. After the town had been taken on 11 June against strong opposition, as Lavarack had expected there would be, he decided to move the 25th Brigade westwards to Jezzine, from where it could support the drive up the coast road by the 21st Brigade. At Merdjayoun Lavarack left a battalion of infantry, a cavalry regiment and some artillery, but made neither of the unit commanders responsible for the Merdjayoun area, trusting instead on their cooperation. Unfortunately the arrangement did not work, and on 15 June the Vichy French counterattacked and the town was retaken. Had Lavarack placed one officer in command a more coordinated, and hence stronger, defence might have been offered with success. Lavarack ordered into the area all troops which could be spared, including some which he had transferred from there to Jezzine a few days before, and placed his artillery commander, Brigadier F.H. Berryman, in command at Merdjayoun.[24]

Fighting to recover the lost ground was still continuing when Lavarack assumed command of the 1st Australian Corps on 18 June, by which time the advance on the coast and in the centre had been brought virtually to a stand-still by the Vichy counterattack. One battalion of the 21st Brigade had been transferred to Jezzine, from where two battalions of the 25th Brigade had been taken to reinforce the Merdjayoun area. Like Blamey in Greece and Freyberg in Crete, Lavarack had been placed in command at a crucial stage.[25]

After taking over as corps commander Lavarack concentrated on taking Merdjayoun, where operations soon came under the control of the headquarters of the newly arrived 6th British Division. Lavarack expected Damascus to fall within a few days and considered it imperative that the right flank and rear of the 7th Division be secured against the possibility of attack from Merdjayoun, so that as soon as Damascus fell the drive along the coast could continue. The drive against Beirut had been uppermost in Lavarack's mind since the campaign commenced and, once resumed, proved to be decisive. At the request of the Vichy French firing ceased on 12 July, by which date Australian troops were north of Damour, only a few miles from Beirut.[26]

The preparations for the campaign in Syria showed that Lavarack had appreciated more keenly the difficulties posed by the country and the enemy than either Wavell or Wilson, who were preoccupied with political considerations. After operations had commenced Lavarack laboured under many difficulties, such as lack of equipment and shortage of troops (although these gradually were remedied), but throughout managed to maintain his aim of delivering a decisive blow against the political and military centre at Beirut. It was not possible to concentrate a force superior in numbers, and so, Lavarack wrote, 'my concentration was one of quality and morale, not of quantity, and the Australians were assembled at Jezzine and Damour, to deal the final blow'.[27]

There were many aspects of the campaign in Syria which were beyond Lavarack's control; nevertheless he conducted operations skilfully, if not faultlessly, under difficult circumstances. Both his decision to move to Jezzine and the plan for the attack on Damour caught the Vichy commander by surprise.[28] However, Lavarack's difficulties did not end with the campaign. Under the terms of the armistice Allied prisoners taken during the campaign were to be freed and Vichy French troops

15 After the armistice in Syria, August 1941. Lieutenant-General Sir John Lavarack inspecting French troops. On the right is his Brigadier General Staff, Brigadier F.H. Berryman. (Imperial War Museum, Negative No. E 5397)

were to be given the opportunity to join the Allies, but the Free French were not to attempt to influence any such decision by personal contact. Both the Vichy and the Free French attempted to circumvent the conditions, and Lavarack was faced with potentially serious problems. These he overcame tactfully, but nonetheless firmly. It was with justification that Lavarack was awarded a knighthood for his efforts in Syria.29

## Java

Although Lavarack did not control operations in the Netherlands East Indies, in early 1942 the appreciations he made there as the commander of the 1st Australian Corps greatly affected his subsequent career. He arrived in Java as the 8th Australian Division and other formations under the command of Lieutenant-General A.E. Percival were retreating to Singapore Island after losing to the Japanese the last hold on the Malayan peninsula. General Wavell, commander of ABDA Command (the American, British, Dutch, Australian command covering South-East Asia) informed Lavarack that the general plan was to hold Burma, southern Sumatra, Java, Timor and Darwin, with an advanced position in Singapore. Lavarack's role was to hold southern Sumatra with one division and central Java with another; in western Java, and thus between Lavarack's two Australian divisions, would be a Dutch formation. Lavarack objected to the splitting of the Australian corps without effect, and decided not to press the matter immediately because of the many difficulties Wavell was facing. Later he 'represented strongly' to Wavell the necessity for maintaining the integrity of the corps if it were to be administered efficiently and to fight effectively. However, Wavell convinced Lavarack that circumstances made concentration impossible.[30]

By 9 February Lavarack had completed his reconnaissances of the areas to be occupied by the 1st Australian Corps, the first formations of which were due to arrive in the Netherlands East Indies shortly. However, on the night 8-9 February the Japanese had landed on Singapore Island and enemy convoys were reported to be moving in the direction of the Netherlands East Indies. Lavarack was being pressed by senior Australian officers to refuse to allow the 7th Australian Division to land in southern Sumatra, but he felt there was time enough to make a decision later. Nevertheless, he was concerned about the fate of some 3000 Australians aboard the *Orcades*, which was due at Oosthaven in Sumatra within a few days.[31]

By 12 February the situation in Singapore had deteriorated considerably and Lavarack was more concerned about the plight of the 7th Division should it be landed in Sumatra, although he noted that at Wavell's headquarters there seemed to be no reservations about taking such a step. On 13 February Lavarack's concern prompted him to prepare an appreciation for the Australian Prime Minister. In it he expressed the fear that the fall of Singapore, which seemed imminent, would free large enemy forces for an invasion of the Netherlands East Indies, and that the 7th Division and the Dutch forces available would delay the Japanese for a short time only. He felt that this delay would be gained probably at the expense of the loss of the equipment of the 7th Division and a large number of its troops, and that even with the 6th Division, which was following the 7th from the Middle East, it would not be possible to hold Java. Lavarack stressed the value of the 1st Australian Corps as the only striking force of any appreciable size within distance of the Far East, implying that it should not be squandered. After reading Lavarack's appreciation, Wavell prepared a similar one for transmission to the War Office. On 15 February the Australian Prime Minister, Curtin, on the advice of the Chief of the General Staff, Sturdee, and influenced by Lavarack's appreciation, informed Churchill that as the Netherlands East Indies could not be held and that as Australia thus had become the main Allied base in the Pacific, he wished the AIF to return to defend it.[32]

Curtin's request of 15 February was reiterated in more detail two days later, and 'there developed a cabled controversy of a volume and intensity unprecedented in Australian experience' concerning the destination of the AIF.[33] Curtin insisted, against Churchill's wishes, that the 7th Division not be diverted to Burma, as Wavell had recommended, and that the AIF return to Australia. Lavarack, too, was 'firmly convinced' that the correct employment for the Australian corps was in

Burma as it was, in his opinion, the best base from which a counterattack could be launched against Japan. At Wavell's request Lavarack informed the Australian government of these views, adding the proviso that such a move should be made only if the defences of Australia would not be jeopardised.[34]

As the controversy over the destination and role of the Australian corps was taking place, Lavarack had a more immediate problem—the fate of the men aboard the *Orcades*, about whom he had first expressed concern on 10 February. Even as Japanese paratroops were landing at Palembang, Lavarack consented to the proposed disembarkation of the troops at Oosthaven, about 250 miles (400 kilometres) to the south. He consented because he believed that without heavy equipment they could be easily re-embarked, and hoped that before they left the ship a Japanese advance would forestall the operation.[35]

With the surrender of Singapore Lavarack persuaded Wavell to order the withdrawal to Java of the Sumatra garrison. Lavarack then began to urge that the Australians on *Orcades* not be disembarked in Java. Wavell would not be moved because, apart from the need to defend airfields, he considered that prestige and the effect on the Dutch if they were not landed justified the disembarkation. Lavarack continued to protest and the unloading of the ship was delayed as long as possible. Also, he informed the Australian government that he could see no possibility that the troops would be employed usefully in Java. Nevertheless, on Wavell's orders the Australians on *Orcades* were disembarked on 19 February.[36]

Lavarack's appreciation that the 1st Australian Corps would have been wasted in the Netherlands East Indies was sound, for time was against the defenders (especially when resistance at Singapore ceased), and even if Lavarack had stayed and won the land battle, he could have been isolated by Japanese naval and air superiority. His recommendation that the Australian corps proceed to Burma, where it may also have been isolated, went heavily against Lavarack, despite the fact that he had been abroad for more than a year and thus out of touch with the military situation in Australia. From that time not only did he have Blamey's dislike to contend with, but also the disapproval of Curtin.[37]

The troops landed from *Orcades* resisted the Japanese invasion of Java, but were quickly subdued and most were taken prisoner. They were landed for reasons more political than military, and were employed to negligible advantage. Once Lavarack realised that the course of events would not prevent the AIF troops being landed, he strenuously opposed any move to have them disembarked. He has been criticised for not being strong enough to prevent the landing, but his Brigadier, General Staff (BGS), Berryman, argued that even if Blamey had been in command he could have done no more than Lavarack. Lavarack was bound to obey Wavell's order that the troops on *Orcades* be disembarked. However, he was successful in his attempts to have diverted another ship carrying about the same number of AIF troops.[38]

After returning to Australia from Java, Lavarack was appointed Acting Commander-in-Chief until Blamey's arrival from the Middle East. In the reorganisation of Australia's military forces which followed he was placed in command of the First Australian Army in Queensland. He languished there for almost two years, his frustration at not being given a fighting command in New Guinea steadily growing. Blamey's animosity and his position as C-in-C ensured that Lavarack's days as a fighting commander were over. Early in 1944, on Blamey's recommendation, Lavarack was appointed head of Australia's military mission in Washington. The appointment was approved by Curtin on the condition that it should not be seen as a promotion. The Washington post had the advantage for Lavarack that it removed him from Blamey's immediate influence, but for the most part it was, as he put it, 'a sadly inactive stay'.[39]

It was not because of any shortcomings in Lavarack's military skill that he had so little opportunity to command troops in battle during the Second World War, but Blamey's fear and dislike of him. Like other generals who had fallen into disfavour, such as Bennett, Rowell and Robertson, Lavarack discovered that Blamey could, and was willing to, keep him on the shelf.

Lavarack's contribution to the Australian Army and to the defence of Australia should not be undervalued. He was an educated, articulate officer well suited to the post of Chief of the General Staff, despite a fiery temperament. Although he was unable to achieve much progress towards a balanced defence policy in the years preceding 1939, in the face of political considerations it is unlikely that any Army officer could have. However, he presented a spirited and coherent criticism of the Singapore strategy. Despite the scarcity of opportunities for training in the command of formations in Australia in the 1920s and 1930s, Lavarack showed after the outbreak of the Second World War that he had a clear understanding of training, administration and command at division and corps level. During the limited time he spent in command, he showed himself to be a determined and competent leader, one who would command troops in battle as well as any citizen officer. He, and other regular officers, such as Vasey, who were given senior command appointments later in the war, exposed the myth of citizen officer superiority.

# Lieutenant-General Sir Vernon Sturdee: The Chief of the General Staff as Commander

D.M. HORNER

The outbreak of war with Japan in December 1941 placed a heavy burden on the already hard worked Chiefs of Staff of the three Australian Services. Until that time, as well as their normal cares for administration, training, equipment and provision of reinforcements, they had advised the government on the strategic deployment and development of the Services. They had also controlled the forces deployed in the Australian area, but they had had no responsibility for or control over the operation of Australian forces deployed in combat zones across the seas and far distant from Australia. Suddenly, with the threat to Australia, the Chiefs now became operational commanders. And, as the only Australian of the trio, none had greater responsibility than the Chief of the General Staff (CGS), Vernon Sturdee (1890–1966).

*The Military Board*

When the Australian Military Board was established before the First World War its main purpose was to facilitate the peacetime administration of the Army. Each of the Board members had their own separate responsibilities, and although the CGS was chairman of the Board (in absence of the minister), he had no authority over the other military members, the Adjutant-General, the Quartermaster-General and the Master-General of the Ordnance. From the beginning it was realised that in the event of mobilisation a commander of the military forces would have to be appointed and it was expected that the Inspector-General, who was not a member of the Military Board, would fill this role with the CGS becoming his chief of staff. Thus although the CGS was responsible for the preparation and revision of defence schemes, the work of the intelligence staff, advice on the raising of units, organisations and establishments, and training, it was not expected that he would command the Australian Army in time of war.[1]

This system continued with minor changes until the outbreak of the Second World War, at which time the Inspector-General, General Squires, became the CGS. Although a Commander-in-Chief (C-in-C) was not appointed there was no doubt about the authority of the CGS who, with the other Chiefs of Staff, frequently attended the War

16 Lieutenant-General V.A.H. Sturdee, on his appointment as Chief of the General Staff, September 1940 (AWM Negative No. 3657)

Cabinet. Nevertheless, it was not clear what would happen if war came to Australia. In 1935 the CGS, Lavarack, had suggested that at the outbreak of war a C-in-C should be appointed with another officer appointed to command the principal field army. The GOC of this field force would be subordinate to the CGS who would become C-in-C.[2] This proposal was raised on a number of occasions during 1940, but General White, who succeeded Squires after the latter died in February 1940, saw no requirement for a C-in-C of the Australian Military Forces (AMF) until war on Australian soil seemed imminent, and in no circumstances could he see a need for a C-in-C for the Home Forces.

White believed that, while Australia could achieve true security only from imperial cooperation, it was vital that Australia should look after her own defence. He had been one of the officers who in 1920 had warned the government that Japan would attack when Britain was occupied elsewhere. His pleas, and those of other senior officers such as

Chauvel, Lavarack and Wynter, had been largely ignored, and during the next twenty years Australia was locked into the Singapore strategy. Under this strategy Britain built a big naval base at Singapore to receive the British main fleet when war with Japan appeared likely. This fleet, to which Australia would contribute, was to ensure British and Australian security in the Far East. Thus, when the Minister for the Navy, W.M. Hughes, stated at a Cabinet meeting on 18 June 1940 that Australia should prepare for invasion, White replied: 'Our strategy is based on the Navy being here to help us when it is needed. Our strategy is based on us holding Singapore; on Japan *not* coming into the war.' The Australian Army was preparing to resist only a 'minor scale of attack'.[3]

The government saw that the main task was to build up the AIF in the Middle East, and as a result few resources were left for the defence of Australia. White and the other Chiefs agreed with this plan, but by August 1940 there was mounting concern that the British forces in Malaya were inadequate for the task of defending Singapore. On 6 August the Joint Planning Committee headed by the Deputy Chief of the General Staff (DCGS), General Northcott, sent a note to the Chiefs of Staff Committee advising that the plan to defend Australia against a 'minor' or 'raid' scale of attack should be changed to the heaviest scale of attack.[4] Before the Chiefs of Staff Committee could consider this note White was tragically killed in an air crash. But it was clear that the new CGS would have to face responsibilities which White had hoped would not eventuate.

## *'A gifted officer of high professional qualifications'*

The foremost military position in Australia, the mantle of Sir Brudenell White, now fell on the shoulders of the 50-year-old Vernon Sturdee, who a year earlier had been a colonel and Director of Staff Duties at Army Headquarters (AHQ). In October 1939 he had been promoted from colonel to lieutenant-general to command Eastern Command, but he had stepped down a rank to raise and command the 8th AIF Division. He had not held this position for two weeks before White was killed. His successor as commander of the 8th Division, Gordon Bennett, who later clashed with Sturdee, farewelled him with an assurance of his loyal support at all times, adding that Australia was fortunate in having him as its military chief.[5]

Fortunate indeed! By training and experience Sturdee was ideally suited for the position. After attending the Melbourne Grammar School he had begun training as a mechanical engineer before joining the Australian Engineers as a sapper in 1908. By 1914 he was adjutant of the 1st Divisional Engineers, and the next year commanded a field company at Gallipoli. Later, as commander of the 5th Field Company he was largely responsible for the construction and maintenance of the ANZAC light railway system in the Somme–Bapaume–Cambrai area in France. For most of 1917 he commanded the 4th Pioneer Battalion as a lieutenant-colonel. Then after four months as chief engineer of the 5th

Division, he joined Field Marshal Haig's staff until the end of the war. This experience gave him a broad view of the British Expeditionary Force and stood him in good stead when in later years he had to cooperate with British and American forces. Since the British Army provided the bulk of logistic support for Australian units, few Australians were in a position to grasp the importance of logistics. Yet during the Second World War Australian forces became responsible for their own administration. For this reason, General Berryman thought that the choice of a commander for the 2nd AIF rested between only three officers, Blamey, Lavarack and Sturdee.

Between the wars Sturdee attended the Staff College at Quetta and the Imperial Defence College, London, and he served for a number of years in England, including a year as the military liaison officer at Australia House. From 1933 to the outbreak of war he was Director of Military Operations and Intelligence (DMO & I), and then Director of Staff Duties at AHQ. General Rowell, who served with Sturdee for part of this time, wrote later:

> To those of us working with him then, Sturdee displayed those characteristics which he retained all his life. He had a very precise mind—he had a great sense of the need for priorities—he saw the problem very clearly—he was able to give orders which left no doubt as to what was wanted and he then left people to get on with the job. When it came to the answer he was kindly and constructive in criticism. But he didn't suffer fools gladly and told them so, while, at the same time, he was unerring in picking out the one who was dragging his feet and who needed encouragement or something stronger. But above all, he knew how to laugh, and thus was a great help to people who...were asked to work beyond the normal.[6]

General Sir John Wilton, who served under Sturdee later in the war, found him to be able, balanced, courteous and fair. He never lost his temper. General Hopkins found him to be 'quite unflappable'. His command of a pioneer battalion in the First World War provided him with more command experience than most of his Regular Army contemporaries and despite his shy, reserved manner, to many observers he presented an aura of command. Indeed, as Rowell observed, he was 'a gifted officer of high professional qualifications with the benefit of the best education and experience that the army of the day could offer'.[7]

As DMO & I at AHQ Sturdee was vitally concerned with the development of the Plan of Concentration for the defence of Australia, which detailed the plan for meeting a Japanese invasion of Australia by deploying two corps each of two infantry divisions, plus two independent cavalry divisions and corps troops. These were, of course, militia formations which would have to be mobilised, equipped and trained in the warning period before attack. Sturdee had quite early discerned that Japan would pose the major threat to Australian security. In 1933 he told senior officers that the Japanese would act quickly, 'they would all be regulars, fully trained and equipped for the operations, and fanatics who like dying in battle, whilst our troops would consist mainly of civilians hastily thrown together on mobilization with very little training, short of

17 The Chief of the General Staff, Lieutenant-General Sturdee, GOC AIF Middle East, Lieutenant-General Sir Thomas Blamey, and the Minister for the Army, Mr Percy Spender, toast the New Year, 1 January 1941, Sea of Gallillee, Palestine. Spender and Sturdee had just arrived in the Middle East for an inspection tour. (AWM Negative No. 4896)

artillery and possibly of gun ammunition'.[8] Sturdee was soon to be faced with the responsibility for handling such a situation.

## Organising for war

From the time of his appointment Sturdee's task as CGS grew in complexity. Although he had no operational control over the AIF in the Middle East and Malaya, he was responsible for its training in Australia and resupply, as well as for the administration and training of the militia, strategic advice to the government, and for the defence of Australia. In February 1941 the government learned that if Japan and the USA joined the war, then the Allied strategy would be to deal with Hitler first. As the Acting Prime Minister, Arthur Fadden, observed, if Australia were to be abandoned by Britain and America 'until the war in Europe was decided, we and our countrymen might well be pulling rickshaws before long'. Soon afterwards, Australia advised Britain that the 8th Division, the first brigade of which was arriving in Malaya, would be retained for use in Australia and the Far East, and that it would not proceed to the Middle East. Furthermore, Australia agreed to provide troops and an air striking force at Darwin to reinforce Ambon and Koepang if required. On 24 February Army Headquarters issued instructions that there was a possibility of war with Japan, the 49th Battalion (militia) was to be raised for service at Port Moresby, the force at Darwin was to be brought up to brigade strength, and militia training was to be stepped up.[9]

But a first necessity was a reorganisation of the command structure. In April 1941 the Minister for the Army, Percy Spender, recommended to the War Cabinet that Sturdee should be appointed C-in-C of the AMF, that the present DCGS, Northcott, should become CGS, and that the Military Board should be abolished. On 9 May the War Cabinet decided in principle to appoint a C-in-C and directed that the necessary proposals should be prepared. In June, however, the War Cabinet could not agree over the necessity for a C-in-C and the appointment was deferred to give the minister an opportunity for further consideration.[10]

It appears that the example of the British system, in which their equivalents of the Military Board and the CGS, namely the Army Council and the Chief of the Imperial General Staff, continued to exist, and a separate GOC-in-C Home Forces was appointed, carried particular weight with the War Cabinet. Despite the reservations of the Army about the efficiency of this system, on 5 August 1941 Major-General Sir Iven Mackay, the commander of the 6th Division in the Middle East, was appointed GOC-in-C Home Forces. His instructions placed him equal in rank to the CGS, but subordinate to the Military Board, and his authority was not clearly defined. In spite of this grandiose title, the responsibility for the defence of Australia still rested with the CGS. After all, Mackay's instructions stated that he was to exercise operational command over the military forces in Australia except for forces withdrawn by the CGS, and he was subject 'to the general responsibility of [the] CGS for the military policy affecting the security of the Commonwealth'.[11] Initially Mackay operated through Army Headquarters rather than from his own headquarters.

Sturdee had his own reservations about this organisation, but at last some headway was being made. Furthermore, at his urgent request Rowell was recalled from the Middle East, and early in September took up an appointment as DCGS, releasing Northcott to command the Armoured Division, which until then he had attempted to command on a part-time basis. Because of the personal attention needed and until Rowell's appointment, much of the work of the General Staff Branch at Army Headquarters had to be left to the various directors.

In the three months to the outbreak of war with Japan, Rowell and Sturdee struggled with every energy to improve the defences of Australia. Meanwhile Fadden replaced Menzies as Prime Minister, and then shortly afterwards there was a change of government, with new, inexperienced ministers to educate. The new Prime Minister was John Curtin and Sturdee advised the government (against Rowell's advice) that Tobruk should be evacuated claiming that he had to support the man on the spot, Blamey. He refused to send the third brigade of the 8th Division to Malaya, holding it in Darwin, and he agitated to have the militia called up for full-time duty. Indeed, he said later that he was almost thrown out of the Cabinet room for insisting on this increase to the militia.[12]

*18* Lieutenant-General Sturdee with Lieutenant-General H.ter Poorten, Commander-in-Chief in the Netherlands East Indies (later commander of the allied land forces in Java) at the handing over of the hospital ship *Oranje*, 28 June 1941. (AWM Negative No. 8031)

## *Island detachments*

Throughout the whole period of Australian military history, 1941–42 marks the time of greatest stress upon the Australian high command. The news that Japan had attacked Pearl Harbor and Kota Bharu on 8 December 1941 came, to most Australians and their government, as a tremendous shock, but initially there was little dismay. It took two months for Australia to come from complacency, or confidence, to the threshold of fear for her own survival. It was not until 23 January 1942, with the fall of Rabaul, that the fear began, and with it the recriminations against Great Britain, whom a number of leading politicians claimed had let Australia down. It was against this background of hysteria that the Australian generals had to make their plans.

From 8 December 1941 onwards for about four months the Chiefs of Staff, Sturdee, Admiral Royle and Air Chief Marshal Burnett, were constantly directed to prepare appreciations for a worried government. The first appreciation was to the point, stating that the attack on Malaya 'might well be a first step in the Japanese plan for a major attack on Australia', and that 'the possibility of a direct move on Australia via the islands to the North and North East must now be considered'. This attack would start, it was argued, by attempts to occupy Rabaul, Port Moresby and New Caledonia; from these bases the attack on Australia could be launched. The Chiefs advised that it was 'necessary to establish and train now the forces that would be required to prevent and to meet an invasion'. One would have thought that this would have been an

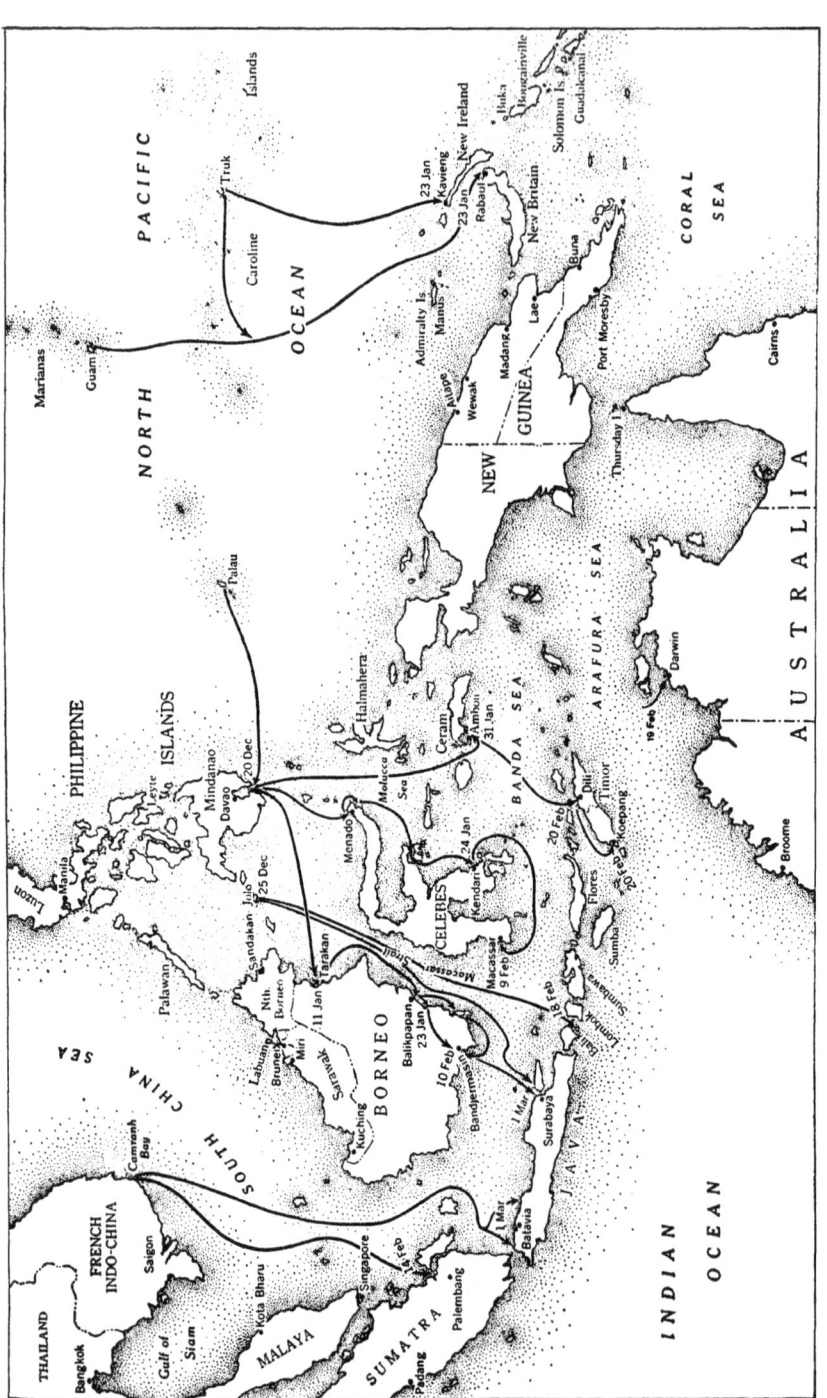

14  The Japanese advance through the Indies and to Rabaul

admirable aim a year earlier, but as late as 4 December the War Cabinet had directed the Defence Committee to consider the possibility of reducing the Military Forces by from 20 000 to 30 000 men to enable additional manpower to be made available for the Navy and Air Force and for munitions production.[13]

Although the Chiefs of Staff recommended on 12 December that the islands to the north of Australia should be held with as strong a force as possible, the Cabinet did not see it that way. Thus in a subsequent appreciation the Chiefs observed that if the Newcastle-Sydney-Port Kembla area were to be secured adequately then there was little that could be done for the outlying islands.[14]

It should be remembered that in the latter part of 1941 the naval and air forces available in Australia were almost non-existent. The only aircraft available for offensive operations were a limited number of Hudsons and Catalinas which in the first few months of operations against Japan were continuously employed, until their ultimate destruction, in daring and effective operations against the Japanese advance to the northeast of Australia. The only remaining force of any magnitude was the Army, largely part-time militia, who in the main were not fully trained, not fully equipped and officered generally speaking by those who had been either too old to be accepted for the AIF, too young or not fit for war, or those who had and continued to have employment in essential industries. Only Mackay and Rowell had seen action in the current war, and while Sturdee and Northcott were men of great ability the remaining senior officers had, on the whole, proved to be too old for service in the Middle East. Immediate plans were made to bring senior officers back from the Middle East, and by the end of January 1942 Generals Clowes, Vasey, Robertson, Savige and Murray had all taken up appointments in Australia.

Meanwhile, during December, as arranged at earlier staff talks in Singapore, and at the request of the Netherlands East Indies, Sturdee ordered one battalion of the AIF brigade at Darwin to Ambon, and another to Timor to protect forward airfields. The third battalion had already been dispatched to Rabaul. An independent company was sent to New Ireland in July, and an element of another had been ordered to New Caledonia. But while these meagre forces formed a thin line to Australia's north, the bulk of Australian forces were assembling in the south of the country.

In view of the crippling limitations of the Australian forces, including the fact that only the AIF could serve outside Australia and there were no trained infantry battalions left, the decision of the Australian Chiefs of Staff not to reinforce the northeastern approaches seems unchallengeable. The movement of large army garrisons to the northeastern approaches to protect advanced air bases, even had trained troops been available, would have served no useful purpose. Whatever the numbers of troops deployed, the effective defence of bases to which they would have been moved would not have been possible without naval and air forces sufficient to support the land forces in this defence, and keep open the lines of communication to these bases from Australia. These naval and air forces were not available and to reinforce the bases would not, therefore, have prevented the investment of the garrisons concerned and their defeat. There is no reason to believe that the reinforcement of the bases would have delayed Japanese operations against the mainland had that been their intention. It would only have reduced the forces available in Australia for defence against a Japanese attack.

That the Australian Chiefs of Staff were fully alive to the need for somehow strengthening the northeastern approaches is indicated by their appreciation of 15 December 1941. It was as a result of this paper that the garrison of Port Moresby was increased to a militia brigade group (the limit of the maintenance capacity) and the decision was taken neither to reinforce nor to withdraw the Rabaul garrison.

While the overall strategy outlined in the appreciation has withstood the test of events, there were inevitable weak links in the defensive chain. With respect to Rabaul the Chiefs of Staff saw three options open to them:

1 to reinforce the garrison to meet the scale of attack likely to be directed against Rabaul;
2 to withdraw the garrison and abandon Rabaul;
3 to retain the existing garrison.

They considered that 1 was not possible, that 2 would have a negative effect on the Dutch in the Netherlands, East Indies, and that option 3 should be pursued because of the importance of maintaining 'a forward observation line'. Colonel Keogh has claimed that there was a fourth option: that there should have been a plan to fight a withdrawal action and in the process inflict damage and delay on the enemy. He pointed out that no instructions to that effect were issued by either AHQ or HQ 8th Military District in Port Moresby. Rowell said later that Army Headquarters had the scale of attack all wrong; the Japanese employed a division against a battalion. It was bad luck for the battalion that the Japanese intended making Rabaul their main base.[15] In any case, the force at Rabaul lacked the supplies, training and air support to conduct a fighting withdrawal.

The decision to send only small forces to Rabaul, Timor, Ambon and New Ireland reveals the desperate position faced by Army Headquarters in Melbourne. It was clear that these forces would be hard-pressed and as General Rowell told the commander of the force on Ambon on 26 December 1941, he was 'to put up the best defence possible with resources you have at your disposal'. Furthermore, despite the great organisational skills of Sturdee and Rowell, Army Headquarters had not been structured to control operations in the field. The operational instruction from AHQ dated 6 December and relating to the forces in Timor and Ambon, did not reach Ambon until 13 January.[16] When, in mid-January, there was doubt about the morale of the Ambon commander, he was replaced by a senior staff officer at Army Headquarters who volunteered for the appointment knowing full well that he would probably be captured.

Sturdee needed all of his strength of character as each garrison surrendered in the face of overwhelming Japanese forces. Meanwhile the government vacillated over whether to withdraw civilians from forward areas such as Rabaul. For example on 15 January the deputy administrator at Rabaul sought permission to evacuate the civil population, hoping to use the cargo ship then in the harbour. The reply on 17 January indicated that government officials should remain at their post. The modified view that unnecessary civilians could be evacuated was received on 20 January after the Japanese had attacked and the ship had been sunk.[17]

In late December 1941 the government directed Sturdee to send reinforcements to Malaya. An AIF machine gun battalion at Darwin plus almost 2000 reinforcements were sent in early January. These reinforcements were largely untrained and some had not even fired a rifle. The official historian, Lionel Wigmore, observed:

> The arrival of such reinforcements in Malaya may be explained partly by the fact that the practice had developed of sending raw recruits to the Middle East where they received their basic training under expert instructors in the excellent training organisation established there. This does not, however, excuse the blunder of sending untrained men forward...to a division then going into battle.

Wigmore added that reinforcements could have been sent from the Middle East, or volunteers could have been sought from militiamen on full-time duty.[18] Sturdee said later that 'subsequent to the commencement of hostilities events moved so rapidly and disastrously that it was not possible to create an organization for training the reinforcements'.[19]

As disasters mounted Sturdee came under increased pressure to change his strategy. For example, on 23 February 1942 Curtin sought views on a letter he had received from Port Moresby claiming that the town had no strategic importance. The Prime Minister's correspondent argued that all of New Guinea should be evacuated and added:

> In Rabaul and New Britain we have the example of a perfectly good force wiped off, because the enemy have been able to deal with it with a much superior force numerically and otherwise and there has been no avenue of escape. Malaya and Singapore are also examples. Port Moresby must follow in the near future, that seems obvious.[20]

Sturdee replied on 4 March that Port Moresby was strategically important because, together with New Caledonia, it controlled the entry into the Coral Sea and their capture was an essential prerequisite for a Japanese invasion of the vital areas of eastern Australia. Furthermore, it was an important flanking position in relation to enemy movement towards Australia and it threatened the Japanese base at Rabaul. Voluntary withdrawal was out of the question not only because of the town's strategic value but also because of the effect on public morale. Sturdee's argument reveals the inaccuracy of General MacArthur's later claims that he decided, when he arrived in Australia in late March, to defend Australia from New Guinea. Indeed Sturdee concluded that provided increased air support and anti-aircraft resources could be provided the retention of the military garrison at Port Moresby was justified.[21]

The overwhelming of the forces in the islands appears to have been caused by a number of factors. First, since the beginning of the war the government and their military advisers had relied upon the Singapore strategy. Second, they had not anticipated the speed and power of the Japanese thrusts. Third, the Army was not ready for war; as a result of the formation, training and support of AIF divisions overseas the forces in Australia lacked trained troops, there were deficiencies in the organisation of headquarters for command, equipment was in short supply and the Army lacked young experienced leaders. Practically none of these shortcomings could be laid at the feet of the army leaders. Indeed the army leaders had not been able to persuade the government of the urgency of the defence of Australia. The policy of continental defence, rather than forward defence, became a reality only after the failure of the latter. But even then, it might be argued that forward defence had bought time for Australia, and still applied so long as the Port Moresby garrison survived.

*The most fateful recommendation*

It is important that the psychological situation affecting the government and the defence planners in early 1942 is understood. W.E.H. Stanner, one of the personal assistants to the Minister for the Army, F.M. Forde, recalled that Forde tended to panic during this period. Stanner had previously been an assistant to Forde's predecessor, Spender, and in the emergency of early 1942 he was impressed by Sturdee's calm approach. Indeed he thought that Sturdee was 'one man who kept his head when the government lost its head after the Japanese attack'.[22]

On 9 December 1941 the Secretary of the Department of the Army, F.R. Sinclair, wrote to Forde suggesting the formation of a guerilla army and urging him to meet a British officer who was supposed to be an expert on guerilla warfare. Sturdee was surprised and told the minister: 'I feel bound to state that it is a matter for regret that so highly placed an officer should, in these critical times, have seen fit to produce a proposal in such despairing terms without consulting any of the government's military advisers'.[23] This incident exemplifies two aspects of Sturdee's character, his sense of propriety and the importance of sticking to the rules, and his determination to face the disasters in a logical and calm manner.[24]

Sturdee's logical approach is shown during the discussions which later led General MacArthur and the Minister for Labour and National Service, Eddie Ward, to claim that the Australian Army planned to withdraw behind a Brisbane Line. On 4 February the GOC-in-C Home Forces, Mackay, produced a memorandum for Forde seeking direction and government concurrence in regard to the area which it considered vital to hold in the event of an attempted invasion. Mackay pointed out that, with the agreement of the government, the Chiefs of Staff had decided that the area most vital to the continuance of the war effort was the Port Kembla–Sydney–Newcastle–Lithgow area. With the decision to develop an important base for US forces at Brisbane, and with the economic and military importance of Melbourne, Mackay noted that it was vital to hold the area stretching some 1000 miles (1600 kilometres) from Brisbane to Melbourne. This area contained a force of barely five divisions. Mackay explained that all other areas had to be looked upon as isolated localities and that Tasmania and Townsville should not be reinforced, although 'for reasons of morale and psychology' it was undesirable to withdraw troops from these areas.[25]

Forde, whose electorate was in northern Queensland, was not happy with this appreciation and he asked Sturdee whether it was possible to move an additional division to Queensland, and also to arm the forces in Queensland more heavily. Sturdee's reply was to the point:

> The problem is...not what can we move out of the vital area to protect say, North Queensland or Tasmania or Western Australia, but whether we may not have to, or should not now, take troops away from these areas to increase our strength at the decisive place.

Sturdee concurred generally with Mackay's views with the qualification that if it became established that the main threat was to the flanks the 'problem would become an entirely different one and the main field army' would be disposed to meet it.[26]

These plans were based purely upon the assumption that only the troops then in Australia could be used for the defence. But Sturdee was now making every effort to ensure that the bulk of the 1st Australian Corps, then destined for Java after having been withdrawn from the Middle East, should be returned to Australia. On 15 February, the day Singapore fell, he produced a comprehensive paper pointing out that the troops in Java would probably be lost, and even if they were not lost Java did not provide a continental base suitable to build up Allied strength for an offensive. Unlike Java, Australia had the shortest sea route to the United States of any large area of land in south-east Asia. He concluded that Australia's northern shores were sufficiently close to Japanese occupied territory to make a good 'jumping off' area for offensive operations.[27]

These views foreshadowed a great allied offensive based on Australia, and indeed the previous day America had decided to send the 41st Division to Australia. Sturdee's paper gives the lie to allegations that the Australian Army planned to give up the north of Australia to the Japanese. What he was saying was that there was a possibility that this plan might be forced on the government if the 1st Australian Corps were not diverted to Australia.

On the morning of 15 February Sturdee telephoned the Prime Minister and urged him to order the convoys to be diverted to Australia. During the previous ten years he had argued consistently that Australia's vulnerability would increase if Singapore were captured and this event had now occurred. At the Chiefs of Staff meeting of 16 February the burden was squarely upon Sturdee, for the other two Chiefs, Burnett and Royle, were British officers. They felt that any advice they might offer could be biased, and consequently they told Sturdee that they would unreservedly support him.[28] This is not to suggest that the other Chiefs were in any way disloyal to the Australian government. Indeed Curtin said that although Colvin, Royle's predecessor, always echoed Admiralty opinion, Royle was more independent. As Royle said of his attitude toward the Admirality: 'Frankly I think that my superior [RN] officers are wrong and I am right'. But the documents show that throughout the period Sturdee was the main spokesman for the Chiefs. Rowell wrote that this was 'the most fateful recommendation [Sturdee] had to make in his service'. When the Chiefs of Staff met the War Cabinet on 18 February to discuss their recommendations, Sturdee informed the ministers that he would tender his resignation if the AIF were not returned to Australia.[29]

Curtin, however, agreed completely with Sturdee. Furthermore, before finalising his paper Sturdee had discussed his concern with the Secretary of the Department of Defence, Shedden, who had prepared a cable to Churchill which followed much the same lines as Sturdee's paper.[30] Eventually, after a night of agonising over the decision, Curtin forwarded the cable to Churchill requesting that urgent arrangements be made for the diversion of the AIF to Australia. The War Cabinet meeting on 18 February confirmed that decision. The events of 1942 were to amply prove the wisdom of this decision; by October four brigades from the Middle East had been in action in New Guinea. Sturdee's manner during the crisis is revealed by the comments of that former school teacher, Sir Iven Mackay, who said that Kipling must have had someone like Sturdee in mind when he wrote the lines:

> If you can keep your head when all about you
> Are losing theirs and blaming it on you ...
> ...you'll be a Man, my son.[31]

The command problems had yet to be resolved. Mackay had eventually opened his Headquarters Home Forces on 22 December 1941, but his command was less than that originally envisaged. While he was responsible for Queensland, New South Wales, Victoria, Tasmania and South Australia, General Plant was responsible for Western Australia, Brigadier Blake for the Northern Territory and Brigadier Morris for Port Moresby. The Sydney *Sunday Telegraph* strongly criticised this arrangement: 'This is no time for finicky politeness. If Mackay is a better man than Sturdee, he should be made GOC—a real GOC—over Sturdee. If Sturdee is the better man, make him GOC with power really to command.' A number of articles in the *Sydney Morning Herald* developed this theme, calling for the appointment of General Sir Thomas Blamey, then in the Middle East, as Commander-in-Chief. On 20 February the Government cabled for Blamey to return 'as speedily as possible'.

Meanwhile Sturdee had been preparing his own proposals. He

recommended that the Military Board should be abolished and that its members should become principal staff officers to the new C-in-C. Once a Supreme Allied Commander was appointed he thought that the Military Board could be reconstituted to look after 'home administration' leaving the C-in-C to devote his time to operations and war training. He proposed that the army should be reorganised into the First Army under Lavarack, who had just returned from Java, the Second Army under Mackay, and independent forces in Tasmania, Western Australia, Northern Territory and New Guinea. With the exception that the Military Board was not reconstituted this was substantially the plan adopted by Blamey when he became C-in-C at the end of March 1942. Sturdee's contribution to this reorganisation has not generally been acknowledged.[32]

## 'We had great admiration for him'

After Blamey returned, Sturdee continued serving loyally as CGS until July 1942 when he was appointed head of the Australian military mission in Washington. This new posting did not signify that Blamey had lost confidence in him; in fact, the reverse was the case. Sturdee had worked exceedingly hard for two years as CGS and needed a rest. General Vasey wrote to his wife at this time that Sturdee 'looked awful'. It was now obvious that the important decisions of the war were being made in Washington, and it was vital that Australia should be represented not only by an officer of high rank, but also one who was thoroughly conversant with the problems of the allied forces in Australia. Rowell wrote that Sturdee 'had an unrivalled knowledge of the strategy of the area and he knew as well as anyone its needs in men and material'.[33] At Blamey's request he accepted the appointment on the condition that after only one year he should return to a command position in the South West Pacific Area.

In essence this promise was kept. His term in Washington was from 10 September 1942 to 28 February 1944, after which he became GOC of the First Australian Army, with his headquarters in Lae. In this appointment which he held until the end of the war in the Pacific, he commanded some 110 000 troops in the campaigns from the Solomon Islands in the east, through New Britain and New Guinea to the Dutch border in the west. On 6 September 1945 aboard the aircraft carrier HMS *Glory* in Rabaul Harbour, he accepted from General Imamura, Commander 8th Area Army and Admiral Kusaka, Commander South East Asia Fleet, the surrender of all Japanese forces in the Solomons, New Britain and New Guinea. At the end of 1945 he succeeded Blamey as Acting C-in-C, before being reappointed as CGS from February 1946 to April 1950. In 1940 he had been appointed CGS by Menzies; now he had been appointed by Chifley—a tribute to the trust he had engendered from both political parties.

So many disasters befell the Australian Army during the period from December 1941 to March 1942 that it would be easy to criticise the CGS. But, as noted earlier, he was not responsible for the AIF in Malaya and he had little option in the deployment of forces to the islands. There was looting in Port Moresby, poorly trained militia were rushed to Papua and many administrative faults were revealed under the pressure of the rapid expansion of the army. Yet Sturdee's subsequent career showed that it was recognised by the relevant authorities that many of these happenings were beyond his control. Years of neglect and parsimony by the government could not be recovered in a few months.

It was Sturdee's lot to fight the first battles, when troops, equipment and training were in short supply and the AIF was overseas. These battles cannot be half as successful or as easy to command as the later avenging battles when equipment and training is more plentiful and heavier weapons can be brought to bear. It is a measure of Sturdee's determination, clear-sightedness, and professionalism that he survived to join in the avenging battles. In the opinion of F.M. Forde, Sturdee had 'performed very well as CGS at a time when there was no glamour attached to the position...he rendered very satisfactory service. We all had a great admiration for him'.[34] More than that, he was the rock on which the army, and indeed the government rested during the weeks of panic in early 1942.

# Lieutenant-General Henry Gordon Bennett: A Model Major-General?

A.B. LODGE

The fall of Singapore marked the end of Henry Gordon Bennett's (1887–1962) active service. Never again would he command troops in battle, despite later efforts to secure for himself an active command. The campaign in Malaya had acted as a bright light played across the surface of his character, throwing into relief strengths and weaknesses far more distinctly than could the pale sheen of peacetime soldiering between the wars. Bennett's performance as GOC of the 8th Australian Division in Malaya must be evaluated on two levels: as the commander of an infantry division, and also as a national commander fighting under the command of a more powerful ally.

*National commander*

After the war Lieutenant-General A.E. Percival (GOC, Malaya Command during the campaign) remarked, without admiration, that 'Bennett cared more for the well-being of the AIF than of anything else'.[1] His observation reflects the essence of Bennett's performance as the Australian commander in Malaya. His determined defence of the rights of his country and her troops, and his low regard for the British, which he did little to disguise, were not always conducive to cordial relations. Like Blamey, who had taken troops to the Middle East and had been given a charter to guide him in his dealings with the British, Bennett was given a directive which, although recognising that the 8th Division would come under the control of the British commander in Malaya, made certain stipulations:

(a) The Force will retain its identity as an Australian force;
(b) No part of the Force is to be employed apart from the whole without your consent;
(c) Should the GOC Malaya in certain circumstances of emergency insist on an extensive operational dispersal of your Force you will, after registering such protest as you deem essential, comply with the order of the GOC Malaya and immediately report the full circumstances to Army Headquarters, Melbourne.[2]

As far as safeguarding the rights of Australian troops was concerned Bennett performed his task well. In August 1941, after there had been trouble between Australians on leave and British military police, Bennett, who believed that his men were being provoked, arranged with

*19* Major-General H. Gordon Bennett, GOC of the 8th Australian Division, 1941. AWM Negative No. 8520)

Percival and Lieutenant-General Heath, of the 3rd Indian Corps, that the AIF would be disciplined only by Australian military police. On another occasion Percival suggested that Australian troops, like British troops in Malaya, should be restricted in their use of army vehicles for recreational purposes to once per month, when they would be required to pay a mileage rate. Bennett refused to agree, explaining that it was the 'policy of the AIF to maintain health and morale at Government expense'. Notwithstanding the value to the Australians of actions such as these, they clearly emphasised the distinction between Australian and British troops and might perhaps have been seen as obstructive. On at least one occasion, however, Bennett also attempted to minimise differences between his men and the British. Percival agreed to a suggestion made by Bennett that a system of exchange appointments be instituted which would allow Australians to serve with British units and vice versa. Although the system was limited by differing conditions of service, it was to Bennett's credit; as Wigmore has noted: 'it was a thoughtful and practical step towards dissipating animosities—often the outcome of misunderstanding—which flourish in static garrison conditions'.[3]

Perhaps the most contentious issue on which Bennett took a stand occurred in May 1941 when he refused to allow his troops to be used to prevent strikes by plantation workers. It was the policy of the Australian government that its troops not be used for such a purpose and Bennett acted correctly as a representative of his government in refusing Percival's request. After the campaign commenced Bennett was determined that the AIF would take an active part in the defence of Malaya and not become just a collection of 'wood and water joeys' for the British.[4] Throughout his time in Malaya Bennett attempted to act at all times in accord with the directive given him by the Australian government. In a similar fashion to Blamey in the Middle East, he resisted the tendency of British commanders to break up Australian formations and use them piecemeal. Whenever possible he ensured that the AIF fought as an integral force and he allowed it to be dispersed only under the pressure of dire operational circumstances, and not before he had made strong representations against the separation.

At times, though, Bennett took the protection of national interests to an extreme, such as on 14 February 1942 when he ordered that Australian artillery would fire only in the defence of the Australian perimeter—an order which deprived support to the 1st Malaya Brigade on his left.[5] As the Australian commander Bennett's task was not only to protect the interests of the AIF; he was also required by his government to cooperate with the British commander in the defence of Malaya, and it appears that at times he fulfilled the former aspect of his duties at the expense of full cooperation with the British. Bennett, correctly, had made it clear that the 8th Division was not another British formation which Percival could dispose at will. However, the effect of this distinction and, more particularly, the undiplomatic character of the man required to preserve it, did not contribute to smooth relations; at best the relationship between Bennett and the British was uneasy.

After the campaign Bennett was eager to minimise any differences he had had with the British. In a letter to the Military Board soon after his return to Australia he wrote: 'The relations between myself as GOC AIF Malaya and my staff with the officers of Malaya Command and all British Staffs in Malaya were most friendly throughout. The AIF found nothing but full co-operation whenever the Staff at Malaya Command were approached.'[6] He was probably concerned, as he had been previously, that his reputation for being difficult to deal with would hamper his chances of being given another active command. Some years later in a letter to the official historian Bennett contradicted his earlier statement that there were no differences of opinion with the British, and in particular with Percival, but still did not implicate himself, claiming that 'our relationship was always harmonious. Any differences that existed between the AIF and Malaya Command being on the staff level.'[7] Bennett was hardly being candid. During the campaign he had openly and officially expressed the views that the senior British commanders were 'not fit for their jobs', that they had let him down in Johore and it was lost for that reason, and that not only could the British not be relied upon, but they could not be trusted.[8] Bennett's criticisms, however, were not always without substance. His opinion that the British had learnt nothing since the First World War was supported by the British official historian, General Kirby: 'The training instructions issued by Command Headquarters were based upon those extant in the United Kingdom and envisaged a linear type of war more suited to Europe in the 1914-18 war than to Malaya in 1940-41'.[9]

Although Bennett distrusted the British, it appears that he did not dislike Percival personally. He recorded in *Why Singapore Fell* that Percival 'was typical of the best [officers] produced by the Staff College at Camberley'. Given Bennett's opinion of regular officers such a statement could be less than complimentary, but he continued on the following page that Percival 'knew Malaya and its problems and was probably the best selection for the appointment to command the land forces there... In spite of all the criticism of General Percival, he did not fail, but he was unable to defeat the system.' Generous words indeed from one so critical, but perhaps like a great part of his book they were included more as propaganda than as genuine appreciation. He had been more critical of his commander in correspondence with Forde during the campaign when he observed of Percival: 'quite a good brain and sound judgement, but very weak. He is dominated consistently by Heath, the Corps Commander. He wants the army to fight and to stop retreating, but lacks the personality to make it fight or even to remove officers who lack the fighting spirit.' Bennett's criticism was similar to that expressed by others who knew Percival at that time, such as the Australian government's representative, Bowden, who had said that Percival was competent, but lacked a strong personality. Wavell, the Supreme Commander in the theatre, was also concerned about Percival.[10]

While Bennett had strong reservations about Percival's ability, he was more critical of other commanders, especially Heath, the commander of the 3rd Indian Corps. On at least one occasion Bennett's antipathy towards Heath influenced Percival's operational plans; in Johore the 3rd Corps Headquarters was withdrawn partly because, in Percival's words, 'the combination of Gen. Gordon Bennett working under Gen. Heath would never have succeeded'. Bennett made it clear to Percival that he thought Heath unfit to command and that he should be replaced as the commander of 3rd Corps; and he also vehemently criticised Heath in his correspondence with Forde, writing that he would not allow the AIF 'to be sacrificed on the altar of Heath's inefficiency'.[11]

The intemperate manner in which Bennett expressed his opinions of Heath should not be permitted to obscure the legitimate foundations of such criticism. There were many points in his vitriolic attacks upon Heath's ability which were shared by other senior officers in Malaya.

Wavell was concerned about Heath, as he was about Percival, and on at least one occasion in early January visited the front in northern Malaya to see for himself if Heath were equal to the task given him. Pownall, Wavell's chief of staff, observed of Heath that he seemed 'all right', but that he was not 'full of fire', although he handled his troops well.[12] Even Percival's opinion was similar to Bennett's. He later related that: 'all the way down the peninsula, I had a feeling that, unless I issued definite orders as to how long such and such a position was to be held, I should find that it had been evacuated prematurely'. Although the official historian, Kirby, did not agree, Percival claimed that Heath's rapid withdrawal from Batu Pahat had 'spelt final disaster'.[13] Percival, like Bennett, gained the impression as soon as operations against the Japanese commenced that Heath suffered from what he called a 'withdrawal complex', which he later argued had manifested itself on several occasions during the course of the campaign. Percival stated that after reaching Johore Heath 'planned daily withdrawals which were...both too long and too fast'; that he issued the orders for the withdrawal to Singapore Island on his own authority; and that he, again without permission, relinquished control of a major portion of Singapore's water supply during the fighting on the island.[14] Percival's biographer has commented: 'There is no doubt...that "Piggy" Heath, though a good soldier and generally popular, was not such a good corps commander for Percival as Barstow would have been'.[15]

Not only did Bennett criticise commanders, but he also turned his attention upon their troops, particularly the Indians. On 30 January he wrote in his diary: 'I thought I could hold Johore—but I assumed that British troops would have held their piece'. While his conclusions as to the effect the poor quality of some troops may have had on operations are not in all cases supportable, and, as at Muar, may conveniently overlook his own part in the defeat, there is little doubt that in the majority of cases his criticisms of the standard of British, especially Indian, troops were valid. And his reports prompted Curtin, the Australian Prime Minister, to cable Churchill in late January to complain that the Indians were not suited to the warfare in Malaya. Although Curtin's cable provoked an immediate rebuttal from London, Percival's opinions were generally in accord with Bennett's. In a note for the Indian war historian Percival commented that while 'morale was never broken...I would not go so far as to say that it "stood up magnificently"'. Pownall, when he visited the 11th Division in late December, observed that: 'They are immature troops and I suppose nobody below CO's had ever before heard a bullet fired in anger'. Percival considered that the poor training of the 45th Brigade was an important factor in the defeat at Muar and reported that: 'This brigade had never been fit for employment in a theatre of war'. During the Muar fighting Bennett recorded that Major-General Barstow of the 9th Indian Division had informed him that Painter (commanding the 22nd Indian Brigade) 'threw in the sponge' because his men were 'unreliable'.[16]

Once again Bennett's criticism was legitimate, if intemperate. He was

20 Major-General H. Gordon Bennett talking to Captain J.A. Collins RAN, on the staff of the Commander-in-Chief Eastern Fleet, 1941. (AWM Negative No. 10103)

arguing from a strong position since his troops were among the best trained in Malaya. As Kirby has observed, this was facilitated by the AIF being a national force, leaving it free to pursue its own policy without interference from Malaya Command.[17]

### British criticism of Bennett

As might be expected, Bennett himself did not escape British criticism. Percival wrote mildly in 1945: 'I had to pull him up once or twice on such matters as direct communication with the Australian Press on defence problems but, generally speaking, we tried to tackle the rather uneasy situation with commonsense and dealing with each problem as it arose'.[18]

Commanding a multi-national force, Percival's task was a difficult one; he recorded of his efforts to weld the separate components of the army in Malaya into a cohesive fighting force:

> From the start I made it my business to try to forge these into a united whole. I cannot claim that I was successful. Whether this was due to my own failings or to an unalterable outlook on the part of some of my subordinates is not for me to say...With the AIF particularly and, to a lesser degree, with the Indian formations one got the impression that their own interests predominated... This lack of unity in the Army of Malaya was certainly one of the chief weaknesses during the campaign...

Later commenting on the problems of controlling Commonwealth forces overseas, Percival elaborated on the Australian and Indian formations: 'In Malaya the situation was never satisfactory...In both cases the commanders had been appointed by, and were to some extent responsible to, their own Governments. It was never made clear to me exactly what my position was vis-à-vis these commanders'.[19]

Percival has described at some length the relationship which existed between himself and Bennett in correspondence with the British official historian, Kirby, in which he alleged that he had been told by Air Chief Marshal Brooke-Popham, the Commander-in-Chief in Malaya, that there were no instructions limiting his authority over the AIF, and that Bennett had never shown him a copy of his directive.[20] Nevertheless, Percival allowed Bennett to communicate directly with the Australian government, which Bennett claimed as his right, and to be independent of Malaya Command in matters of promotions and appointments within the AIF. This would have been unusual behaviour if Percival had not seen any evidence permitting such procedures. Kirby was surprised at Percival's remarks and replied: 'Gordon Bennett had instructions from his own Government...There was no secret about them and it is most surprising that you never saw them or never insisted on seeing them. Brooke-Popham I am quite sure had them or at any rate could have got a copy for you.'[21]

It was in Percival's nature that he would not insist upon seeing the directive, and would accept Bennett's claims despite the displeasure he felt at Bennett's high degree of independence from Malaya Command, which probably prompted him to attempt to prevent Bennett communicating directly with the Australian government in late December 1941. Had he been successful, telegrams such as the following from Curtin to Churchill probably would never have been despatched: 'The General Officer Commanding the Australian Imperial Forces Malaya reports that part of his force has been cut off without the possibility of relief. It would appear from information received regarding the disposition of the AIF and its operations that support for it has not been forthcoming.'[22] The political consequences of having a relatively independent subordinate must have been the cause of many misgivings for Percival, no doubt as was Bennett's penchant for press conferences—at one he informed journalists that Percival was 'clever but weak'. Not surprisingly, nor without cause, Percival criticised Bennett in *The War in Malaya* for his outspoken comments and for his communications with the Australian government of which Percival had not been informed.[23]

The fact that he had no influence over appointments made in the 8th Division also rankled with Percival, as his comments on Maxwell's appointment to command the 27th Brigade indicate: 'Brig. Maxwell, who was a doctor by profession, was a personal friend of Gordon Bennett's and was appointed to the command by him. His military experience and training was, in my opinion, quite insufficient for a difficult operation...' In another place he wrote: 'There were wheels within wheels here. Maxwell was a great friend of Gordon Bennett's—in fact G.B. had got him appointed over somebody else's head. Taylor, the other Brigadier, was a much better chap, but never got on with Gordon Bennett. Indirectly, Gordon Bennett was responsible for a lot of our troubles on the Island.' Apparently commenting on the confusion concerning the 27th Brigade on Singapore Island on the 11 February, when it was uncertain whether Maxwell's formation was under Heath's or Bennett's command, Percival wrote: 'Maxwell was, of course, Gordon Bennett's nominee for command, which probably accounts for a good deal'.[24]

As far as operations were concerned, Percival held Bennett at least partially responsible for the failure to hold the Muar River because of his disposition of the 45th Indian Brigade: '[A] considerable part of that Brigade was dispersed, some on the wrong side of the river and some towards Lenga'. Percival could see no excuse for Bennett's failure:

> As regards the defence of Johore, General Bennett therefore, in addition to the pre-war period, had four weeks in which to study the problems and formulate outline plans after the outbreak of war, though of course he was not in a position to draw up detailed plans at that stage because the forces that would be available when the time came were not known. He should however have been thoroughly conversant with the ground and the strategical and tactical problems.[25]

Percival also made a more direct, and more serious, allegation against Bennett's ability:

> Gordon Bennett's personal bravery is not in question, but he was an intensely ambitious man...It is my opinion, from information which has since come into my possession and from my personal contacts with Gordon Bennett at the time, that from about the 25th January onwards his personal ambitions dominated his outlook to the detriment of his duties as Comdr. AIF. He was not a man to contemplate with equanimity the possibility of spending the rest of the war as a POW and he probably knew that his rival, Blamey, was due to return shortly to Australia from the Middle East.[26]

The extent to which Bennett's plans for escape may have affected his duties is difficult to determine (and will be discussed later). However, Kirby seems to have accepted this assertion uncritically and, in *Singapore: Chain of Disaster*, supports Percival almost verbatim, adding that Bennett 'began to make arrangements with a couple of trusted staff officers, through the Sultan of Johore, to escape...' It was to Kirby that Percival made his most scathing criticism of Bennett when he wrote in 1953: 'The Australian Govt. must be held responsible for putting

Gordon Bennett in command of their troops. They presumably were satisfied that he was fit to command'.[27] Obviously, Percival was not.

There is no doubt that Bennett was a difficult subordinate. As early as August 1941 it had been suggested by Sturdee, the Australian CGS, that Bennett be replaced in Malaya because of this. Colonel Thyer [General Staff Officer, Grade 1 (GSO1) of the 8th Division] has observed that Bennett was 'incapable of subordinating himself to Malaya Command or of co-operating whole-heartedly with other commanders', and Brigadier Taylor related to Gavin Long:

> Bennett says that he didn't get cooperation from British authorities and commanders. What happened was that Bennett would go in and tell them how to do their jobs. You don't approach British regular officers that way. Duncan [Maxwell] and I were given everything we ever asked for by the same British officers. They were good and they were prepared to collaborate in every way.

Thyer conceded after the war that: 'In fairness to General Bennett it must be stated that his attitude may have been entirely impersonal and that he was at all times battling for the rights and entity of the AIF. That may be perfectly true, but the result was the same, a denial of unity of command and integration of effort.' Percival simply wrote that Bennett's outspoken criticism 'did a lot of harm'. It would be unrealistic to disagree with him.[28]

## Division commander

An infantry division is a weapon just as a sword is a weapon, and like a sword it must be honed to a fine edge to achieve best results. To evaluate Bennett as a commander is not simply to consider his tactics, although that is of course of great importance, but it is also to examine how well the sword he wielded in Malaya had been prepared for battle. There is little argument that the 8th Division had attained a high standard of proficiency by December 1941, although Maxwell has argued that this was achieved more through the efforts of Taylor than Bennett.[29] However, the degree to which troops have been trained is of little consequence if they are not efficiently directed by senior commanders and staff, and it is here that Bennett failed.

It is beyond the capabilities of one man to exercise personal control over all the administrative and operational aspects of a division. Of necessity the divisional commander is the leader of a team which, if it is to function properly, must be comprised of members who as far as possible understand and support one another. It rested largely upon Bennett's shoulders to foster a spirit of cooperation at the higher levels within his command. It was especially important that a close working relationship be established between Bennett and Thyer, who as GSO1 was most likely to be required to act in Bennett's name, and who should have been confident of support after the decision had been taken. It is clear that this desirable situation did not exist in the 8th Division; the headquarters operated within an atmosphere of intrigue and distrust emanating from Bennett himself and readily reciprocated by his staff, both citizen and regular officers alike.

The suspicion and lack of trust which had been evident from the time Bennett assumed command of the division until the outbreak of war with the Japanese, attacked the very roots of efficiency within the headquarters when it affected the relationship between the commander and his GSO1. Bennett's deep dislike of Thyer had alarming ramifications. On a number of occasions Bennett allowed his personal feelings towards his chief of staff to override purely military considerations. His failure to consult with Thyer, and his decision at times to exclude him from conferences, was extraordinary behaviour. If Bennett's opinion of Thyer's military abilities was as low as such actions indicate, he should have replaced him with someone in whom he had confidence; although Bennett's high opinion of himself and the correspondingly low value he placed on the staff may have convinced him that he did not need a GSO1 at all:[30] 'Thyer is constantly pressing me,' Bennett wrote in late December, 'to amend my plans to suit imaginary dangers—illogical dangers. Generally, it involves dissipating the force,—a company here, a company there.—He seems to worry a lot unnecessarily [sic].'[31] Clearly Bennett saw such suggestions by Thyer as evidence of weakness. Yet as GSO1 it was Thyer's responsibility to make such suggestions; it was up to Bennett to consider them and act upon them if he thought fit. Whatever Bennett's assessment of Thyer, events at the crossing of the Causeway, when he took control in Bennett's absence, showed him to be at least competent. Percival accurately assessed the situation at 8th Division headquarters when he wrote that 'the relationship was never cordial and ... it deteriorated as the campaign progressed'.[32] Thyer wrote: 'I considered it imperative that I should subordinate myself completely to the GOC in order to achieve smooth working. It in no way indicated that I agreed with the tactical and strategical decisions of the Commander. On the contrary, I disagreed in a major way.'[33]

It appears that complete subordination to his views was the only attitude which Bennett could accommodate in a subordinate, as he wrote during his earlier confrontation with Taylor: 'A junior must be subservient to a senior and should not clash with him'. After the confrontation with Bennett in early 1941, Taylor 'rarely, if ever, visited Div HQ'. The rift was so great that the two men neither understood nor fully cooperated with each other. During the fighting on Singapore Island Bennett relieved Taylor of his command indefinitely and to replace him promoted Varley, whom he had earlier regarded as 'poor as [a] Bn. Comd. in battle'. It is difficult to imagine that all that had passed between the two men since the division's arrival in Malaya did not greatly influence Bennett's action.

It would be unjust to Bennett to imply that the people with whom he had to work were perfect colleagues, for they were not, and were unlikely to be since Bennett's staff had been largely not of his own choosing (although he had chosen Thyer to replace his original GSO1, Rourke). However, Bennett's prejudice against regular officers and his dogmatic personality, which seem to have fostered reciprocal attitudes among some of his subordinates, largely prevented the reconciliation of differences within the 8th Division, and hampered its effective operation.[34]

In relations with his senior staff and commanders, Bennett clearly did not fulfil one of his most important tasks: to support and encourage his senior officers; instead he fought openly with them and actively undermined them. Bennett apparently had such faith in his own abilities that he required from his senior staff and commanders not advice, but compliance. This over-confident and unrealistic attitude led him to make at least one serious error during operations against the Japanese.

*Conduct of operations*

Bennett's ideas on how best to deal with the Japanese were aggressive, unlike those of either Heath or Percival.[35] On 16 December 1941 he had circulated a letter to all ranks under his command in which he outlined the methods to be adopted:

> The recent operations in northern Malaya have revealed the tactics adopted by the Japanese in their offensive movements. It is simply that they endeavour to infiltrate between posts, or if that is difficult, to move small parties via the flank to threaten the rear or the flank of our position... This is not a new system; it is as old as war itself... Our training during the past twelve months has been to outflank any enemy position which is being held; similarly in any attack, the main attack should come from the flanking party. All units in defence will hold a small reserve in hand which will have the duty of moving around the enemy flanks and creating despondency and alarm by firing into their rear elements. Should it be possible for a small party of the enemy to penetrate between two posts and open fire on the rear of posts, arrangements must be made for alternate sections in a post to face the rear and deal with this enemy party by fire. At the same time a patrol must be sent forward to capture or destroy the enemy which has been successful in penetrating the position. It is imperative that the offensive spirit be maintained... There will be no withdrawal; counter-attack methods, even by small parties, will be adopted.[36]

Bennett's observations were similar to those of Stewart, the commanding officer of the Argyll and Sutherland Highlanders, who was sent out before the surrender to convey information about Japanese tactics. Stewart's opinion of the best method of dealing with Japanese tactics was the same as Bennett's—to avoid a withdrawal it was essential to attack: 'Static defence has no hope of success. It will be walked around, infiltrated, the road in rear cut... There are only two alternatives—to attack or delay by gradual withdrawal to avoid the encircling move.'[37]

It was the second method, withdrawal, which had allowed the Japanese to reach Johore in five weeks, and it was Bennett's vehement advocacy of the first which appealed to Wavell as the most likely way of halting the Japanese. Consequently, Wavell shelved Percival's orders for the defence of Johore, and substituted a less practical organisation to allow Bennett to fight the battle.

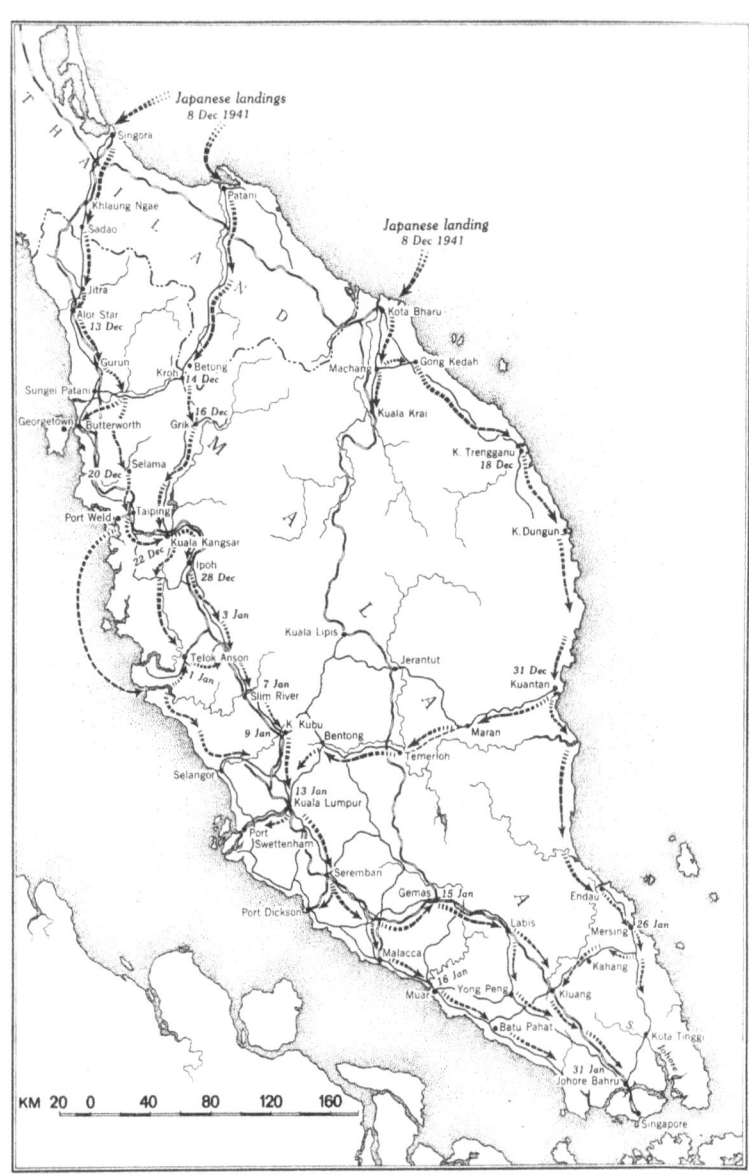

15 The conquest of Malaya, December 1941–January 1942

Aggression and determination, not least among military virtues, were Bennett's greatest attributes in Malaya, and must have been refreshing for Wavell to encounter, for, as Leasor has observed of Heath and Percival: 'Their ideas lacked aggression and novelty. They were aging men cast, for the most part, in moulds too large for them... Their deliberations, like their plans, lacked conclusion and decision.' Nonetheless, Wavell's confidence in Bennett was misplaced. 'The test... does not lie so much in the general conception as in the execution of a plan', wrote Belloc, and in northern Johore Bennett failed this test; his dispositions at Muar suggested that his pronouncements on the tactics to be employed against the Japanese were merely lip-service to an ideal only dimly perceived. He spread two battalions along a frontage of 25 miles (40 kilometres) in difficult terrain, which left what should have been his primary concern, the road leading south from Muar, extremely vulnerable. It was not breadth of defence, but depth which was paramount. In the light of his observation that Japanese tactics were 'to threaten the rear or the flank of our position' and the successful employment of such tactics at Telok Anson in circumstances similar to those at Muar, Bennett's dispositions were surprising, and clearly exposed his weakness as a commander in this instance. He failed to recognise the vital ground and to concentrate his forces accordingly. On the other hand, he should be given credit for the outstanding success of the ambush at the Gemencheh River near Gemas, probably the most severe setback inflicted upon the Japanese during the campaign.[38]

Accusations that Bennett's conduct of the ensuing withdrawal to the Causeway resulted in the loss of the 22nd Indian Brigade cannot be substantiated. However, at the crossing of the Causeway Bennett was at fault for being absent and, therefore, being unable to deal with the difficulties which arose and were handled by Thyer.

Once on Singapore Island the odds were heavily against Bennett. He had little chance of conducting a successful defence because of Percival's appreciation that the Japanese would attack from the north-east, which seriously deprived the western area of a large amount of men and equipment. Nevertheless, there are several points on which Bennett can be criticised. First, he seemed to have neglected the possibilities of several defensive positions, in particular the Kranji–Jurong Line. Second, he injudiciously issued Percival's instructions for a defensive perimeter around Singapore, which led to the abandonment of the line by Taylor. Third, he failed to countermand Taylor's order to withdraw. Furthermore, it is possible that although it was not under his command, he ordered the 27th Brigade to attack Bukit Panjang village on 11 February, thus exposing the left flank of the 11th Indian Division which led directly to the abandonment of the Singapore naval base. After the war there was confusion as to who had actually ordered the attack.

The assertions made by Percival that Bennett's interest in the campaign waned during the last weeks of fighting appear to be correct. As mentioned earlier, Bennett had been making preparations for an escape even before the withdrawal to Singapore. After he had been ordered to counterattack to retake the Kranji–Jurong Line on Singapore Island he took a pencil in his fist and in an off-hand manner drew three lines upon a map—these represented the three stages of the assault. The line denoting the assembly area for the attack was then noticed to be behind the enemy lines. His lack of interest and frustration are also clearly shown by his willingness to surrender at a conference on 13 February and again on 15 February, on both occasions supporting

Heath, whom he had hitherto disparaged and disagreed with at every possible opportunity.[39]

Bennett had already shown that he was prepared to subordinate operational considerations to other factors, whether national, personal, or a combination of the two; in December 1941 he had opposed the despatch of any large force of Australians to northern Malaya to assist the British. On 27 January 1942 he criticised Forde, the Australian Minister for the Army, for not promoting him in early January and threatened resignation if he were not promoted to command an Australian corps in Malaya after more divisions had arrived, as was planned at one stage. He felt destined to command the whole Australian Army and apparently admitted as much to Sturdee and, most surprisingly, Blamey, after his return to Australia at the end of February. Bennett was both intensely ambitious and patriotic, and it is most likely that a mixture of these traits turned his mind more frequently to escape as his hopes for a successful defence faded, perhaps to the detriment of operations.[40]

From observations made before and during operations against the Japanese, Anderson, the commanding officer of the 2/19th Battalion, has remarked that although Bennett was receptive to new ideas he 'never really understood jungle warfare'. In general, Bennett had advocated appropriate tactics, but was unable to put them into practice. Sometimes this was because of the handicaps under which he laboured, such as the lack of air support and the large area he had to defend on the Island, but at other times it was because of defects in his ability as a commander; his aversion to planning in detail, which prompted Maxwell to observe that Bennett 'tried to fight the war off a map five miles to the inch'; and his failure to go forward to view the battle for himself, even when communications were badly disrupted as they were after fighting commenced on the Island.[41]

There are several reasons for Bennett's shortcomings as a divisional commander. The first is the system of military training which existed in Australia during the years following the First World War when Bennett had returned to civilian life, but retained a commission in the militia. Thyer saw this period as the root of Bennett's problems. It is true that he had attempted to remain up to date in military affairs between the wars, but under the poor system of militia training it was a difficult task. Percival placed the blame not so much on Bennett himself, but on the system of which he was a product. He wrote that Bennett's handling of operations in Johore had 'shown that his technique in the conduct of operations was very much out of date. He had of course had little opportunity of keeping himself up to date in such matters in the years between the wars.'[42]

By itself, Bennett's study of tactics between the wars could not confer upon him experience. Although he was promoted to command the 2nd Division of the militia in 1926 and held that command for five years, it does not necessarily follow that he was a thoroughly trained divisional commander in 1939. The scope which the citizen forces provided for the training of both officers and men was extremely limited. The militia was very much understrength, divisions often being at a mere quarter of their full establishment, and the time allotted for training—six days in camp and four days at the local centre per year—was totally inadequate. A reasonable knowledge of current theory could be obtained through wide reading and discussion, but the overall result was virtual stagnation in the practical aspects of the training of senior officers like Bennett. As Sir Granville Ryrie, the commander of the 1st Cavalry Division, observed in 1926: 'We have no training for brigadiers or divisional commanders...' After the Second World War the Directorate of Military Training confirmed that the militia system had not been effective and that its 'formations and units were too weak to provide useful experience for the leaders and not much more than elementary training for the troops'. Even in regular armies it had been difficult during the 1920s and 1930s to provide senior commanders with adequate experience in the handling of large formations, so it is not surprising that the task was almost impossible with only a skeletal citizen force which trained for an insignificant fraction of the year.[43]

The problem was aggravated by the lack of imagination in British tactical doctrine during the inter-war years, with the result that by 1939 British armies were inferior to the Germans and Japanese in both theory and practice. 'Nothing is more dangerous in war', wrote Fuller prophetically in 1933, 'than to rely upon peace training; for in modern times, when war is declared, training has always been proved out of date'. Not surprisingly, Bennett's First World War experience was not of great benefit: the task of a brigade commander under the largely static conditions of the Western Front was vastly different from that of a divisional commander fighting a mobile defence in Malaya. The problem was not peculiarly Bennett's, but one which presented itself to the army as a whole. Horner's comments on the difficulties facing the Australians in New Guinea in 1942 are applicable to Malaya:

> With the exception of the small expedition to German New Guinea in World War I, the army was faced, for the first time, with war in the tropics...The paradox of this new warfare was that it required a flexibility of thought and a standard of leadership not always found in the professional soldier, but it also required a technical skill which could be gained only by considerable study and application.[44]

Perhaps given more favourable circumstances, such as those enjoyed in the Middle East by another citizen general, Mackay, Bennett would have been able to make good some of the deficiencies in his expertise. However, Mackay, unlike Bennett, was prepared to rely on his regular staff officers, especially Berryman (his GSO1), to guide him. This Bennett did not do; he allowed his deep-rooted antagonism for the Staff Corps to prevent him from drawing upon the knowledge of the regular officers under his command—knowledge which could have compensated for the paucity of his inter-war experience.

Finally, Bennett was over-confident in Malaya, as he had been in 1939 when he expected to be appointed to command the 2nd AIF. His statements that he knew how to stop the Japanese influenced Wavell to place him in command in northern Johore, which under the circumstances was a reasonable decision, but Bennett deluded himself as to his own infallability and so ignored factors which pointed to an enemy assault on two fronts. This is, as Lewin has noted, 'the too frequent response of commanders and staffs who, while involved in a campaign, are inclined to discount intelligence which might lead to diversions from what, in their eyes, is the main front'.[45] Thus, these three factors reacted one upon the other with disastrous results.

Ian Morrison, a war correspondent who spent some time with Bennett in Malaya, described Bennett as 'a rasping, bitter, sarcastic person, given to expressing his views with great freedom. As a result he quarrelled with a good number of people. But he did have a forceful personality. He was imbued with a tough, ruthless, aggressive spirit.'[46] A 'forceful personality' and an 'aggressive spirit' are great assets for a general, but in Bennett's case they were more than offset by his limitations. Although he was diligent in protecting his country's interests and the welfare of his troops, his forthright personality and professional prejudices did not make him especially suited to a semi-diplomatic command, requiring tact and a readiness to compromise, such as GOC AIF Malaya. Bennett's dislike for the British was clear and had affected not only relations, but also operations. Although his criticisms were frequently shared by others, the blunt manner in which he expressed them, both to his superiors in Australia and to his commander in Malaya, did not contribute to a spirit of cooperation during the campaign, and seriously questions his suitability for such an independent command.

Bean recorded that Bennett, as a brigadier-general, was 'a young front-line leader';[47] in 1941 Bennett was neither a young nor a front-line leader, but even so was still more suited to the command of a brigade under the controlled conditions of the Western Front of 1918 than he was the command of a division fighting the Japanese in Malaya more than twenty years later. He was probably the best of the mediocre collection of generals in Malaya, but it is doubtful whether his performance against the Japanese and the information he conveyed to Australia justified his departure from Singapore, or that his responsibility to the Australian government required, or even excused, his action. What is not in doubt, is that he was hardly (to borrow W.S. Gilbert's phrase) the very model of a modern major-general.

# Lieutenant-General Sir Leslie Morshead: Commander, 9th Australian Division

A.J. HILL

The First AIF threw up a remarkable galaxy of fighting commanders at unit, brigade and divisional levels. Some, whose names were famous in the Force, such as Humphrey Scott, T.P. McSharry, Philip and Owen Howell-Price, may well have risen high had they survived. Among the more fortunate were William Glasgow, Charles Rosenthal, Talbot Hobbs, L.C. Wilson, C.F. Cox, R.L. Leane, H.E. Elliott, Iven Mackay, H.W. Murray, Blair Wark and Norman Marshall, representatives of a brilliant company. Some had been private soldiers on Gallipoli and many a commanding officer of 1917 and 1918 was in his twenties. Such a leader was the CO of the 33rd Battalion, Leslie James Morshead (1889–1959).

When Australia went to war in 1914, Morshead was a young schoolmaster not quite 25, with such military knowledge as can be acquired by a keen lieutenant of cadets. He was commissioned in the 2nd Infantry Battalion under G.F. Braund who in the first few days on Gallipoli was to show 'every quality of a really great leader'.[1] Morshead landed at Anzac on 25 April 1915 as a platoon commander and was in the hard fighting for Baby 700. Promoted major in June he led a company at Lone Pine where he was wounded. He was invalided first to England in October then to Australia, where in April 1916 the 3rd Division was being formed. Appointed second in command of the 33rd Battalion he became its commander after only a few days and was to lead it until demobilisation in 1919.

Morshead trained his battalion under the eye of Major-General John Monash, one of the most meticulous and imaginative trainers the Australian Army has produced. It is hardly fanciful to suppose that almost two years under such a commander was for Morshead as formative an experience as his five months in the front lines of Gallipoli. Morshead led the 33rd with distinction in the great set-piece attack at Messines in June 1917 and on through the agonies of Third Ypres. In the critical days of the German spring offensives of 1918, he won new laurels especially in his counterattack with the 12th Lancers at Lancer Wood on 30 March.[2] When the war ended he had been awarded the CMG, DSO and French Legion of Honour and had been six times mentioned in despatches. Monash, having been impressed by his powers as an organiser, took him into his demobilisation staff in London.

On his return to Australia Morshead tried unsuccessfully to establish himself as a grazier. Turning from the land to the city he began work as a clerk. In 1924 he joined the staff of the Orient Line in Sydney and then rose quickly, becoming manager of the Sydney Branch in 1938. Except for a break of four years while establishing himself with the Orient Line, Morshead served in the Citizen Forces commanding successively the 35th, 19th and 36th Battalions and the 14th, 15th and 5th Infantry Brigades. It was a lean time for the services which suffered sharp retrenchments in 1922 and again during the Depression years, 1929–31, when the compulsory system was suspended and the pitifully inadequate training time was reduced to a token level. In the army modern equipment was non-existent; only the dedication of the handful of regular officers and instructors and of veterans such as Morshead kept the army in being and prepared the cadre of young leaders who were to provide the framework of the Second AIF.

What these years contributed to Morshead's development as a commander is not easy to determine. Commanders of Citizen Military Forces (CMF) units and brigades were chiefly concerned with training their officers and administering their commands with the help of regular adjutants and brigade majors. The brief annual camp, increased from six to eight days in 1924, and an occasional TEWT [tactical exercise without troops] supplemented by such studies as a busy man chose to pursue, were all that was possible; there was no Australian Command and Staff College and militia officers were not eligible for the courses at Camberley and Quetta, even had they been able to take two years off to attend one of them. It is probable that Morshead and others of his generation, while maintaining the habit of command and a theoretical acquaintance with tactical and other developments, did more for the Army than it could do for them. In particular, they passed on and kept alive the traditions of the AIF and they succeeded, within the stringent limits imposed by successive governments, in preparing a useful officer corps for the next great emergency. However, no officer of Morshead's experience had any illusions about the usefulness of the CMF as a fighting force.

When the 6th Division was raised in October 1939 Morshead was one of its original brigadiers, being appointed to command the 18th Infantry Brigade. However, chance intervened early; when France was invaded in 1940 his convoy was diverted from Egypt to the United Kingdom. The 18th Brigade never joined the 6th Division but, after solid training, went to Palestine in November where it eventually became part of the 7th Division. When the illness of Major-General H.D. Wynter made it necessary for General Blamey, commanding the AIF in the Middle East, to appoint a new commander for the infant 9th Division, his first choice was Major-General J. Northcott, Deputy Chief of the General Staff in Melbourne, but as the CGS was unwilling to release him, Blamey chose Morshead on 27 January 1941.[3]

*Tobruk*

In January 1941 Morshead left one of the best trained and equipped brigades in the AIF—perhaps the best trained—for an incomplete

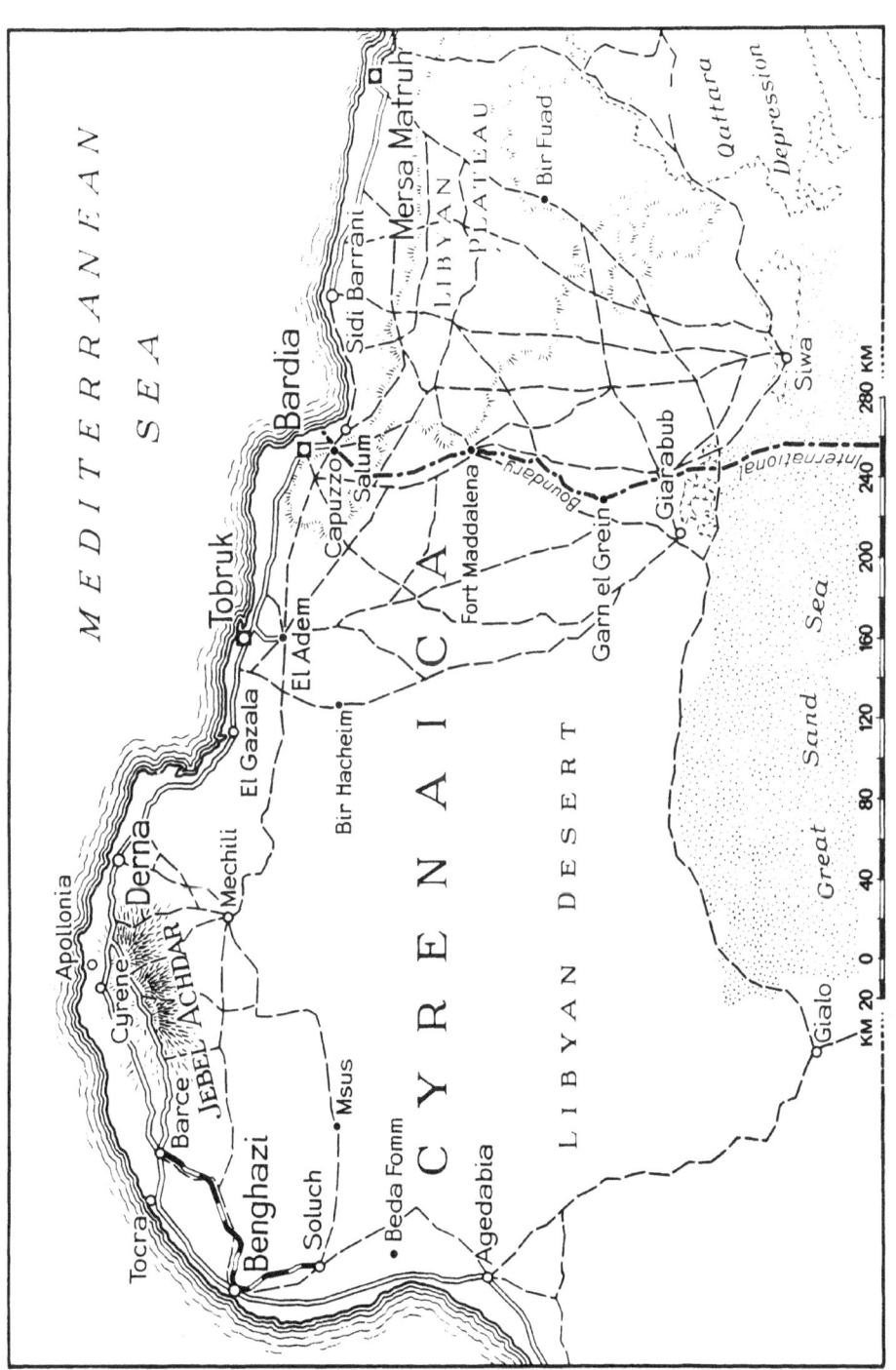

16  Cyrenaica 1941

assemblage of brigades and units called the 9th Division. Only three weeks later the AIF was regrouped to prepare for the campaign in Greece; his more advanced 18th and 25th Brigades were transferred to the 7th Division in exchange for the part-trained 20th and 26th Brigades, and Morshead was ordered to relieve the 6th Division beyond Benghazi in Cyrenaica and continue training there. The newly arrived 24th Brigade, with only two battalions, remained with him but Morshead had to begin again the process of getting to know his commanders and troops while in the midst of a major move into an operational area.

March 1941 was dominated for Morshead by two parallel struggles, the equipment of his division and its concentration on sound tactical principles. His diary records the demands he pressed on Lieutenant-General Sir Philip Neame, GOC Cyrenaica Command, and Brigadier Harding, his Brigadier, General Staff (BGS): the dire need of anti-tank guns, transport, artillery ammunition, signals equipment and troops. When could the 26th Brigade come forward to Benghazi from Tobruk? The divisional signals, except a party of twenty, were at Tobruk and without transport; he needed his field park company (mobile ordnance supply depot), infantry tanks, vehicles and every kind of weapon for the infantry who as late as 15 March had only seventeen brens per battalion![4] Morshead succeeded in obtaining more brens but the movement of vital units, including field ambulances, was beyond the means of Cyrenaica Command, if only because of the immense shortage of transport. The western flank had been stripped to equip the force going to Greece.

Perhaps even worse than the equipment problem was the tactical. The 20th Brigade had relieved the 17th Brigade in the Marsa Brega position from which armoured cars patrolled to keep in touch with the enemy. When Morshead reached the forward area he was alarmed to discover that Neame proposed to place the 20th Brigade under command of the 2nd Armoured Division, a scratch force of worn-out British tanks, captured Italian tanks and 'ersatz' armoured cars. His headquarters would move back to Gazala to train the rest of the 9th Division in the Tobruk area. He also quickly discovered that the widely dispersed and immobile 20th Brigade was in open country, leaving the one area of defensible ground around El Agheila to the enemy whose forces were already known to include German armour. When he had seen enough Morshead put his views on paper for Neame but the arrival, on 17 March, of General Wavell, C-in-C Middle East, enabled him to press his demand for the withdrawal of the 20th Brigade 'to a defensive position where they would have an opportunity of stopping and fighting the enemy'. Wavell agreed and supported Morshead's demand for immediate withdrawal; he later wrote that Neame's dispositions were 'just crazy'. Why Wavell did not remove Neame forthwith and recall General R.N. O'Connor, the victor of the first Libyan campaign, is a mystery.[5]

The withdrawal of the 9th Division to Tobruk during the period 4-9 April 1941 was a stiff test for a new general who had not been granted the time to know his command and who lacked the signal communications and transport essential to fast-moving operations. Besides, he had little reason for confidence in Neame's handling of the situation and could only observe, helplessly, the disintegration of the 2nd Armoured Division. A nice beginning to the desert war! The fact remains that Morshead and his staff, by their calmness and by ceaseless effort, managed to acquire or borrow enough transport in successive crises to lift the two brigades from one rearguard or blocking position to another and the green Australian infantry, strongly supported by British artillery and machine gunners, and by the Air Force, always presented a solid front to any Germans who collided with them.[6] Perhaps the bloody check at Regima, administered by the 2/13th Battalion and 51st Field Regiment to the 3rd Reconnaissance Battalion and their supporting armour, imposed a degree of caution on the Germans.[7] In any case, the Afrika Korps of the First Benghazi Handicap was not the veteran force of the Crusader battles of November 1941 and of the summer of 1942. If the 9th Division was finding its feet and learning on the job, so were the well-equipped and mobile Germans, who also suffered shortages of fuel and supplies, breakdowns and fatigue.

Every commander needs luck and Morshead was well served by this fickle friend in April 1941. One of the first flashes of good fortune was Rommel's decision to advance on three axes so that when his leading elements ran into the 9th Division there was little weight in their thrust. The capture of General Neame on 6 April removed one of Cyrenaica Command's main handicaps, if rather too late, so that control of the withdrawal in its final stage fell to the outstandingly able BGS, Brigadier John Harding, a future field marshal. It must also be counted as luck that the little port of Tobruk with its defences and dumps, lay squarely in the path of the 9th Division at a time when the weight of the pursuit was increasing and the makeshift condition of the division was beginning to tell. Moreover, Morshead had watched the 6th Division's assault on Tobruk in January and, after its capture, had spent three days inspecting its defences.[8] Therefore, when he drove into the fortress and found Harding's headquarters on 7 April, the two were able to agree quickly that Tobruk would be held. Morshead found much to help him towards this decision. His 24th Brigade, still short of a battalion, was there as he expected but he was delighted to find his old 18th Brigade had been sent up and the two, under Brigadier George Wootten, were already moving into the defences.

Wavell, who had also decided that Tobruk must be held came up, at Harding's request, on 8 April, bringing Major-General J.D. Lavarack, commander of the 7th Australian Division, but now to take charge of Cyrenaica Command. There is a story in John Connell's admirable *Wavell: Scholar and Soldier* of Morshead and his GSO1, 'Colonel Loyd' [sic] meeting Wavell on El Adem airfield: 'They were dog tired, unshaven, and conscious that they looked and smelt of the desert, of defeat and of retreat. Wavell's presence gave them back their confidence—in him and in themselves. Tobruk, he said, was to be held; he had merely come to settle the method by which this was to be done.'[9]

*21* Major-General L.J. Morshead with his GS01, Colonel C.E.M. Lloyd, at Torbruk. Both appear "unaffected by their gruelling time". (Imperial War Museum Negative No. E 2840)

This tale is supported neither by the official historian, Barton Maughan in *Tobruk and El Alamein* nor by Morshead's own diary. Maughan states that Morshead 'visited Cyrenaica Command headquarters and found that General Wavell had just arrived—Wavell was already in conference...Morshead, the senior military commander of the field force, had not been invited but was at once shown in.' Anyone who served under Morshead, while surprised to learn that he was unshaven, could only be amused at the idea that it took Wavell's appearance to restore his confidence. After all, he had successfully withdrawn in the face of a better equipped force and one of his battalions had stopped the Germans dead at Regima; he had two brigades well posted at Acroma and two brigades in the Tobruk defences. Lavarack, who saw Morshead and Lloyd at the conference, noted afterwards: 'The GOC of 9 Australian Division [General Morshead] and his G1 [Colonel Lloyd] were unaffected by their gruelling time'.[10]

Lavarack appointed Morshead commander of Tobruk Fortress with the bulk of the fighting troops, leaving certain mobile elements in a harassing role outside and keeping the 18th Brigade and the available armour as his reserve. It was Morshead who set the tone for the defence of Tobruk. This was to be 'static' warfare, an affair of patrols, wire, mines and defences dug or in concrete, of which the former CO of the 33rd Battalion had once been a master, as were his brigadiers, G.F. Wootten, J.J. Murray, A.H.L. Godfrey and R.W. Tovell. Morshead told them on the eve of their move into Tobruk: 'There'll be no Dunkirk here. If we should have to get out, we shall fight our way out. There is to be no surrender and no retreat.'[11] His operation instruction of 12 April laid down a tactical policy which derived from his own experience and from what he had learnt while training in Britain after Dunkirk. He emphasised the importance of hitting the enemy by night with strong fighting patrols while he was probing the defences: posts were to be developed for all-round defence and were to hold out in the event of penetration by tanks; enemy infantry who made gaps for armour were to be counterattacked. In an earlier order he had insisted that where tanks had penetrated, posts were to hold out and engage the enemy to their flanks and rear.[12] It was this system, backed by aggressive artillery and Lavarack's armour held in reserve, which smashed Rommel's hastily mounted attack on 14 April and gave the garrison more than a fortnight's respite.

The most important feature of Morshead's defensive tactics was their thoroughly offensive spirit, as shown above. This was carried far beyond the perimeter by ceaseless patrolling; as Morshead told the war correspondent, Chester Wilmot: 'I determined we should make no-man's land *our* land'.[13] In the Red Line, everyone patrolled. It was exhausting but it built up expertise as well as information and denied information to the enemy. What was accomplished is best shown in the words of the CO of 1st Royal Horse Artillery in a report to the fortress artillery commander: 'It was simply through the fearless and meticulously thorough investigation of the terrain out of view and often deep inside the enemy defended localities, that we gradually built up a clear knowledge of his defences and organization'. He added that the 'continuously brilliant patrolling' had enabled the gunners 'to strike deeply and accurately' and had 'persistently impressed' them as 'regular soldiers'. The 'astounding accuracy' of certain patrol reports was later confirmed by aerial photography.[14] These patrols were supplemented by frequent fighting patrols and raids sometimes in company strength with tank and artillery support. As Morshead put it: 'We set out to besiege the besiegers' (11 June 1941).

Rommel's next attack, beginning on 30 April, was a better planned operation of much greater weight than that of Easter. Lessons had been learnt by both sides in the Easter Battle and Morshead could now expect a sterner struggle. He prepared accordingly, laying new mine fields within the fortress, one of them in the path of Rommel's planned penetration. The attack on the 2/24th Battalion opened with dive-bombing and artillery bombardments. This time the panzers worked with the assault troops, methodically subduing the Australians post by post before attempting an armoured drive on the port. For Morshead, 1 May was a classic example of the fog of war, reinforced by a thick mist which did not clear until about 7.30 am. Communication with the posts on the Red Line had been lost from the start and neither the 2/24th Battalion nor the 26th Brigade could provide firm information for many hours. The extent of the enemy penetration was very difficult to establish, nor was it possible to determine which posts were still in Australian hands.

17 Defence of Tobruk. Dispositions, afternoon 5 May 1941

During the night the GSO1, 'Gaffer' Lloyd, had moved an anti-tank company up to the 2/24th Battalion and had ordered the 3rd Armoured Brigade to move tanks up behind the 2/24th under cover of darkness. He had also kept the fortress reserve, the 18th Brigade, informed. Morshead was unwilling to counterattack until the position was clearer and he could be sure that the thrust against the 2/24th was the main effort and not a feint.[15]

No plan survives contact with the enemy. In this case the Axis attack lost momentum owing to the resistance of the 2/24th Battalion, the efforts of the British armour and the ceaseless fire of the artillery after the lifting of the mist. Yet it was not until 4.20 pm on 1 May that Morshead ordered the 26th Brigade to counterattack with its reserve, the 2/48th Battalion, and recapture the perimeter posts before dark on a front of about 4500 yards (4150 metres)! It is unlikely that Morshead expected to seize Point 209 and re-establish the perimeter with one reinforced battalion; as an experienced infantryman what he must have expected was to upset preparations for a renewal of the attack and to show the enemy that the fight was still on. Hence his remark to the CO of the 2/48th: 'Listen, Windeyer, it is important that this be done and done today'.[16] The 2/48th failed to take their objectives but so had Rommel, apart from Point 209 and the adjacent posts. His losses in men and tanks had been heavy and there was no sign that the garrison was giving in. He wrote afterwards: 'But next day [2nd May] it became obvious that we were not strong enough to mount the large-scale attack necessary to take the fortress and I had no choice but to content myself with what we had achieved...[17]

Morshead, on the other hand, was not content that the high ground around Point 209 and fifteen posts of the Red Line should be in Axis hands. He began to prepare a more deliberate counterattack with his reserve brigade for the night 3–4 May. Again, too much was expected of the infantry. In a complicated operation with two battalions assaulting on converging axes, artillery support was necessarily divided, nor was there time for the thorough preparations needed for success against Germans in position. The attack ran into trouble from the start and Morshead called it off at 3.40 am on 4 May. Three months later he tried again with the 24th Brigade in a carefully prepared operation but repeating the tactics of 3–4 May. A battalion attacked from each end of the Salient with the object of capturing two or three of the perimeter posts. What had not worked in May was a costly failure in August, yet a week later Morshead ordered the 18th Brigade to prepare a plan for an attack on the right shoulder of the Salient over most unfavourable ground.[18] One can only be grateful that General Auchinleck's decision to relieve the 18th Brigade in August obliged Morshead to cancel this attack, as it is difficult to see how it could have been other than a bloody disaster. The Salient was an irritation to both sides but Rommel showed greater realism than Morshead by deciding to hold it defensively until he could launch a major assault on Tobruk. Yet the justification of Morshead's policy may well be seen in Rommel's concentration of scarce

German (rather than Italian) troops in the Salient, including three infantry battalions, armour and artillery, all needed elsewhere.

Tobruk provided not only tactical problems. Morshead, still something of the martinet, as in Bean's much quoted description, banned the playing of two-up as a potential source of trouble just as he banned the private use of light AA guns for which there was abundant ammunition in old Italian dumps.[19] These guns enlivened the night sky but were dangerous to the garrison. Only constant pressure by Morshead through Blamey and Air Chief Marshal Tedder, brought improvement in the supply of much needed air photographs. On the other hand, his protest against the withdrawal of the remnant of No.73 Squadron, RAF, from Tobruk, was unavailing. In a strong letter to GHQ Middle East, Morshead argued the need for a much larger number of periodical awards and mentions in despatches for the garrison; he won his case.[20] It was Morshead who, in early July, told Blamey that the garrison's capacity 'to resist a sustained assault was diminishing'. He watched the troops' condition closely and, when the time came, backed Blamey's stand on the relief of the force. He also kept an eye on his staff, acting quickly when he discovered that two senior staff officers had not visited brigade or unit headquarters.[21]

Morshead's aggressive defence of Tobruk, according to General Auchinleck, was largely responsible for the 'freedom from embarrassment' of the weak force on the Egyptian frontier for four and a half months:

> Behaving not as a hardly pressed garrison but as a spirited force ready at any moment to launch an attack, they contained an enemy force twice their strength. By keeping the enemy constantly in a high state of tension, they held back four Italian divisions and three German battalions from the frontier area from April to November.[22]

There was no let up for the besiegers because there was no let up for the besieged—they patrolled, attacked, dug, wired, stood to their guns and hit the enemy by day and by night. They did this by grace of the Navy, British and Australian, who nightly delivered men, food, ammunition and stores and took away the wounded. They did it also by grace of the 4th AA Brigade, Royal Artillery, including one Australian light battery, who defended the port against the Luftwaffe, as the four British and one Australian field regiments defended the perimeter and struck at every target offered. (The latter was the 2/12th Field Regiment which had arrived on 17 May.) Morshead and the 9th Division knew what they owed to their comrades in arms, not forgetting the 1st Royal Northumberland Fusiliers, machine gunners without equal, 18th Indian Cavalry and the 3rd Armoured Brigade.

Morshead commanded a division which, however much it lacked higher training and the flexibility and speed that come with it, nevertheless wanted to fight. Defending Tobruk gave it the opportunity, while robbing the Germans of some of the advantages of their better training, equipment and mobility. What was said of the American Civil War remained true: 'Put a man in a hole and a good battery on the hill behind him and he will beat off three times his number even if he is not a very good soldier'.[23] Morshead had all the ingredients of a stout defence with his own indomitable spirit to stiffen it: the KBE conferred on him early in 1942 was very moderate recognition of a great feat of arms. Furthermore, the defence of that miserable Cyrenaican harbour and fortress from April until the last elements of the 9th Division (except for one battalion) were relieved by the British 70th Division in October, was one of the few gleams of light in a year darkened by withdrawals and disasters.

*First Alamein*

The period in Palestine and Syria after Tobruk restored the health of the 9th Division but the demands of work on the Ninth Army fortress areas, raised in February 1942 to six days a week, were resisted by Morshead who was anxious to have a trained division ready for whatever lay ahead. he also resisted the demand of the army commander, General 'Jumbo' Wilson, for one of his brigades as army reserve.[24] He continued to fight the same old battle for equipment as he had when he took the division into the desert. On 31 May 1942 he had only 50 per cent of his motor vehicles; he had no mortars and, on 12 March, only 5 per cent of his medium machine guns.

On 25 June Morshead was ordered to move his division to Egypt as quickly as possible. Tobruk had fallen to Rommel on 21 June and on the day that Morshead received his orders Auchinleck, at last, had taken command of the battered remnants of Eighth Army at Mersa Matruh. Morshead's first task was the defence of Cairo where the writer had the agreeable role of reconnoitring the Gezira Club for Div. HQ but was rudely interrupted by a change of orders. This task was soon changed to defending Alexandria. There Morshead showed that he was prepared to act drastically, directing the engineers to prepare channels to flood areas with seawater, ordering the removal of civilians from defended areas, the cutting of fields of fire, and even demolitions. Fortunately the situation eased before the decision to let the Mediterranean in had to be taken.[25]

In this emergency, equipment for the 9th Division was eventually forthcoming and Morshead could now feel confident that he would launch a complete Australian division into battle. However, General Auchinleck had other plans. The Commander-in-Chief had taken the remarkable decision to change the organisation and tactical doctrine of his army in the midst of operations; he wanted only enough infantry to escort artillery batteries which were now to be the tactical units apart from armour. So spare infantry were to go back to the Delta after the new 'battle groups' had been formed. When on 3 July Morshead was ordered to form the 9th Division into such battle groups and to move a brigade group less one battalion to 30th Corps immediately, he flew up to the Eighth Army and confronted Auchinleck with a blunt refusal. He came away with Auchinleck's agreement that the 9th Division would operate as a division under his command and he, in return, agreed to send the 24th Brigade forward as a temporary detachment. It appears to have been an unpleasant interview but this was a battle Morshead had to win.[26]

Morshead came up to the desert when the first great crisis at El Alamein was over and Rommel, having been stopped on 1–3 July, had gone over to the defensive. The 24th Brigade Group now fully equipped, but less the 2/28th Battalion, moved into 30th Corps area on 4 July. Next day Morshead visited Wall Group, a collection of battle groups assembled from many different units. He was not impressed; perhaps he was even less impressed when he was visited on 7 July by the new commander of the 30th Corps, Major-General W.H.C. Ramsden. Theirs was to be an unhappy relationship, not merely owing to Morshead's senior rank, for he was now a lieutenant-general as GOC AIF Middle East, Blamey having returned to Australia.

The Eighth Army had already passed the nadir of its fortunes when Morshead led the last fresh division in the Middle East into battle at El Alamein. In one respect the scene had not changed much since March 1941 when Neame had wanted to place the immobile 20th Brigade under the ramshackle 2nd Armoured Division; this time Morshead had to save the 9th Division from fragmentation at the hands of the Commander-in-Chief himself. On the other hand the division was complete, experienced, equipped and better trained although not to the level that Morshead would have liked. It was going into action in another great crisis but confident now that it could do anything that would be asked of it. One significant change in the summer of 1942, noted with delight by every soldier in the 9th, was the constant presence of the RAF and other air forces and their increasingly powerful support.

The positional warfare which Auchinleck had succeeded in imposing upon Rommel in the El Alamein bottleneck deprived the Germans of much of the advantage they enjoyed through their skilful and rapid prosecution of mobile operations. To Morshead and the 9th Division this was a godsend after their long experience in Tobruk, shared by no other division in the Eighth Army. Moreover the British, with heavy American support, could handle far better than the Axis the logistics of positional warfare with its great demands for material and its higher casualty rates, a point of which Rommel was acutely aware.[27] Nevertheless the July fighting was, for Morshead, unsatisfactory in spite of some notable successes because at no stage from 7 to 27 July did he fight a true divisional battle. The Eighth Army then—as it would be for the October battle—was short of infantry, which was hardly surprising after the heavy losses of Gazala, Tobruk and Mersa Matruh. Divisions were required to hold wide frontages defensively while mounting attacks, so that their operations were usually at brigade level with one or two battalions in the assault supported by armour (in the operation order though not always in the battle) and the divisional artillery.

First Alamein, as some like to call it, began for Morshead when the 24th Brigade raided the Germans on Ruweisat Ridge on 7 July. Although (or because) the brigade had been detached to 30th Corps, Morshead

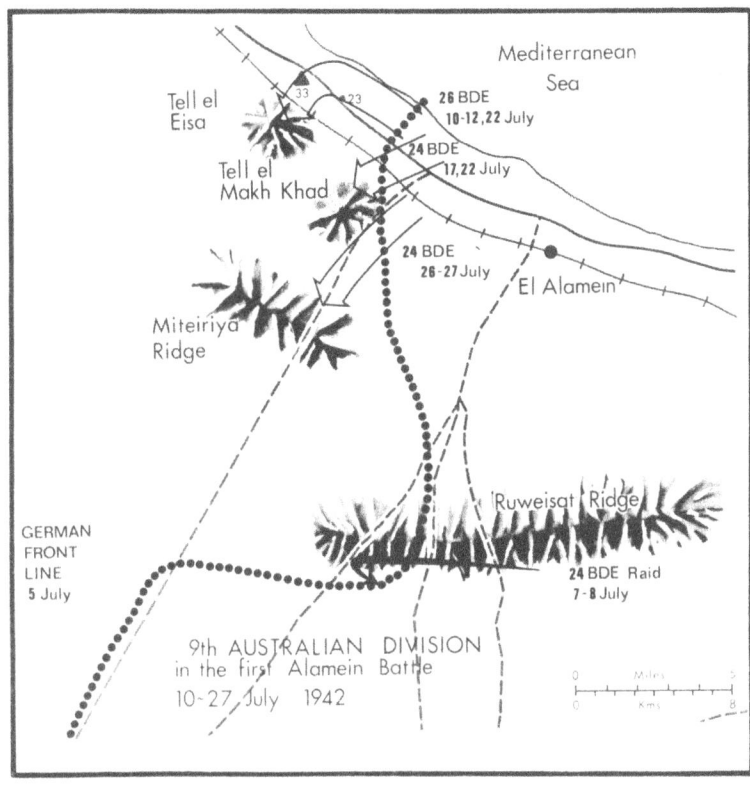

18  9th Australian Division in the first Alamein Battle, 10-27 July 1942

kept a close watch on the preparation of the raid, making two reconnaissances of the area on 5 and 6 July. Being dissatisfied with certain parts of the artillery plan he made his objections known to Auchinleck in person. From the British point of view he must already have appeared to be a difficult subordinate but, as he was to point out to Auchinleck a fortnight later, his job was also to minimise casualties.[28] The incident is one of a number which reveal the care he took to see that his men were put into battle with the best possible chance.

Another characteristic of his method of command was his frequent visits to brigade and unit headquarters. While the records throw little if any light on his planning, Morshead's diary and the War Diary of his general staff show how closely he kept in touch with his commanders. For example, on 17 July he went forward on three occasions to HQ 24th Brigade; the day before he visited the three commanding officers of 26th Brigade; on 19 July he was up twice with 24th Brigade, going to each battalion on his second visit.[29]

In less than three weeks Morshead was required to mount four attacks as part of 30th Corps operations on 10, 17, 22 and 26–27 July. The capture of the ridge Point 23–Trig 33 parallel to the coast road and of Tel el Eisa was a brilliant success. Although Tel el Eisa was soon lost, the persistent defeat of the German counterattacks during the period 10–15 July, was no less an achievement. On 17–18 July more useful ground was gained when part of Makh Khad Ridge was seized by 24th Brigade and Rommel was again forced to react to a serious threat. The two later attacks were more complicated and the last ended in disaster.

Auchinleck's plan for the operations of 22 July required the 30th Corps to attack in the north to prevent the enemy from concentrating against the main effort of 13th Corps on Ruweisat Ridge. Ramsden's plan gave the 9th Division so many and such diverse tasks that Morshead tackled him in a two-hour conference on 21 July 'during which I objected strongly to scope of my attack tomorrow and [to] several changes in timings'. Ramsden, no doubt exasperated, reported this to Auchinleck who sent for Morshead and came up to 30th Corps, accompanied by his *éminence grise*, Major-General Dorman-Smith. Morshead's diary for 21 July states:

> Commander-in-Chief explained plan of 13 Corps' attack. I did not like our plan because of wide dispersion and difficulty to support and pointed out that our immediate objectives were much more difficult than realised by Army and Corps. Commander-in-Chief according to Ramsden was very annoyed and perturbed but he did not show it. He stressed that he realised he must have a willing commander. I stressed that my concern was a task which was reasonably certain of success and could be held and supported, and that my job too was to minimise casualties. Altogether it turned out to be rather like a family party.[30]

It appears that Morshead accepted Auchinleck's explanations but the dispersion of his own attack—westwards and southwards—with its consequent dispersion of artillery support, remained.

The attacks by the 26th and 24th Brigades hit the enemy hard but won little ground while the 13th Corps on Ruweisat suffered disaster largely at the expense of the New Zealand Division which only a week earlier had lost over 1400 when attacking in the same sector.

Auchinleck was determined to keep up the pressure. He called for a new thrust by 30th Corps to seize Miteiriya Ridge with armoured raids into Rommel's rear areas. In another complicated plan, Morshead was ordered to seize the ridge where it was known as Ruin Ridge then exploit northwestwards along Miteiriya, while the 1st South African Division was to clear paths through the minefields southeast of Miteiriya to pass through the 69th Brigade. They would clear paths for two armoured brigades which would then drive northwestwards.

When the time came on the night 26–27 July, every phase of the operation broke down. Morshead's attack, carried out by the 2/28th Battalion reached its objective on Ruin Ridge but was cut off by unsubdued German posts to its rear. Eventually the force was surrounded and overrun by German tanks; 554 of the 2/28th Battalion were lost with their supporting detachments of anti-tank gunners, engineers and machine gunners.[31] It was a bitter blow. The 69th Brigade had a similar number of casualties while the armour took little part in the battle.

In the three dominions divisions there had developed a bitter distrust of the armour which, for its part, had good reason to believe that the infantry had little idea of the employment and limitations of tanks. Morshead expressed the general view when, on 4 August, he wrote to HQ 30th Corps: 'It is vital that on the next occasion our armour restores our lost faith in them...Until we can be certain about our armour we must have more limited and less exposed objectives than those in recent operations. The only justification for recent objectives was that our armour would effectively operate.'[32] He had already made the point forcefully when visited by Ramsden on 29 July.

The July battles cost the 9th Division 2552 casualties in just three weeks but the division emerged from them more united, confident and stronger than ever before. At last all its arms and services had fought and worked together and with marked success, under the eye of a hard commander who was becoming widely known to his troops although not without help from the German radio—their references to 'Ali Baba Morshead and his twenty thousand thieves' delighted the division and were duly celebrated in verse by Hugh Paterson, Banjo's son.[33] This German identification of Morshead with his troops was also given public circulation in the *AIF News* on 15 August 1942.

*Second Alamein*

The period between July and the battle which opened on 23 October can hardly be called a breathing space. Defensive positions had to be improved, reinforcements had to be absorbed and the case for more to be argued and sent off to Melbourne, while consideration had to be given to the next battle. Morshead was deeply uneasy about the high command. He told Maughan in 1955: 'Auchinleck thought in terms of brigade groups and Jock columns and was continually alternating between optimism and depression. I used to ask Wells [his GSO1] "How is the barometer this morning"'. He may have been recalling his diary entry of 6 August 1942 which, after using the same metaphor, continues: 'No stability, a wealth of plans and appreciations resulting in continual TEWTS. Fighting always in bits and pieces and so defeats in detail [Rommel's own criticism of Eighth Army!]. Formations being broken up automatically—it has been difficult and unpleasant keeping 9th Division intact.'[34]

Morshead's unease had long been felt in far more influential circles until, at last, Winston Churchill made sweeping changes in the high command of the army. General Sir Harold Alexander became Commander-in-Chief and the field command went, although not at first, to Lieutenant-General Bernard Montgomery. Montgomery visited the divisions the day after taking command so that it was on 14 August that Morshead said to Brigadier W.J.V. Windeyer at HQ 20th Brigade: 'This man is really a breath of fresh air. Things are going to be different soon.' Indeed they were. Nevertheless, Morshead's quick enthusiasm for the new commander was moderated, if only temporarily, by his appointment of Sir Oliver Leese to command 30th Corps. This vexed question of

22 Lieutenant-General Sir Leslie Morshead with Winston Churchill at headquarters 9th Australian Division, August 1942 (AWM Negative No. 24769)

senior commands for dominions officers, which stirred General Freyberg, commander of the New Zealanders, as it did Morshead, was taken up by General Blamey, the Commander-in-Chief of the Australian Army, in Melbourne and passed into the political realm with an exchange of cables between Curtin and Churchill. Blamey also exchanged cables with Morshead who had discussions with both Alexander and Montgomery. While the former indicated that the highest commands in the Middle East were open to dominions officers, the latter, 'who has revitalised Eighth Army and is quite friendly just doubts capacity any general who has not devoted entire life to soldiering'. The upshot of it all was that Montgomery, who by early October must have formed his own opinions about his generals, told Morshead that, if Leese were to become a casualty, he would command the 30th Corps.[35]

Morshead's chief preoccupation in the period between July and October was the training of his division for the next offensive. The Battle of Alam Halfa, Rommel's last bid for Cairo and the Canal, 30 August–7 September, did not involve the Australians in the coastal sector except for a raid near Tel el Eisa which was carried out on 1

September. Operation Bulimba had no diversionary effect on the battle to the south but its value as a reconnaissance of the Axis defensive system was immense; it bore the same relation to the battle of Alamein as the Dieppe raid of 19 August 1942 bore to the Normandy landings.

The task of initiating the 51st Highland Division into the mysteries of desert warfare was given to Morshead and welcomed by the Australians. D.N. Wimberley, their commander said of Morshead: 'He gave me a higher feeling of morale than anyone else I had met so far'. He must also have been given a frank briefing on life in the Eighth Army for he recalled the warning: 'The staff here are mad on breaking up divisions. They'll——you about for a dead cert.'[36] Perhaps Wimberley remembered this when called upon to hand over a brigade for Operation Supercharge at the end of October.

The battle of Alamein was being won by the quartermasters, the planners and the trainers in the few weeks between Alam Halfa and 23 October. Few if any soldiers could have been unaware of the new tanks and guns, the dumps of ammunition and the many preparations for a major battle. The planners, under the impulse of Montgomery's plan (a personal, not a staff paper) issued to corps and divisional commanders on 14 September, were busy getting the details right while the trainers, instructed and supervised by Montgomery himself, had the soldiery hard at work.[37] Morshead, who would not have seen himself as second even to Montgomery as a trainer, drove the two assault brigades of the 9th Division, especially in practising the attack with armoured support on highly organised defences. The positions attacked were laid out in rear areas in accordance with the enemy positions that each battalion would be attacking on 23 October.

When Morshead briefed his unit commanders on 10 October the final words of his address summed up his own feelings about the fight ahead of them: 'We must go into battle with our heads high and the will to win...it will be a killing match...If you have anyone you are not sure of then don't take the risk of taking him in. Give him some other job than fighting...We must regard ourselves as having been born for this battle.' On 22 October he visited both assault brigades and went forward to all their battalions to speak to the COs. Similarly, when he came up to his newly dug Tactical HQ about 2000 yards (1.8 kilometres) from the infantry start line on 23 October, he went round his staff chatting with each one in turn.[38]

A fortnight before the battle, Leese, the new commander of 30th Corps, had proposed an earlier H hour, 9.30 instead of 10 pm. Morshead, with his memories of Lone Pine and many another infantry assault in France, strongly opposed the change and wrote:

> The troops will have laid 'doggo' in slit trenches all day. I cannot conceive anything psychologically worse than such a solitary confinement in a tight-fitting, grave-like pit awaiting the hard and bloody battle. There must be some relaxation before the fight.
>
> To avoid giving the show away to the air these troops will not be able to emerge until close on 1900 hours. Then they have dinner—in the semi-

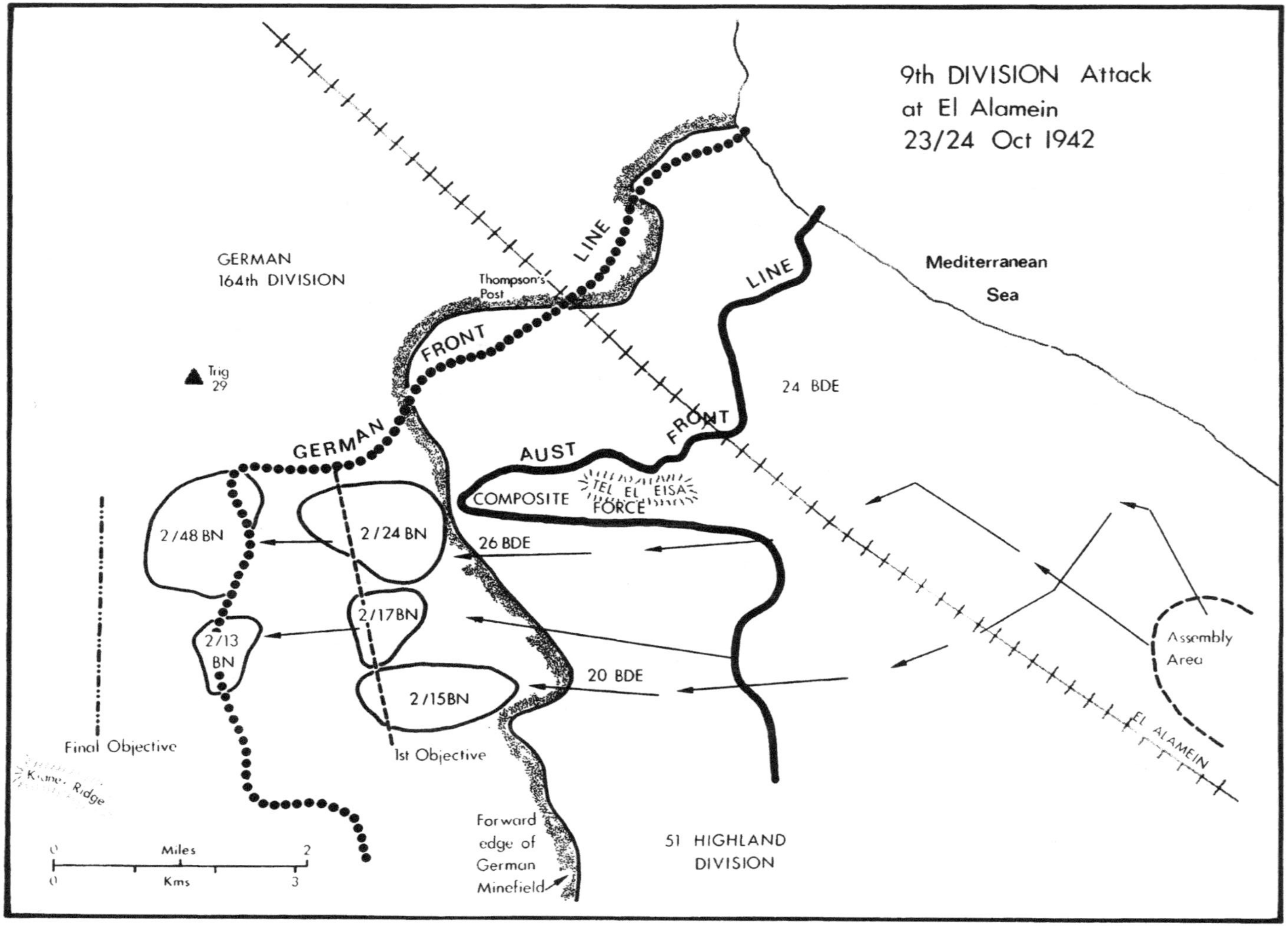

19  9th Division attack at El Alamein, 23/24 October 1942

darkness—and they do not want to be rushed off as soon as they have eaten. During the day they will have been ungetatable for final instructions and advice—however long and thorough the preparations there are inevitably those very last instructions which a platoon commander gives to the whole of his platoon. All this could be done by 2030 but 2100 would be preferable as it would avoid any chance of last minute rush and excitement. Then we have the approach march of two miles and it will take up to an hour by the time the battalions are in position on their start-lines.

To sum up, we will be demanding much of our men and we want them to start off in a proper frame of mind; we could be ready by 2120 hrs but I would much sooner have the extra 20 mins up my sleeve. This would further ensure getting A Echelon vehicles forward from Alamein in time.[39]

The writer was required to supervise the typing of this most secret letter and has an indelible memory of the impact made on him by Morshead's plea for the men in the rifle companies. Other divisional commanders may have argued similarly: Leese kept to the original time.

The Eighth Army and 30th Corps plans left little room for the initiative of divisional commanders. Thus Morshead was required to hold the coastal position with one brigade which would simulate an attack while the division assaulted on a front of 3300 yards (3 kilometres) to a depth of 6000 yards (5.48 kilometres). The other three divisions of 30th Corps would carry out similar attacks with a final objective for the whole corps of about 16000 yards (14.6 kilometres). Montgomery's revised plan issued on 6 October involved, after the 'break-in', a 'dog-fight' or crumbling process to be carried out by the infantry divisions behind the armoured shield of the 10th Corps. For Morshead this meant that he must be ready to attack northwards towards the coast or westwards, depending on how the battle went.

Although the final objective of the 'break-in' phase was reached in the course of the first two days, there was no 'break-through'; the Germans quickly established a new front, the British armoured thrusts faltered and it was clear that the 'Master-Plan' had failed.[40] Montgomery, however, was equal to the occasion. He closed down the southwestwards attack on Miteiriya Ridge, withdrew the 10th Armoured Division from the battle and ordered the 10th Corps to continue operating westwards from the bridgehead won by 1st Armoured Division. The 30th Corps was to attack northwards using the 9th Division. As this change of axis would threaten to cut off the German 164th Division opposite Morshead's 24th Brigade, the enemy could be expected to react fiercely. The battle might now turn on what Morshead and his men could achieve, as in the south the 13th Corps attack had also run into trouble.

Morshead was prepared for this situation and on 24 October had warned the 26th Brigade to be ready to capture Trig 29. This operation began at midnight on 25 October. The rapid capture of the high ground around Trig 29 and to the east of it was the first of three major attacks, all at night and supported by 360 guns under the 9th Division artillery commander, Brigadier A.H. Ramsay. The others were on 28 and 30 October. Tired as the infantry and all the supporting arms were by 25 October, they answered every call made on them. 'These attacks', wrote

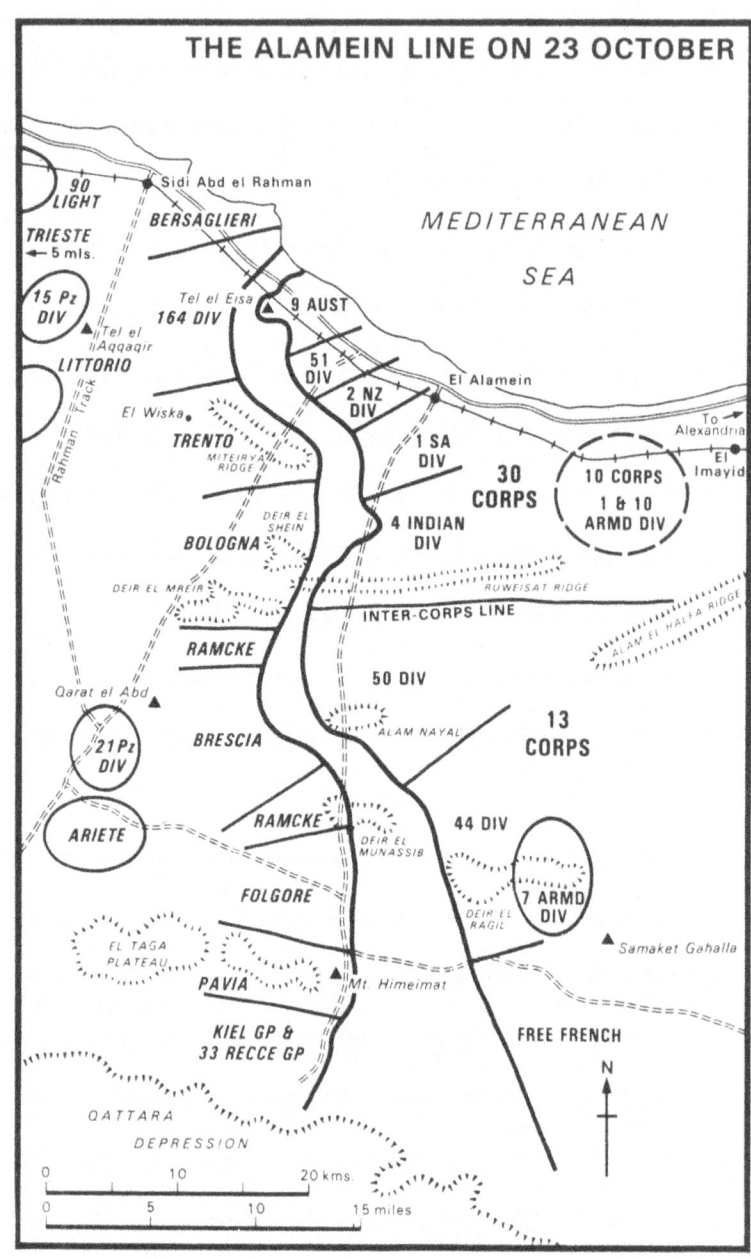

20  The Alamein Line on 23 October 1942

Morshead referring to the second and third, 'were magnificently carried out and were a vital contribution to the success of the whole battle. They were the Division's greatest achievement.'[41] Their growing threat to

164th Division and to the Axis communications caused Rommel to concentrate the Afrika Korps, 90th Light Division and a large part of the Italian mobile troops against the 9th Division, thus unbalancing his army and helping to create a situation in which Montgomery could launch a decisive blow.

It was vital now to the outcome of the battle and to the renewed attempt to smash through the Axis minefields and armour which Montgomery was preparing, that the 9th Division should be able to withstand the intense pressure of the almost ceaseless German counter-attacks. The third night attack had been only partly successful as the 2/24th and 2/48th Battalions, after withdrawing, had between them less than the strength of one company, while the 2/3rd Pioneer Battalion which thrust through nearest to the coast, was overrun by tanks and its survivors had fallen back. This had been an ambitious and a highly complicated operation made possible only by the high efficiency and resolution of the infantry and the great weight of artillery support under the hand of the CRA, 9th Division.[42] In the light of the near destruction of two battalions of 26th Brigade and the disaster to the Pioneers, one may reasonably ask whether Morshead asked too much of the troops employed. Certainly the remnants of three of the four battalions grouped under 26th Brigade were in no condition to fight a protracted defensive battle; Morshead ordered their relief by the 24th Brigade at 7.30 pm on 31 October and this was completed by 3.30 am 1 November. (He had warned 26th Brigade of this relief at his final Orders Group early on 30 October.) It was one of the decisive moments of the battle, ranking with Montgomery's order to 30th Corps to start attacking northwards. When the full fury of the Germans fell next day on the Australians in the Saucer area, it was held by the 24th Brigade with two fresh battalions (the 2/32nd was already there having been under command of 26th Brigade for the third attack) and a fresh commander and staff.[43] That night, at 1.05 am on 2 November, Operation Supercharge was launched and the last phase of Alamein began.

The momentous events just south of Tel El Eisa soon reduced the pressure on the Australian salient. Morshead saw both brigadiers in the Saucer area that day but there were no counterattacks by the Germans. These visits were a key part of Morshead's method of command, helping him to get the 'feel' of the battle and giving confidence and encouragements right down to the battalions. He also went to the main dressing stations of his field ambulances to talk to the wounded.

For the 9th Division the battle was over apart from patrols; the last casualties were suffered on 4 November and the last prisoners were captured next day. On 4 November both the Commander-in-Chief and the Army Commander called on Morshead to congratulate him and express their appreciation of the division's achievements. Messages from abroad began to pour in and General Horrocks, commanding 13th Corps in the south arrived to congratulate Morshead 'on the magnificent fighting carried out by his division'. Horrocks records with pleasure

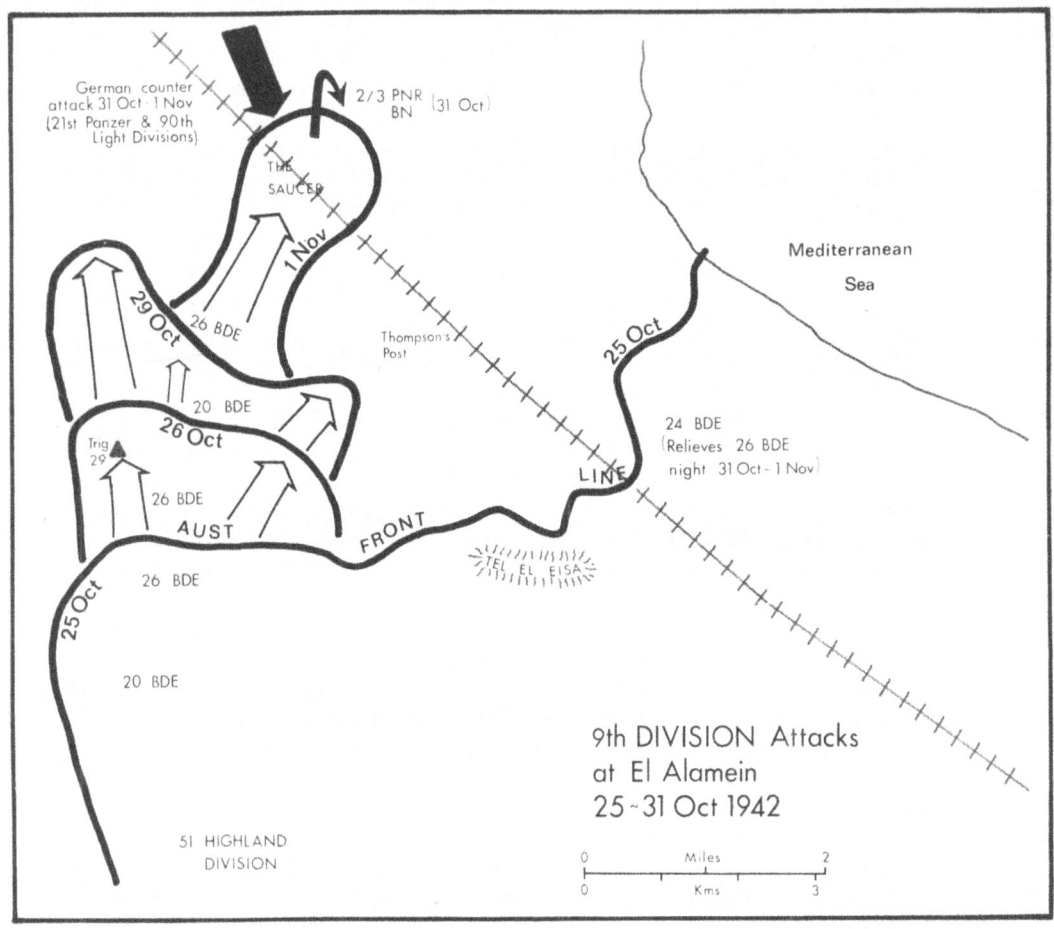

21  9th Division attacks at El Alamein, 25–31 October 1942

Morshead's reply: 'Thank you, general. The boys were interested.'[44] Sir Oliver Leese, commanding 30th Corps, in a long and highly laudatory letter said:

> The final break was...a very bold conception...and one which he [Montgomery] could never have carried out unless he was certain of the valiant resistance that would be put up by your Division. If the Germans could have broken your Division, the whole gun support of the attack would have been disorganized and its success vitally prejudiced.[45]

The KCB awarded only a few days after the battle was an appropriate recognition of the achievement of Morshead and '9 Aust Div'.

When the pursuit began there was no role for the exhausted Australians whose fate was being decided in areas remote from Eighth Army and 30th Corps. In the course of the battle they had suffered 2694 casualties of whom 620 were killed or died of wounds. When essential clearing of the battlefield and salvaging had been completed, Morshead characteristically held a commanders' conference to lay down his training policy. (Among his opening remarks about the battle was the sage advice: 'And don't forget a good word to the cooks'.) He emphasised the need to keep a firm grip on the troops after their terrible experience in battle; there must be no slackness and to counteract the natural tendency after such great efforts and four months in the desert, he prescribed a good dose of close order drill. Weapon training, the preparation of officers and NCOs for later collective training were included; nor was recreation forgotten. Morshead also added desert navigation, probably as a neat piece of camouflage given his knowledge of his government's desire to bring the 9th home.[46] Later he asked Blamey for his instructions on the kind of training he wanted carried out having regard to the division's future employment and was told to work on combined training, opposed landings and jungle warfare, a tall order for a division in Palestine.

That Morshead understood the importance of the outward signs of achievement in the military world was shown by two of his decisions in December 1942. He instituted a new T-shaped colour patch to replace the variety of shapes worn by the division; as he later explained it, 'The T stood for Tobruk'.[47] On 22 December he held a parade of the 9th Division, 12 000 strong, on the airfield at Gaza when General Sir Harold Alexander inspected and addressed the parade. For all who were in the ranks or watching that day, this was a moving and unforgettable spectacle. A month later the division and others of the AIF embarked for home. The Australian epic in the Middle East was over.

## Corps commander

Morshead's command in the Middle East was the zenith of his military career. His two years as a corps commander were in some respects an anti-climax as his command of the 9th Division had provided his most demanding tests. Yet he still had important responsibilities. It was not until October 1943 that he took the field again when he relieved General Herring in command of the 2nd Corps in New Guinea. George Wootten had succeeded him in command of the 9th Division. At this time the 7th Division had occupied Dumpu in the Ramu Valley, Finschhafen had been captured by the 20th Brigade and Wootten was about to move to Finschhafen with another brigade group of the 9th. He took command there in time to handle the counterattack which the Japanese hoped would regain Finschhafen.

Morshead's contribution to the defence was the movement of the 26th Brigade from Lae to Finschhafen which he initiated on 17 October 1943, having earlier obtained assurances that shipping would be ready. Fortunately both New Guinea Force in Moresby and the American naval commanders had learnt from an earlier crisis and this time the shipping was immediately forthcoming. The brigade began to disembark near Finschhafen at midnight on 20 October. Three days later Morshead himself went to see Wootten and on 24 October visited the 20th Brigade whose War Diary recorded: 'Wherever General Morshead went the troops made it evident that they were glad to see him. News of his visit rapidly spread amongst all troops and produced a general feeling of confidence which was most noticeable.' That afternoon Morshead also visited HQ 24th Brigade whose brigadier, Bernard Evans, had drawn upon himself the severe censure of his commander for having given up vital ground, as some said, unnecessarily, although he had, in fact, defeated the Japanese counterattack. Wootten wanted to remove Evans and Morshead agreed, which must have troubled him as he had appointed Evans at Alamein when Brigadier Godfrey was mortally wounded.[48] This was a rare occurrence under Morshead; certainly he had his senior administrative staff officer replaced in Tobruk, while at Alamein in July he removed the acting CO of the 2/28th Battalion when an attack was stopped some 2000 to 3000 yards (approximately 1800–2800 metres) short of its objective. Nevertheless it is true to say that Morshead, in spite of the nickname 'Ming the Merciless', was not associated with purges and the rolling of heads. After this unhappy affair, he went on to visit the 7th Division in the Ramu.

The secondment of Lieutenant-General Sir Iven Mackay to be first Australian High Commissioner in New Delhi and the appointment of Lieutenant-General Sir Edmund Herring as Chief Justice of Victoria caused important changes in field commands. Blamey's selection of Morshead to administer command of New Guinea Force and be GOC Second Australian Army raised misgivings in some quarters where it was regarded as one more Machiavellian move by Blamey to side-track a rival. More likely was the fact that Blamey, who was soon to accompany Curtin to London, 'knew he could count on Morshead...and Berryman, who had taken over the field command from Morshead, to give the enemy no rest...'[49]

Morshead gave up command of New Guinea Force in May 1944 after the fall of Madang. In July he received the best command in Blamey's gift, the 1st Corps with the three original AIF divisions now training around Atherton. This was the beginning of a long and frustrating period of preparation which ended in the futility of the Borneo campaign. Morshead now found himself dealing direct with MacArthur's headquarters as a result of arrangements made between that officer, the Australian Prime Minister and General Blamey who would, however, be kept informed by receiving copies of correspondence between MacArthur and Morshead.

In the three landings in Borneo—Tarakan, Brunei Bay and Labuan, and Balikpapan—Morshead's role was somewhat remote but he was able to intervene effectively on two occasions in the planning stage. The D Day fixed by GHQ for Oboe Six (Brunei Bay and Labuan), 23 May 1945, was impossible because large numbers of troops, guns and vehicles would not arrive in Morotai until after D day. GHQ's insistence that the date could not be changed provoked Morshead into suggesting that the Brunei landing be reduced to one battalion and that *ad hoc* units be

23 Rear-Admiral Forrest Royal (USN), Lieutenant-General Sir Leslie Morshead and Air Vice-Marshal W.D. Bostock on Morotai for a conference to plan the Australian landing at Tarakan, April 1945. (AWM Negative No. RAAF OG 2415)

created to support it. Morshead sent Brigadier Wells to Manila with this outrageous plan; the Americans changed their minds and D Day for Brunei was postponed to 10 June.[50]

During the planning of Oboe Two, the assault on Balikpapan by 7th Division, the US naval planners objected strongly at first to the area chosen for the landing by General Milford, the divisional commander. Although they eventually agreed, the commander of the US naval attack group, Rear-Admiral Noble, renewed the objections but Morshead decided in favour of General Milford's plan.[51] On 10 July 1945 Morshead joined the assault force by flying boat to watch the landing. The Catalina came down on dangerously rough water and broke up, but this had been foreseen and boats were standing by to pick up the corps commander and his party. There could easily have been a tragic beginning to the last major Australian operation of the war and only five weeks before its end.

*An indomitable tactician*

Morshead, like other Australian generals, never handled large forces as an independent commander. Whether as GOC of a division or a corps he was never required to take a strategic decision but rather to operate as a tactician. At that level, when commanding the 9th Division, he showed a complete understanding of the infantry battle; the tactics he laid down for dealing with armoured and infantry penetrations in Tobruk were the basis of his success in the Easter battle. His aggressive policy, especially in patrolling, 'besieging the besiegers' as he called it, paid dividends in the defence of Tobruk and had a favourable influence on the general situation in the Western Desert.

As both Tobruk and Alamein were essentially static infantry battles, not unlike those of the Western Front, should Morshead be regarded simply as a lucky general who was working in a familiar medium? Would he have shown the speed and flexibility necessary to cope with the Germans in mobile operations? One can only speculate on these interesting questions. That he withdrew to Tobruk in good order in spite of every difficulty argues something more than mere luck, and he was the only Australian general who, when presented with the opportunity of fighting the Afrika Korps in two successive campaigns, beat them on both occasions.

It may be argued that in 1941 he tended to repeat his tactics in the fighting in the Salient and ask too much of his infantry, some of whom were not fully trained. Nevertheless, the Germans from Rommel down, held the Australians in respect and the pressure exerted in the Salient tied down German units much needed elsewhere. At Alamein, with an experienced and better trained division, his three night attacks launched at intervals of about 48 hours stretched his division to the limits of endurance and won the necessary breathing space for the mounting of Operation Supercharge. He then made the crucial decision to relieve 26th Brigade and saved his defensive battle with the Afrika Korps and 90th Light Division. It was a brilliant, if costly performance, 'a magnificent piece of fighting by a great division led by an indomitable character', as Sir Oliver Leese called him.[52] Only a division solidly trained, well led and with good staffs at every level could have carried out those attacks and the swift relief of the night of 31 October–1 November.

Morshead was in his early fifties in 1941–42, not young for divisional command, yet he lacked nothing in energy and driving power; he also had the will and the moral courage without which energy may be fruitless. At Alamein especially, when every day was fraught with its own crisis whether in July or October, he moved around his division. By the end of that battle he was known to an extent far beyond Tobruk days and his name was associated with success without any gimmicks or seeking for acclaim. If the 9th was proud of its achievements it was also proud of its leader.

Morshead did not command at a level at which he dealt with his government; even when he was GOC AIF Middle East 1942–43 he corresponded with Blamey or the CGS. However, in that role and as divisional commander, he was required to deal with the British Commander-in-Chief, first Auchinleck then Alexander. In spite of his confrontations with the former or, more likely, because of them he won his respect. Under the new regime it was easier. In the Pacific, his relations with his own high command and with MacArthur in 1945 appear to have been harmonious but, as shown above, he was prepared to take on an American admiral when it came to the selection of a landing beach.

With the soldiers Morshead was relaxed and informal. The CO who played the piano in the 33rd Battalion mess and played in the battalion's football team did not lose the common touch when he became a general and famous. He was hard on some officers, he 'never forgave' as Maughan has written,[53] especially where he considered there had been failure; he was always a stern disciplinarian but he defended the good name of his newly organised division in March 1941 against General Neame. He also saw to it that there should be little chance of further trouble. Such was his grip on the division that by the end of 1942, his men were praised by the British authorities in Cairo for their conduct when on leave.[54] But the ultimate test of a commander, of his nerve and skill and his capacity as trainer, is the test of battle. That test Morshead passed triumphantly at Tobruk and at Alamein.

# 11 Field Marshal Sir Thomas Blamey: Commander-in-Chief, Australian Military Forces

D.M. HORNER

Thomas Albert Blamey (1884–1951) was one of the most controversial figures in Australian military history. Indeed he inspired such passions that for some 40 years after the Second World War his detractors have tended to discount automatically the admiration in which he has been held by a score of senior officers. As A.J. Sweeting observed, these detractors seem to believe that 'by some mysterious power the senior officers still dance to the beat of the dead field marshal's baton'.[1] While not overlooking his meritorious work as GOC AIF during the early years of the war, this chapter concentrates on Blamey's achievements as Commander-in-Chief of the Australian Military Forces (AMF) during the last three and a half years of the war. It was this crucial task which consolidated his high reputation; but how well does this reputation stand up to critical evaluation?

When Blamey was appointed C-in-C of the AMF on 26 March 1942, the Army, and indeed the whole nation was in deep crisis. In the four months from the outbreak of war in the Far East Japanese Forces had overrun South-East Asia, had defeated and captured the Australian Forces in Singapore, Java, Ambon, New Ireland and New Britain, and had landed on Timor and New Guinea. Darwin and Broome had been bombed and an assault against the mainland was expected any day. The government and some sections of the population seemed close to panic.

Yet, although the country stood largely defenceless, military and political authorities had not stood idle while the Japanese had come steadily closer. In December 1941 there had been less than a brigade of adequately trained and equipped troops in Australia. Now the 7th Division and one brigade of the 6th Division were returning from the Middle East. Some six infantry and two cavalry divisions of militia had been mobilised and were undergoing intensive training, albeit with limited equipment and facilities. In addition, a number of senior commanders had been recalled from the Middle East to take command of many of these newly mobilised formations; they brought new energy as well as recent battle experience. Then, a few days before Blamey's appointment, General Douglas MacArthur had been appointed Commander-in-Chief of the South West Pacific Area. Australia no longer stood alone, but seemed assured of American help. Until this help

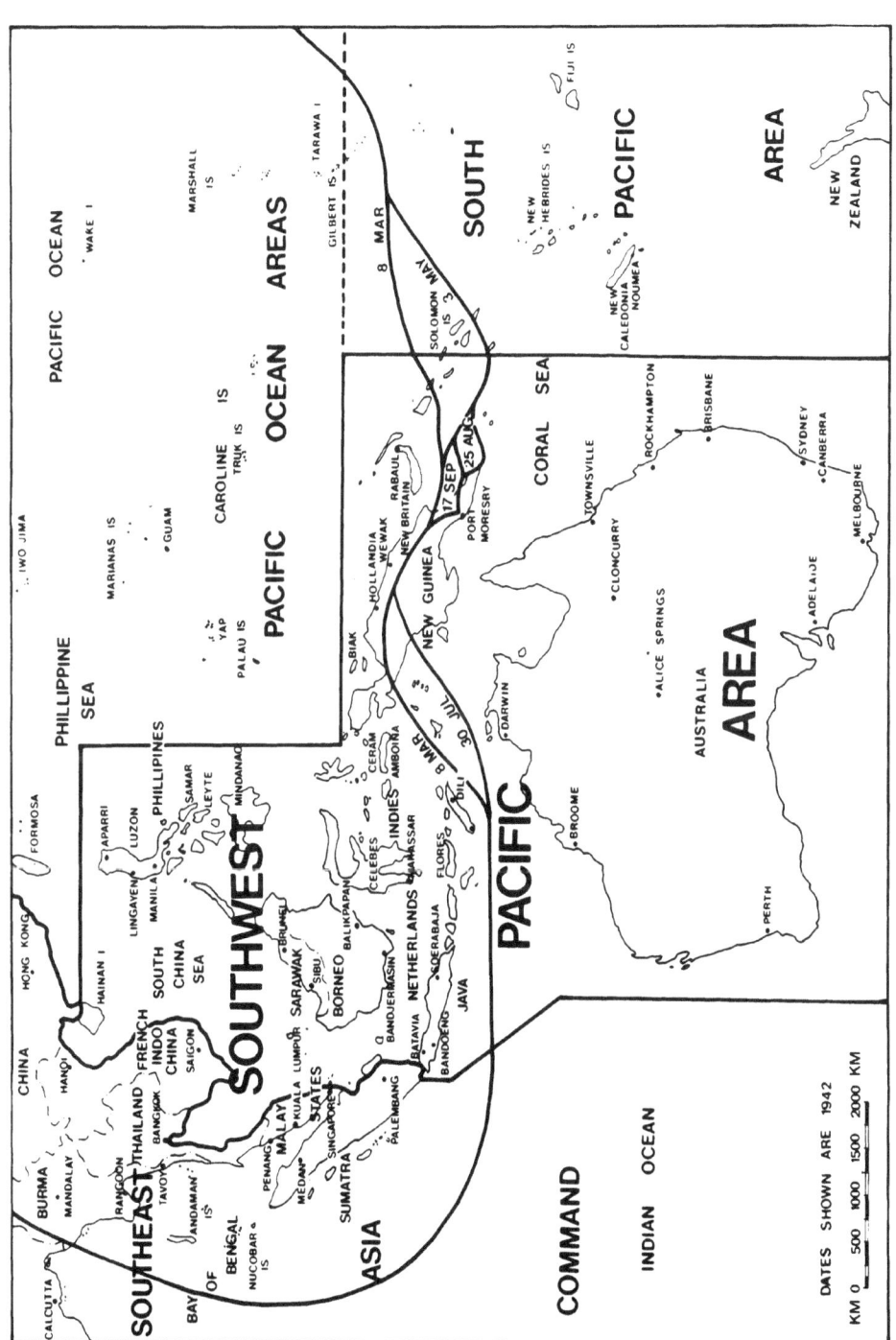

22 The Boundaries of the South-West Pacific Area and the Extent of the Japanese Advance

materialised, however, and until the troops could be trained and equipped, the Japanese threat would continue to grow.

In these circumstances Blamey was faced with an immense and complicated task—a task which overshadowed in its complexity and importance to Australia that presented to any military leader in Australian history, including Monash in the First World War. As Commander of the Allied Land Forces, under MacArthur, Blamey was responsible for the land defence of Australia and for offensive operations planned by MacArthur. As C-in-C of the AMF he was responsible for the training, development and administration of an Army soon to exceed twelve divisions, plus a multitude of training and base establishments. As the government's chief military adviser he was responsible to advise the Prime Minister on high-level defence policy. And he accepted a further, self-imposed task of ensuring that Australian interests were safeguarded against the wider interests of her more powerful allies, particularly the United States.

*Early experience*

To his new appointment Blamey brought wide experience and knowledge, but his career had not been without difficulties. Before the First World War he had attended the Staff College at Quetta, and during that war he had proved to be a brilliant staff officer. For the last six months of the war he was Brigadier-General, General Staff (BGGS) under Monash, and had won strong praise from that demanding soldier. Unfortunately, he had commanded a battalion for only three weeks and a brigade for only six. After the war he had been appointed Second (Deputy) Chief of the General Staff before retiring in 1925 at the age of 41 to become Chief Commissioner of the Victoria Police.

In 1936, following several earlier controversies, he was forced to resign after he had issued an untrue statement in an attempt to protect the reputation of the police. The next year he relinquished command of the 3rd Division (Militia), and he was on the unattached list until September 1938 when, on a suggestion from the Secretary of the Department of Defence, F.G. Shedden, he was appointed chairman of a recently established Manpower Committee and Controller-General of Recruiting. Shedden later explained that: 'The aim was twofold. His military experience and organizing ability would be most valuable to the Committee, and he would be brought back into the Defence Organization as the most probable Army Commander in the event of war.'[2]

Despite the damage to Blamey's reputation following his resignation from the Victoria Police, at the outbreak of the Second World War the Prime Minister, Menzies, supported by the senior ministers, R.G. Casey and Sir Henry Gullett, had little hesitation in appointing him to command the 6th Division of the Australian Imperial Force (AIF). The previous Prime Minister, Joseph Lyons, who felt that Blamey lacked the moral qualities to lead Australian soldiers in battle, had to agree after meeting him that, 'He's really somebody'. The raising and training of the 6th Division at a time when there was doubt about its service overseas needed all Blamey's resilience and experience, and when it was decided in early 1940 to form an Australian corps by raising an additional division, he became the corps commander. Major-General L.E. Beavis, Blamey's chief ordnance officer throughout the war, has written:

> I had the greatest admiration and respect for him as a commander. He had his shortcomings, such as aspects of his personality which affected some of the personal clashes in which he became involved, evidenced in a degree of ruthlessness when he felt sure he was right.

Gavin Long noted that:

> He had a mind which comprehended the largest military and politico-military problems with singular clarity, and by experience and temperament was well-equipped to cope with the special difficulties which face the commander of a national contingent which is part of a coalition army in a foreign theatre of war.[3]

Yet at the outbreak of the Second World War he was out of touch with recent developments in military technology. Although he had an excellent grasp of the corps level command and staff techniques in operation at the end of the First World War, he had had no opportunity to study the problems of joint warfare, particularly the coordination of air and armoured units which became the hallmark not only of the German blitzkrieg, but also of the Japanese advances in China and Malaya. As the Second World War approached he had concerned himself with world affairs by delivering a weekly radio broadcast. Although his work with the Manpower Committee had given him an insight into the senior policy-making level of the Department of Defence and the government, this background prepared him more for high command appointments such as command of the AIF in the Middle East, or of the Army, than for a battlefield command, such as command of a division or a corps.

This experience of the policy-making level stood him in good stead during the eighteen months he spent in the Middle East as Commander of the Second AIF. In particular, he fought long and hard to maintain the integrity of the AIF, beginning with his arguments with Wavell in 1940 and culminating in his showdown with Auchinleck in 1941 over the relief of Tobruk. His chief of staff, Brigadier Rowell, summarised Blamey's achievement in notes prepared for the Army Historical Section on 25 January 1941:

> It is not my business to give the Commander bouquets, but I think anybody with less force of character and sense of responsibility both to the AIF here and his Government at home, would probably have split the show, and we would have had a lot of difficulty in restoring it again. But he never ceased to make it clear that he owed his duty to his Government. He made it clear that the British could not simply order Australian units about as they liked.[4]

Because of Rowell's later disagreement with Blamey, particular weight should be attributed to these comments.

Views on Blamey's performance as Commander of the ANZAC Corps in Greece are mixed. On the one hand the withdrawal down the Greek

*24* The Prime Minister, R.G. Menzies, with General Blamey in the Middle East, February 1941. On the left is Mr. F.G. Shedden, Secretary of the Department of Defence. (AWM Negative No. 5868)

peninsula was handled with great skill, and the evacuation was organised from beaches reconnoitred earlier by Blamey in expectation of just such eventuality. On the other hand Rowell has claimed that at the height of the withdrawal Blamey was 'physically and mentally broken'. While supported by some other officers, this accusation has been strongly refuted by a number of senior officers, such as Herring and Savige who retained tremendous regard for Blamey, and Blamey's ADC recalled that in Greece he never saw him 'fearful or abnormally troubled'.[5] Whatever the truth of Blamey's performance as commander, he lost much support when he chose his son to fill the one remaining seat on the plane carrying him out of Greece. He has also been criticised for not warning the government earlier that he doubted the wisdom of the Greek campaign.

Thus while in many quarters Blamey enhanced his reputation in the Middle East, he failed to win the unanimous support of a small but influential group of senior officers. There were many tensions alive in the upper echelons of the AIF as ambitious, and at times disloyal, regular and militia officers vied for commands. Perhaps Blamey could never have won all their support. Admirably, it was not his style to curry favour with subordinates, but at times he seems to have fueled the antipathy towards himself, rather than try to dissipate it. Rarely able to inspire complete loyalty and trust among the soldiers, he enjoyed life to the full in the manner which soldiers understood, but which they did not expect of their commanders. Sensitive to criticism, relentless in pursuit of personal enemies, perhaps his greatest failing was his lack of understanding of the importance of public relations; in short, he lacked tact.

*Military administrator*

Blamey's appointment as C-in-C of the AMF had a profound effect on the organisation and administration of the Army. As a first step, the Military Board was abolished and he established Land Headquarters (LHQ) with five principal staff officers, the CGS, Sturdee, who retained his position, the Lieutenant-General in charge of Administration (LGA), Wynter, the Adjutant-General, Stanke, the Quartermaster-General, Cannan, and the Master General of the Ordnance, Beavis. The day-to-day control of operations was in the hands of the DCGS, Vasey, and in July 1942 he moved to Brisbane to establish Advanced Land Headquarters. From that time Blamey had two headquarters, one devoted to operations, and the other to the training and administration of the Army. As well as these changes, the old, State-based geographic command system was disbanded and the Army was divided into the First Army (responsible for the defence of NSW and Queensland), the Second Army (Victoria, SA and Tasmania), the 3rd Corps (Western Australia), Northern Territory Force, the forces in New Guinea, and LHQ units. The old military districts became Line of Communication Areas supporting the field forces in their areas.

Although the differences between the militia and the AIF were never completely resolved, Blamey's task was to build and train the Army (both militia and AIF), initially for the defence of Australia, and then for offensive operations. Under his drive, and also that of the other senior officers who returned from the Middle East, the Army was organised and prepared for battle. Brigadier Field of the 7th Brigade at Milne Bay remembered meeting Blamey at Port Moresby in mid-1942. Blamey asked Field what his officer position was, and Field replied that it was fair but that he would like some AIF captains. Blamey said, 'Very well', and the next day four captains from the 7th Division arrived. Blamey ordered, 'Promote corporals to lieutenants if necessary', but these measures needed time to take effect.[6]

Blamey realised that once all the militia divisions were mobilised, along with the AIF divisions, the country would not be able to sustain the force, and at the beginning of July he began to reduce the number of divisions. This process was to continue throughout the war. But Blamey's letter to the Prime Minister Curtin, shows his blunt approach. After outlining the proposed changes he concluded:

> I would be grateful, therefore, for an early conclusion of the determination of the powers requested by the Commander-in-Chief. In the meantime the circumstances demand that I proceed with the reorganization policy with all possible speed, and I propose to do so unless you direct to the contrary.[7]

Despite the very fine work of Sturdee, and his deputy Rowell, in the first months after Japan entered the war, in the opinion of General Berryman, Blamey's appointment as C-in-C was vital in resolving the problems facing the Army. Blamey's signals with the prefix 'Z' ensured immediate action, and the expansion and training of the Australian Army in mid-1942 was an impressive feat, facilitated by the rapid decisions from Land Headquarters.

In exercising his responsibilities as C-in-C of the AMF Blamey faced a number of difficulties. First, he failed to win the unanimous support of the government. Blamey tended to ignore the Minister for the Army, Forde, and dealt directly with the Prime Minister on many matters. And while he retained the confidence of Curtin, other ministers were critical of his administration, particularly his selection of commanders. Indeed Blamey's ADC, Norman Carlyon, believed that after Curtin died in July 1945 Blamey would have been dismissed as C-in-C had not the war ended so soon afterwards.

Second, Blamey and General Wynter were involved in a bitter contest with the civilian staff of the Department of the Army over financial administration. Advised by Wynter, Blamey was led to believe that the Secretary of the Department, F.R. Sinclair, was attempting to restore the Military Board. In a letter written in March 1944 Wynter described the 'bastardry' of the civil staff: 'They take any and every opportunity to oppose the C-in-C'.[8] Perhaps Blamey took too much notice of Wynter, but whatever the rights and wrongs, this internecine war made Blamey's job more difficult.

Third, Blamey was faced by personal antagonism, and indeed disloyalty, from a small group of senior officers who thought that they had been overlooked, side-tracked or victimised by the C-in-C. It is true

25 The Minister for the Army, F.M. Forde, and General Blamey at the 1st Armoured Division review, Puckapunyal, June 1942, (AWM Negative No. 12692)

that Blamey was a great hater—ruthless in pursuit of his enemies—but he was intensely loyal to those who had supported him in his earlier days. These undercurrents provided additional fuel to the criticism of Blamey's administration.

Fourth, Blamey had a constant battle to balance the demands for manpower from other sections of the economy against the needs of the force he considered necessary to conduct operations. He progressively reduced the number of divisions until at the end of the war the Army consisted of six divisions, all of which were in action. His critics accused him of demanding an excessive level of manpower in an effort to maintain his own prestige. In Blamey's view the men were required so that Australia could play its full part in the war. In this respect the government failed to give Blamey clear directions, tacitly accepting his strategic policy. Nonetheless many ministers criticised the drain on manpower resources.

But by far the greatest difficulty to be overcome in achieving the smooth administration of the AMF was that Blamey could devote only part of his time to that task, for he also had extensive responsibilities as Commander of the Allied Land Forces. The dual responsibility worked tolerably well before MacArthur's General Headquarters and Blamey's Advanced Land Headquarters moved to Brisbane, but after that move Blamey had to fly constantly between Melbourne and Brisbane, not to mention the requirement to visit both operational and training units. Notwithstanding the fact that the CGS (Sturdee and then Northcott) handled everything that did not need the personal attention of the C-in-C, once Blamey became involved closely in operations, some urgent administrative matters had to wait.

In September 1942, soon after arriving in New Guinea, Blamey recommended that Northcott be made Deputy C-in-C as well as CGS so that he could take over some of the administrative responsibility. The government was unwilling to take this step as it felt that Morshead, still in the Middle East, would make a suitable Deputy C-in-C. Blamey was adamant that the Deputy C-in-C should be the same person as the CGS, otherwise 'such an appointment is like placing a fifth wheel on the coach'.[9] The matter was not resolved at this time, and later the government, on advice from MacArthur, denied Blamey's request.

Both during the war, and in books written since the war Blamey has been criticised for trying to wear his two hats as C-in-C AMF and Commander, Allied Land Forces. Before discussing the merits of Blamey's two appointments, however, it will be of value to describe his role as Commander, Allied Land Forces.

*Commander, Allied Land Forces*

From the time of his arrival in Australia, MacArthur was determined that American and Australian ground troops should remain separated and should operate under their own commanders. It was only under pressure from Washington that an Australian was given command of both American and Australian forces, but MacArthur never intended that Blamey should exercise command over American forces. Indeed MacArthur planned to act as his own land forces commander with operations conducted by task forces operating under his personal control.

The fiction that Blamey was commander of the Allied Land Forces lasted until there was the likelihood of American troops being committed to battle. Thus, following the Japanese landings in Papua, MacArthur suggested on 1 August 1942 that the 32nd US Division should be sent to New Guinea to operate directly under the control of GHQ. This move would have led to an impossible command structure with two separate superior headquarters in Australia controlling separate national forces in one operational area. Apparently Blamey talked him out of that folly. Again, on 14 September, MacArthur ordered the US general, Eichelberger, to prepare to command a task force alongside that of Rowell in New Guinea, but again he was persuaded to drop the scheme.[10]

MacArthur, however, was soon to get his own way, for when things appeared to be going badly in New Guinea he persuaded Curtin to send Blamey there to act, in effect, as a task force commander. This mission was against Blamey's wishes and resulted in the unfortunate dispute with, and eventual replacement of Rowell, the Commander New Guinea Force. Nor would MacArthur allow Blamey to return to Brisbane, even when Herring arrived to take over from Rowell.[11]

While there is no doubt that MacArthur aimed to become his own land forces commander, and to operate through subordinate task force commanders, Blamey's dual role gave MacArthur a ready-made excuse for doing so. For example, even before his headquarters opened in Brisbane, Blamey had decided that he would not be located there permanently. As Sturdee explained on 20 July: 'He combines the functions of operational commander with head of the housekeeping side and will therefore spend a considerable period of time in Melbourne in addition to travelling about Australia generally'. Between 1 August, when his headquarters opened in Brisbane, and 23 September when he left to take command in New Guinea, Blamey spent barely half of his time in Brisbane. This absence might have been acceptable later in the war, but during the South West Pacific Area's first campaign, with Port Moresby threatened, and with MacArthur's staff 'like a bloody barometer in a cyclone—up and down every two minutes', as General Vasey exclaimed, it seemed Blamey was inviting trouble by being away.[12]

MacArthur was far from satisfied when Blamey took command in New Guinea, for he realised that when the 32nd US Division, which had begun to arrive in New Guinea in September, was finally committed to battle, Blamey would, in effect, resume his position as Commander of the Allied Land Forces. On 19 October MacArthur commented to his British liaison officer, Wilkinson, that it 'would not do to leave [Blamey] in Supreme Command'. Wilkinson suggested that another American commander would 'have to be put in over' Blamey. MacArthur was not sure how that 'should be handled' but added that he 'could handle Blamey'. Significantly, on 6 November MacArthur moved his own advanced headquarters to Port Moresby. It is not surprising, therefore, that before leaving for New Guinea MacArthur told Shedden that 'it would be necessary for General Blamey sooner or later to make a decision as to whether he was going forward in command of the advanced forces in any offensive operations, or was remaining in Australia to command the forces left there for the defence of the base'.[13] And this was to be one of MacArthur's constant themes for the next two and a half years.

General Berryman is probably right in his assertion that when Blamey journeyed to New Guinea he was fighting for his military life. From the time when he returned to Australia from the Middle East in March 1942 Blamey would have been aware of criticism in some quarters of his suitability for high command. There had been the so-called 'Revolt of the

26 General Sir Thomas Blamey and General Douglas MacArthur, with Australian troops, Port Moresby, New Guinea, 2 October 1942. (AWM Negative No. 13422)

Generals', and Generals Robertson, Lavarack and Bennett had all coveted his position. But for a while Blamey's position seemed secure. After the war MacArthur described Blamey as 'a veteran soldier of highest quality', and before the operations in New Guinea he was lavish in his praise.[14] Blamey's performance was raised at the Prime Minister's War Conference in Canberra on 17 July 1942. Shedden's notes recorded that MacArthur said that he considered Blamey:

> to be the best of all the Australian Generals. He was above the average military ability of Generals. In his considered professional opinion, General Blamey was a first-class Army Commander.
>
> General MacArthur added that as Commander-in-Chief he was not concerned with any personal idiosyncrasies which General Blamey might possess. He judged him by results, and he considered that he had effected great improvements in the Australian Army since his return to Australia. Furthermore, he had the confidence and respect of the United States Army Staff.
>
> General MacArthur said that he had heard much loose talk from some people about General Blamey and he regretted to say that much of it had originated from officers in the Australian Army. Other Australian officers coveted the post of Commander-in-Chief and had made representations against General Blamey. He had also received anonymous letters on the subject.

After the conference, Curtin confidentially told newspapermen that MacArthur had commended Blamey: 'He praised his organisation of the evacuation of Greece as one of the most outstanding events of the war, greater even than Dunkirk where the evacuating troops had the protection of almost unlimited aeroplanes'. Curtin added that Blamey's private life had nothing to do with his military office: 'He said when Blamey was appointed the Government was seeking a military leader not a Sunday School teacher'. On 22 September 1942, the day before Blamey flew to Port Moresby to relieve Rowell, MacArthur described him as possessing a 'sensual, slothful and doubtful character, but a tough commander likely to shine like a power-light in an emergency. The best of the local bunch...' After the war, Blamey recalled that during these early months a good relationship was established with MacArthur 'which ripened into a deep regard and friendship'.[15]

Despite these comments, there is plentiful evidence that as the crisis in New Guinea deepened, Blamey's position became increasingly precarious. On MacArthur's part, Blamey could provide a convenient scapegoat if Port Moresby fell. After all, MacArthur knew that if he suffered another defeat, following the fall of the Philippines, his own position would be unsteady, hence the reports to Washington of the poor fighting quality of the Australians. MacArthur also advised General Marshall, in Washington, that Blamey's 'entire methods and conception differ so materially from ours that his actions during my absence would be unpredictable'.[16]

Nor were all the politicians happy. The Advisory War Council on 17 September 1942 attempted to determine responsibility for the deteriorating situation in New Guinea, and it must have been obvious to Blamey, who was present, that some members were blaming him. That evening MacArthur told Curtin, who already thought, incorrectly, that the place for the C-in-C was in the front line, that Blamey should proceed to New Guinea 'to save himself' and 'to meet his responsibility to the Australian public'. Aware of the government's decreasing confidence in his ability, Blamey said later that he 'raised no question'. William Dunstan, the General Manager of the Melbourne *Herald*, wrote soon afterwards (to Rowell) that, at two background conferences for senior newspapermen, Curtin had discussed Blamey's position 'more or less freely—off the record, of course':

> They are not satisfied with him...His every move is watched....He was sent to New Guinea for no other reason than to give him one final chance....To this extent we must sympathise with him....He has had to submit to the MacArthur holiness. Just how much do we know of what he has suffered under the set-up? Granted he is GOC Land Forces, just how much does it mean if he took strong directive over US Forces? Would he have to go cap in hand to MacArthur?

Dunstan continued that there had been a canvass of names to find a replacement.[17]

But if some observers could accuse Blamey of a lack of moral courage in not challenging the government at this time, others might have seen it as the pragmatic approach of an experienced soldier who was convinced that he was the only man for the job. By strong and effective action in New Guinea, helped by the fruits of Rowell's careful planning, Blamey re-established his reputation to a point in December 1942 where he was able successfully to challenge MacArthur over the employment of an American regiment. And if the Advisory War Council had its doubts about Blamey on 17 September, on later occasions he left them in no doubt. Air Marshal Sir George Jones, who was present at one meeting, recalled that the Prime Minister criticised Blamey at length over a minor administrative matter. When Curtin paused Blamey said, thumping the table: 'Prime Minister, I want you to know that I have no ambitions in this war. If you don't like what I have been doing you damn well get somebody else.' What was an attempt to censure Blamey almost turned into a censure of the government.[18]

Throughout the war Curtin maintained his loyalty to Blamey, prompting Blamey to write later that he had 'no need to bother about rear armour'.[19] But the events of September 1942 show that Blamey's ability to influence Allied strategy through his dual positions of power rested on a fragile structure of prestige and performance and the various perceptions of these by other key actors.

Furthermore, the Papuan campaign showed that MacArthur was determined that his operations should be controlled by task force commanders, rather than by Blamey as Commander, Allied Land Forces. Indeed MacArthur told Shedden in January 1943 that Blamey was unable to perform adequately his duties as Commander, Allied Land Forces as well as those of C-in-C of the AMF. Thus on 11 January 1943 MacArthur asked Marshall to send General Walter Krueger from America 'to give the US Army the next ranking officer below General Blamey in the Allied Land Forces which is not now the case and is most necessary'. Soon after Krueger's arrival MacArthur formed Alamo Force to conduct the operations of Krueger's Sixth US Army. There were not yet enough troops to form a US Army in Australia, but Krueger, who also commanded Alamo Force, 'realised that this arrangement would obviate placing Sixth Army under the operational control of CG Allied Land Forces'.[20]

There is other evidence to support this view. General Dewing, the British Army representative in Australia, wrote that MacArthur was 'working steadily to exclude the Australians from any effective hand in the control of land or air operations or credit in them, except as a minor element in a US show'. Krueger's deputy chief of staff commented later that Alamo Force was created 'to keep the control of Sixth Army units away from General Blamey'. In the opinion of General Eichelberger, Krueger was sent to Australia because the Americans were disadvantaged by the high rank and experience of many of the Australian officers: 'Whether Walter's rank will help solve the problem only time can tell...It reminds me of the poker games in Shanghai...where the cuspidor was put on the centre of the table because no one dared look away to spit'.[21]

Colonel Keogh has written that in forming Alamo Force to avoid placing any large body of American troops under Blamey's command, MacArthur needed 'to be congratulated for his skill in extricating himself, his own Government and the Australian Government from an embarrassing situation'. This may well have been the case, for towards the end of 1942 Curtin stated that he intended to discuss Blamey's role when MacArthur returned from New Guinea. Perhaps Shedden advised Curtin that the formation of Alamo Force made a change to the organisation less urgent. Such a change, coming at the same time as an upheaval in the command of the RAAF, might have been difficult to institute, for Blamey was advising the Government that the results achieved in Papua 'would have been quite impossible had he not been' C-in-C.[22]

But if Keogh was right in his suggestion that MacArthur needed to be congratulated for his skill, he overlooked another aspect of MacArthur's motivation. Keogh claimed that if the Commander, Allied Land Forces had not had the additional responsibility of administering the Australian Army, then 'the subsequent command arrangements would have adhered more closely to the original plan'.[23] What Keogh failed to realise was that whether the nominal Commander, Allied Land Forces had other responsibilities or not, MacArthur had no intention of allowing him to operate as such. Blamey's 'two hats' provided a convenient excuse for MacArthur's rearrangement of his forces; they were not the real cause of the change.

There is no doubt that MacArthur's new system worked, although Krueger doubted whether it would have done so if the land forces had been faced with defeat. Nevertheless, MacArthur's method of achieving it was, in the words of Gavin Long, 'by stealth and by the employment of subterfuges that were undignified, and at times absurd'. These subterfuges revealed a lack of consideration by MacArthur towards a subordinate who, to date, had shown outstanding loyalty. Indeed on 17 May 1943, MacArthur's Chief Public Relations Officer, Colonel L.A. Diller, wrote to MacArthur from Melbourne: 'I found no question from any source of General Blamey's loyalty and fidelity'. F.M. Forde commented later that Blamey 'worked very hard in order to give satisfaction to General Douglas MacArthur and to render a maximum of assistance'.[24] It was to be almost two more years before Blamey lodged a formal complaint with Curtin over MacArthur's degradation of the role of Commander, Allied Land Forces.

The incident which spurred Blamey to complain took place in February 1945, when during the planning for the attack by Morshead's 1st Australian Corps on Borneo, GHQ announced that it intended to eliminate Advanced Land Headquarters completely from the chain of command. Morshead's 'task force' would answer directly to GHQ. Blamey wrote a long letter to Curtin explaining how his authority had been gradually reduced by MacArthur. Curtin wrote to MacArthur seeking further information and MacArthur replied that he intended to operate through task force commanders. He made no mention of Blamey's position as Commander of the Allied Land Forces. Yet at the same time, MacArthur told his British liaison officer that 'Blamey's position was an intolerable situation'. Unless Blamey were assigned entirely to MacArthur's command, the latter was 'not prepared to take him as commander of the Australian Forces for any future operations which might arise'.[25] As Blamey had expected, Curtin took no further action.

*Operational commander*

While Blamey was progressively edged out as Commander, Allied Land Forces, he nonetheless retained wide responsibility for the operations of the Australian Forces. Initially he was concerned with the defence of Australia, which he organised with energy. With MacArthur's agreement, in April 1942 he deployed the First Army, the largest of his formations, in New South Wales and Queensland with the task of defending the east coast of Australia. The Second Army was based initially on Melbourne, and soon afterwards, on Sydney. One division (4th) was sent to Western Australia, where the 3rd Corps was to be formed, and the 19th (AIF) brigade was ordered to Darwin. In addition, US anti-aircraft and engineer troops were sent to Darwin, a squadron of heavy bombers and a US anti-aircraft regiment were ordered to Perth, a US anti-aircraft regiment and an additional Australian infantry brigade went to Townsville, and the remainder of the US anti-aircraft troops were grouped in Brisbane. The air force concentrated most of its striking force in the Townsville–Cloncurry area, where airfields were becoming available.

During this period only a limited number of units were deployed to Port Moresby. These units included a light anti-aircraft battery, US engineer and anti-aircraft units and an Australian independent company. But until the Battle of the Coral Sea in early May 1942, the Australian Army stood on the defensive. Indeed, the Australian official historian noted that:

> So hesitant had General MacArthur and General Blamey been to send reinforcements to New Guinea that on 10th May, the day on which the Japanese planned to land round Port Moresby, the defending garrison was not materially stronger than the one which General Sturdee had established there early in January.[26]

After the Battle of the Coral Sea MacArthur decided to reinforce Port Moresby, and it is in executing this order that Blamey has been criticised for his decision to send a militia rather than an AIF brigade. This error was compounded by his decision to send a militia brigade to Milne Bay. On 19 May the Advisory War Council expressed the view that well-trained and experienced troops should be sent to Port Moresby, but the CGS said that the 7th AIF Division had to be kept for training for overseas operations contemplated later. In a report for the Prime Minister Shedden observed that 'though this may have fitted in with

27 Lieutenant-General R.L. Eichelberger, Lieutenant-General E.F. Herring and General Sir Thomas Blamey examine a three inch AA gun captured from the Japanese at the Government Gardens, Buna, January 1943. (AWM Negative No. 14099)

projected offensive plans, the security of such a vital place should have had priority'. MacArthur supported this view, and later claimed that he had asked Blamey to send his best troops to New Guinea. He did not think that the militia were adequately trained. General Rowell recalled that the decision not to send the AIF made his 'headquarters weep at the time', and Gavin Long wrote in his diary that the 'decision to keep the best troops to last was "criminal"'.[27]

Shedden concluded that the reason Blamey did not send the AIF to Port Moresby was that he and the Australian Chiefs of Staff had decided that two militia brigades would be sufficient to repel a seaborne attack, and that there was no chance of an overland advance on the town. Shedden stated that it was 'probably not an unfair surmise that the Owen Stanley Range and the difficulties of communications on the southern side induced a 'Maginot Line' complex that there was an easily defensible barrier to a Japanese advance beyond Kokoda'. After reviewing Shedden's paper, Rowell said that he fully agreed.[28]

During the Papuan campaign Blamey can be criticised for bowing to MacArthur's demands by replacing Allen as Commander of the 7th Division on the Kokoda Track, and he unnecessarily sacrificed troops in using bren gun carriers as tanks at Buna. But on the other hand Blamey treated his senior field commander, Herring, with sensitivity. He tried to protect him from MacArthur's whims and unlike MacArthur, he visited the battlefield at Buna and Sananda. The campaign became a slogging match but there was probably little alternative.

By contrast, the capture of Lae and the attack on Salamaua to draw the Japanese away from Lae, was a brilliant orchestration of sea, air and land resources. During the campaign which began in March 1943 with the defence of Wau and ended a year later with Australian forces advancing along the coast to Madang, Blamey rotated commanders and formations as troops became exhausted from battle, terrain and sickness. Relentless pressure was maintained on the enemy by five divisions, three of which were militia. All divisions were well trained for jungle warfare—a tribute to Blamey's foresight in establishing training areas in Queensland, particularly on the Atherton Tableland, and to his organisational ability and unparallelled knowledge of how an army should be formed and put to work.

With respect to the campaigns at Wewak and on Bougainville during late 1944 and 1945, Blamey has been criticised for employing an offensive policy when, it has been argued, it might have been better to protect the allied bases by patrolling and waiting for the war to finish. Blamey defended his policy in an appreciation prepared for the Advisory War Council, arguing that the troops' morale and health would deteriorate if they were not involved in an offensive. He said that the natives had to be liberated from the Japanese. His aim was to eliminate the Japanese garrisons so that the army could be reduced from six to three divisions, one of which would be made available to MacArthur. Blamey could hardly have expected that the war would end abruptly in August 1945, and he wanted to make sure that he could provide troops to take part in the invasion of Japan. After all, it was government policy that Australia should win recognition at the peace table by the important role played by her troops. Perhaps Blamey was neither authorised nor competent to decide on a policy with such wide political implications, but if he had ignored the political aspect and in absence of advice from the government had made his plans purely according to military facts, he would surely have been condemned by many critics. Given the paternalistic attitude of the government and its advisers, the aim of the operations as stated by Blamey, to 'liberate the natives from Japanese domination', seems unchallengeable.[29]

While Blamey had little opportunity to display his ability as a field commander in the Pacific War, he quickly grasped the nature of the war, the need for the use of sea and air resources, the debilitating effect of the climate and terrain, the need for thorough training and fitness as well as frequent reliefs of commanders and soldiers, the importance of logistics and the value of accurate intelligence. He did not immerse himself in detail which he left to his first-rate chief of staff, General Berryman, yet he had a clear grasp of the details. Importantly he was not wasteful of Australian lives (apart from his miscalculation of Buna), and was always protective of Australian interests. Brigadier Sir Kenneth Wills, Controller of the Allied Intelligence Bureau, wrote after the war: 'Few people realise how much of the credit of the successful Australian operations, both in the Middle East and in New Guinea, was due to the Chief's personal control and personal planning'. Wills thought that Blamey was ahead of Mackay, Lavarack, Morshead and Herring in 'brainpower, leadership and drive'.[30]

## The Government's chief military adviser

The mainspring of the machinery for the higher direction of the war was the Prime Minister's War Conference, which consisted of the Prime Minister, MacArthur, and any ministers or officers whom the Prime Minister wished to summon. The Defence Department provided the secretary, and this was nearly always Shedden himself. The first Prime Minister's War Conference was held on 8 April 1942, and it was here that MacArthur made his wishes known with respect to the higher direction of the war. He was adamant that he should deal directly with Curtin alone and vice versa, and as a result on 14 April Curtin produced a memorandum formalising the arrangements.[31] Subsequent letters between Blamey and Curtin made it clear that in his capacity as C-in-C of the Australian Military Forces Blamey also would have direct access to the Prime Minister on matters of broad military policy.

Thus the whole high command structure was revised. Whereas previously the Chiefs of Staff had been the government's principal advisers on strategy, they were now replaced by a foreign general. Furthermore, individually, if not collectively, the Chiefs of Staff previously had been responsible for the operations of the forces in defence of Australia. Now this responsibility rested with MacArthur. It is true that Curtin was still able to receive advice from the Australian Chiefs of Staff, as well as Blamey, but the Prime Minister, supported by Shedden, looked to MacArthur as the main source of advice. It might be argued that Curtin had no option but to rely on MacArthur because only the latter had access to information from the Combined Chiefs of Staff through the Joint Chiefs of Staff. (The Combined Chiefs of Staff consisted of the British and American Chiefs of Staff meeting together. The Combined Chiefs delegated responsibility for the South West Pacific Area to the American Joint Chiefs of Staff.)

This problem would have been overcome if Australia had been represented on the Combined Chiefs of Staff, but that did not prove possible. Nevertheless, it is difficult to see why Blamey, in a capacity such as Chairman of the Australian Chiefs of Staff, could not have attended the Prime Minister's War Conference on a regular basis. Curtin would then have had the benefit of Blamey's military experience in his negotiations with MacArthur, rather than having to rely solely on Shedden. The explanation may well be that Shedden preferred Blamey to be excluded, thus strengthening his own position.

Yet from April 1942 until the end of the war Blamey fulfilled a vital role as the government's chief Australian military adviser. Given the weakness and inexperience of the government, Blamey realised that if he was to provide some degree of a check to the almost complete power of MacArthur he needed to maintain his dual positions of C-in-C of the AMF and Commander, Allied Land Forces. By virtue of his role as Commander-in-Chief, Blamey had direct access to the Prime Minister, even if he was not present at the vital discussions between MacArthur and Curtin. The other Chiefs of Staff could only present their views either through their respective ministers or when asked at the Advisory War Council. By virtue of his role as Commander, Allied Land Forces, Blamey had more direct access to MacArthur's strategic plans than would have been possible had he been merely the commander of the Australian Army's operational forces or the CGS. The importance of Blamey's dual role was heightened by the Prime Minister's lack of military knowledge.

For most of the war Blamey enjoyed close relations with Curtin and he offered sound advice on a wide range of military matters. But Blamey knew that if his advice was contrary to that offered by MacArthur the latter would prevail. Moreover, he could not rely on Shedden's assistance. When there was a conflict in views, for example over Blamey's role as Commander, Allied Land Forces, Shedden did not seek an independent solution but deferred to MacArthur. And when Blamey complained to Curtin in mid-1943 of excessive American control over Australian supplies, equipment and services, he received little assistance from the Prime Minister, who readily accepted MacArthur's explanations.[32]

In April 1944 Curtin journeyed to Washington and London to seek approval for Australian strategic plans, and he was accompanied by Shedden and Blamey. This journey marked a decline in Blamey's influence as chief military adviser. Whereas previously Curtin had dealt with Blamey on a professional basis, they were now thrown together socially. Curtin, a teetotaller who in earlier years had had a drinking problem, was not impressed by Blamey's rowdy parties on board the supposedly 'dry' ship which took them to America. After the overseas trip Shedden commented about Blamey: 'Though good as a Commander-in-Chief, he is not suitable as a member of a Prime Minister's party'.[33]

While in London Blamey became attracted to the possibility of a separate British–Australian offensive into the Netherlands East Indies, which he hoped to command. Curtin preferred to retain the existing command organisation in the South West Pacific Area. When he returned to Australia Blamey threw much of his resources into the plans and preparations for the new British command. Furthermore, it seems that he told MacArthur that the Australian assault divisions would not be ready for some time, thus denying them to MacArthur and leaving them available for the British Commonwealth force. This was the opening which MacArthur wanted, for he now went to Curtin and privately advised him of Blamey's disloyalty to the present command arrangement in the South West Pacific Area.[34] Since Blamey had said that the assault divisions could not be ready in time MacArthur then said that it would no longer be possible to use them in the Philippines invasion, and in the following months he used all means available to him to ensure that the proposed British operation did not come off.

Curtin pleaded with MacArthur to make one last attempt to include the Australians in the Philippines campaign, and in due course

*28* Blamey with Lieutenant-General Walter Krueger of the US 6th Army, Leyte, Philippines, 14 December 1944. (AWM Negative No. 17892)

MacArthur produced a plan which was highly unacceptable to Blamey. It involved splitting the 1st Australian Corps and equipping the Australian divisions with American arms and equipment. Blamey objected and another operation, scheduled to land at Aparri in Northern Luzon in January 1945 and involving the Australian Corps, was concocted, but it never came to pass.

When the Allied Conference at Quebec in September 1944 decided that the British contribution to the Pacific would be the British Pacific Fleet, Blamey's plan at last collapsed. But by this time American troops had been assigned to their Philippine tasks. The Australians were now kept on tenterhooks for the next five months before they received their orders for the Borneo campaign.

During the nine months before the American landing at Leyte on 20 October 1944, Blamey had made strenuous efforts to influence Allied strategy in the South West Pacific Area. It appears that he miscalculated severely in pressing for the use of British forces. But although this proposal might have given MacArthur an excuse to exclude the Australians from the Philippines, the evidence shows that MacArthur's staff had determined to exclude the Australians anyway. MacArthur's personal views are more difficult to pin down, but it seems that he was not willing to go out of his way to ensure Australian participation.

It seems that MacArthur misdirected Curtin in March 1944 when he said the Australians would be involved in the Philippines, for his plans at that time show no evidence that this was his intention. Blamey, too, was misled by MacArthur about his plans to include the Australians. Yet even when Curtin found that MacArthur's claim that the Australians were not ready was false, he applied no pressure to the Americans.

This incident also reveals something of the decision-making machinery in Australia, for Curtin and Blamey were shown to have been working at cross purposes. Blamey attempted to play a lone hand in controlling strategic policy. He kept his plans to himself, telling few of his senior staff. One reason for this attitude was so that the CGS, Northcott, and the Adjutant-General, Major-General C.E.M. Lloyd, would not inadvertently reveal the plans to the Minister for the Army, Forde.[35]

While both Curtin and Blamey were concerned to preserve Australian national interests they differed markedly on how this aim was to be pursued. Lacking military experience Curtin placed his faith in MacArthur and was determined to maintain the command relationship established in early 1942. Blamey saw clearly that MacArthur was less concerned with Australian and more concerned with American and his own interests. In May 1945 the Australian Defence Committee prepared a list of subjects concerning the development of the Australian defence forces which they suggested MacArthur should be asked to advise upon. In Blamey's opinion these affairs had nothing to do with MacArthur, and he wrote that it was 'entirely wrong in principle that any foreign officer should be invited to advise upon matters which are entirely Australian'.[36]

In 1942 Blamey had made it clear that he would resign as C-in-C at the end of the war, and in September 1945, after representing Australia at the surrender ceremony in Tokyo, he offered to resign. Meanwhile he advised that rather than reintroduce a Military Board the government should establish an Army Council comprising the Minister as Chairman, the GOCs of Eastern and Southern Commands, the CGS, a militia representative, the Secretary of the Department of Army and the Chief Finance Officer. For a while the government chose to retain Blamey as C-in-C informing him in early November that the complexity of problems confronting the Army made it desirable for him to remain on the job. Then in mid-November Forde advised Blamey that Sturdee was to become Acting C-in-C from 1 December until the Military Board could be reformed.

This peremptory dismissal of the government's top military adviser without any recognition or reward for his service shows the depth of feeling he had aroused in some quarters of the Labor Party. Clearly many members of the government were anxious to dismantle the wartime high command organisation and the first step was to remove Blamey. Other advisers, such as Shedden, were anxious to strengthen the role of the Department of Defence and Blamey had some unacceptable ideas.

As Commander of the Allied Land Forces and as the government's chief military adviser Blamey suffered such serious frustrations and disappointments as would have broken most men. His strength of personality and determination showed that he was well equipped for the task of safeguarding Australian interests. He had strong views on the need to maintain Australian sovereignty, and unlike some Australian generals, realised that wars are fought for political purposes. Blamey's critics have

29 General Sir Thomas Blamey, Commander-in-Chief, Australian Military Forces. (AWM Negative No. 107532)

assigned personal motives to all of his actions. To them he was a self-seeking, devious manipulator who cared little for Australian lives and who struggled to retain his powerful position and to fuel his own ego. But to many others he was Australia's greatest general. To them he revealed a deep experience of military and political affairs and proved a wise and forceful administrator. He fought relentlessly to maintain Australian independence in military matters and he had a genuine concern for the welfare of his troops. Without his efforts MacArthur would have more easily disregarded Australia's wishes.

The most credible evaluation of Blamey's character lies somewhere between these two views, probably closer to the second view than the first. He walked a tightrope between maintaining his own position and protecting Australian interests, between risking his own replacement and risking the distrust of his subordinates. He made few concessions to his critics. He advanced his own point of view ruthlessly—a course which, like MacArthur, he saw as one identical with the best interests of his Army and nation.

Blamey's record contains no outstanding peak, rather it is marked by year upon year of wise decisions, stubborn determination to further the interests of Australia, and a deep concern for the well-being of his soldiers. He had many weaknesses, but greater strengths. He was Australia's senior soldier for the full period of the Second World War during which time the Army fought with skill and bravery in a score of campaigns. His appointment as field marshal in 1950 was justified recognition of his achievements.

# Lieutenant-General Sir Sydney Rowell: Dismissal of a Corps Commander

### D.M. HORNER

Lieutenant-General Sir Sydney Rowell (1894-1975) is the only Australian corps commander to have been dismissed from his command. Yet just over three years later he was returned to the rank of lieutenant-general as Vice Chief of the General Staff, and after four more years, on his appointment as CGS, he reached the most senior position in the Australian Army. Rowell's dismissal had repercussions throughout the Army, and the incident is a focal point for a wide range of problems affecting Australian security. These problems include Australia's lack of preparedness for war, the split between the volunteer AIF and the compulsorily enlisted militia, political faint-heartedness and inexperience of military and strategic affairs, and Australian subservience to the United States.

There has never been any doubt about Rowell's professional ability, intellectual qualities or strength of character. Yet the question of whether he failed as a commander in New Guinea, as General Blamey later claimed, or rather was the victim of the coming together of long-term national security weaknesses, as well as other influences beyond his control, has never been and perhaps never will be determined to the satisfaction of all involved.

*Inexperience in command*

When Rowell and his staff of Headquarters 1st Australian Corps arrived in Port Moresby on 11 August 1942 they were faced with a crucially important task. As early as March Rowell, as DCGS, had discerned that if the Japanese were to disrupt Australia's communications with the United States then they would need to seize Port Moresby. In May 1942 the Japanese invasion convoy bound for Port Moresby had been forced to turn back after the Battle of the Coral Sea. Then on 22 July Japanese forces had landed at Buna on the north Papuan coast. By 13 August the Japanese had advanced beyond Kokoda along the Kokoda Track to Isurava where their advanced units waited for the remainder of the force to join them. Rowell had thus taken over a campaign which had already begun, and in the six weeks of his command in New Guinea he was to face a series of crises which would have tested the most experienced of commanders.

Yet for all his fine qualities, Rowell lacked experience as a commander; indeed the corps command to which he had been appointed in

April 1942 was his first since he had commanded a light horse troop and, for a short while, a squadron at Gallipoli some 27 years before. He had been a member of the first class to graduate from Duntroon in 1914 and after service at Gallipoli he had been invalided home to Australia. He had missed, therefore, the opportunities for command and battle experience at senior levels that had come to his contemporaries in France and Palestine. During the heart-breaking inter-war period he had devoted himself to professional study, had attended the Staff College at Camberley and the Imperial Defence College, and had developed into one of the outstanding staff officers in the Army.

At the beginning of the Second World War Rowell had been selected as GSO1 (chief operational staff officer) of the 6th Division, and had then progressed to BGS (chief operational staff officer) of the 1st Australian Corps in the Middle East, where he had performed with distinction in the campaigns in Greece and Syria. But the appointments brought him into close contact with General Blamey, the GOC of the 6th Division and then of the 1st Australian Corps until he became Deputy Commander-in-Chief of the Middle East in late April 1941. Rowell conceded that Blamey had performed well before the campaign in Greece, but after the campaign relations became sour, and he wrote to the CGS that he would never again serve in the field under Blamey, whom he thought lacked both moral and physical courage. To another officer he wrote: 'we are now not tuned to the same wave-length and are never likely to be'.[1] Events were to prove him right.

Despite Rowell's disagreement with Blamey the latter still found much to praise in his BGS and he wrote to the CGS:

> Rowell has very great ability; is quick in decision and sound in judgement. There can be no question of his personal courage, but he lacks the reserves of nervous energy over a period of long strain. I found him difficult in the last few days in Greece and, as Commander, had to exercise considerable tact. Rather a reverse of what it should be.[2]

Rowell had disagreed with Blamey over the composition of the small group of officers who had accompanied the commander when, on orders, he had left his corps and had flown out of Greece. Blamey had included his son in the party of seven men.

Rowell was not required to continue as Blamey's chief of staff for the latter was appointed Deputy C-in-C Middle East, and Lavarack commanded the 1st Corps in Syria. Then, in August 1941, after the successful conclusion of the Syrian campaign, Rowell was recalled to Australia at the urgent request of the CGS, Sturdee, to become DCGS. This appointment probably came as a disappointment, for as Brigadier Vasey observed: 'Isn't it just too awful. I consider it a tragedy for the AIF and for Syd too. As you know he didn't see much of the last show [First World War] and here he is being defeated this time.' But Rowell had a vital role to play, and for seven months he and Sturdee struggled to improve the defences of Australia.[3]

When, in March 1942, the government decided to reorganize the

30 Lieutenant-General S.F. Rowell, GOC 1st Australian Corps, soon after his arrival in New Guinea, August 1942. Laverack held him in 'the greatest esteem for his strength of character, intelligence, judgment and loyalty' (AWM Negative No. 13174)

Army, the Minister for the Army, Forde, asked the Acting Commander-in-Chief, Lavarack, for the names of officers suitable for promotion to senior positions. Lavarack placed Rowell at the top of his list and wrote to Forde:

> This officer has had no experience in actual command during this war, mainly because his services as a staff officer, to both Blamey and myself, have been indispensable. I should have no doubt myself about taking him either for command or high staff work, as I hold him in the greatest esteem for his strength of character, intelligence, judgement and loyalty. I should be glad to accept him in any capacity including a Corps Command or Lieutenant General on the General Staff.[4]

Obviously Blamey, when he returned to Australia as Commander-in-

Chief later in the month, agreed with Lavarack, for Rowell was soon afterwards promoted to command the 1st Australian Corps, which by this time had returned to Australia from the Middle East.

Proud, very austere and sensitive, Rowell was highly principled to the degree that one senior officer remarked, 'the trouble with Syd is that he expects everyone to act like a saint'. Throughout the Army he was respected for his integrity and competence. Aged 47, he was at the peak of his ability. Less than three years before he had been a lieutenant-colonel. Now, as a lieutenant-general and a corps commander the campaign in New Guinea was to be the first real test of his leadership.

*Early problems*

From the beginning, before Rowell arrived, the Papuan campaign had been shaped by inaccurate strategic assessments, for despite accurate signals intelligence, MacArthur's staff at General Headquarters (GHQ) in Brisbane did not believe that the Japanese would advance over the Owen Stanley Ranges. Indeed MacArthur had spoken confidently of seizing positions around Buna before the Japanese arrived. The Japanese landings sent a shiver of apprehension through the War Cabinet, but MacArthur and Blamey reassured the politicians that 'a satisfactory defensive position had been established'. MacArthur's communique of 6 August confirmed that the New Guinea defence line 'was along the almost impassable Owen Stanley Range'.[5]

From his headquarters near Port Moresby Rowell was already facing considerable problems. At Milne Bay the garrison consisted of one militia and one AIF brigade (from the 7th Division) all under his Duntroon classmate, Major-General Cyril Clowes. Around Wau there was a force of commandos, and at Port Moresby he had two partly trained militia brigades, of which the 39th Battalion already in action on the Kokoda Track, was part. Soon another brigade of Major-General Allen's 7th Division (AIF) would arrive.

His chief initial problem was one of logistics. Despite the fact that he had warned the US Air Force commander of the danger of an air attack, on 17 August all five available transport planes were destroyed or damaged by a Japanese raid. At the same time Brigadier Potts, commanding the 21st Brigade on the Kokoda Track, informed him that instead of the 20 days' supplies he thought were available at the forward base of Myola, there were in fact only two days' for two battalions forward of Myola, and five days' at Myola. Offensive operations, therefore, had to be deferred. Already Rowell was feeling the pressure of disapproval from Brisbane, and he later wrote to Blamey's chief of staff, Vasey, that he 'quite expected to be recalled after my hard luck signal about the supply situation on the Ranges'.[6] In Brisbane both MacArthur and Blamey had failed to grasp the magnitude of the problems faced by Rowell.

Even before the Japanese landing at Milne Bay in late August, GHQ had demonstrated a lamentable lack of understanding of the situation in New Guinea. For example, on 13 August GHQ suggested that the 'pass' through the Owen Stanley Ranges could be blocked by demolitions. Rowell replied that since parts of the track already had to be negotiated on hands and knees, the explosives which would have to be man-handled up the track could hardly block it. GHQ believed that a Japanese overland operation against Port Moresby should be 'discounted in view of the logistical difficulties of maintaining any force in strength' on the Kokoda Track.[7] But at the same time GHQ was urging Rowell, with similar logistic problems, to advance over the mountains. Thus Rowell's performance as a commander must be seen in the context of the lack of understanding displayed by GHQ in Brisbane and the government in Canberra. Furthermore, as the principal area of fighting in the South West Pacific Area, the campaign in Papua became a focal point for the tensions in the inter-allied command structure.

## Milne Bay

These early misunderstandings were to pale beside the lack of grasp shown by GHQ following the Japanese landing at Milne Bay on the night of 25-26 August. Rowell found himself caught between a series of hysterical orders from Brisbane, and his firm belief that Clowes, commanding Milne Force, was best placed to conduct the battle. Before the end of 26 August, Vasey signalled Rowell to pass on MacArthur's instructions to attack the Japanese landing and destroy it as soon as possible. Rowell realised that Clowes, reliable and unperturbed, was in the best position to judge the necessity for action, and did not pass on this signal to Milne Force. Rather, he sent another message to Clowes: 'Confident you have situation well in hand and will administer stern punishment'.[8]

Although Blamey was confident that Clowes would be able to defeat the Japanese, MacArthur was becoming increasingly apprehensive. His original strategy had been to establish bases in New Guinea from which to strike at Rabaul. He had now to alter this strategy. At a conference that evening his chief of staff, Sutherland, declared that if the Japanese succeeded in getting a footing and then attempted to exploit it, General MacArthur's intention was to send additional troops to Milne Bay and to 'fight it out in New Guinea'.[9]

This comment explains why the 25th Brigade, the remaining brigade of the 7th Division, had not been sent earlier to New Guinea. Until now MacArthur had not been convinced that he would have to fight a major campaign in Papua. His mind had been already leaping forward to his spectacular island-hopping campaign, overlooking the possibility that the Japanese might attempt to wrest control of his bases from him. Despite the fact that his naval adviser said that he had insufficient ships and no escorts, it was decided eventually that the 25th Brigade would be sent to New Guinea.

For the next few days Rowell was bombarded with demands for action and information. Eventually, after one signal Vasey wrote privately to Rowell:

> If you took a lousy view of our message about offensive action, I hate to think of your reactions to my letter of this morning... It boils down to the question of who is commanding the army—MacArthur or T.A.B. [Blamey] and it seems the sooner that is settled the better...
>
> I am now awaiting the result of Cyril's activities yesterday. I'm dying to go to these bastards and say I told you so, we've killed the bloody lot.

Vasey then added a postscript: '...this is a war of nerves isn't it. Let's give it to our own people too.'[10]

MacArthur had already started to look for scapegoats and on 30 August sent a long message to General Marshall in Washington:

> This is the first test of Australian troops under my command. The Australians claim that the commander is excellent and rate half his troops as good. The other half from the 7th Australian Division they rate excellent... With good troops under first class leadership I would view the situation with confidence unless enemy reinforcements are landed but, as I have previously indicated, I am not yet convinced of the efficiency of the Australian troops, and do not attempt to forecast results.[11]

It was, in effect, the first step in shifting any blame for a disaster to either the 'inefficient' Australians or the 'short-sighted' Joint Chiefs of Staff.

MacArthur's worries were now placing increasing strain on Rowell and on 28 August he wrote to Vasey:

> I'm personally very bitter over the criticism from a distance and I think it damned unfair to pillory any commander without any knowledge of the conditions...
>
> I suppose there will be heresy hunts and bowler hats soon. I hate to think what would have happened with our Allies in charge up here.

Two days later he followed with another note:

> I realise the position down there and I know how you are faced [sic]. I do hope that there is a showdown between Blamey and MacArthur. Taking it by and large, we do know something about war after three small campaigns; it is idle to assume that we can clean up damned good troops (as the Japs are) in five minutes.
>
> Sorry I was a bit abrupt but I've got three wars [that is, Milne Bay, Kokoda and Wau] on my hands now. (I was going to say four, but that would be uncharitable.)[12]

These letters indicate the complexity of the pressures affecting the commanders. Blamey and MacArthur were subject to more diverse influences than Rowell in New Guinea. Difficult though Rowell's situation was, he had to satisfy only one master—Blamey. So long as the correct chain of command was maintained, he should have been shielded from the direct influence of politicians, public opinion, arguments over Allied cooperation and the fears and exhortations of MacArthur.

There is an immense strain imposed on army commanders at the political–military interface, but there is another less obvious interface—that between the generals who have to balance political and military decisions, and those who merely have to follow military instructions and whose problems are those of execution. Rowell and his subordinates in New Guinea were in the latter category. It was important that Rowell recognise the different pressures borne by Blamey. His failure to do so may have been a result of Blamey's unwillingness to explain his difficulties to Rowell. This delicate task was left to Vasey, who could explain his own problems, but not those of Blamey. Rowell was undoubtedly well equipped to understand Blamey's difficulties for he had been Blamey's chief of staff. Yet at no time did Blamey take his main operational commander into his confidence. The result was that Rowell felt that he was fighting 'four wars'.

*Kokoda Track*

Rowell's confidence in Clowes was well placed, and by the end of August it was clear that the Japanese landing had been soundly defeated, but the situation on the Kokoda Track was now causing real concern. The Japanese had begun their offensive along the Track at the same time as the Milne Bay landing and they were slowly driving the Australians back. The Japanese advanced remorselessly for the next fortnight as Rowell struggled to move sufficient forces forward to hold them. Brigadier Potts of the 21st AIF Brigade had assumed command of the militia battalions originally sent forward. Then, on 8 September, Rowell relieved Potts of command of the brigade and replaced him with Brigadier Porter.

The recall of the brave, solid, inspiring Potts underlines a serious fault in command, for neither Allen nor Rowell had been able personally to inspect the conditions in the mountains. No man could have done more than Rowell in the busy fortnight between the time he took over and the Japanese attacks. He was desperately short of transport aircraft, he lacked much of his signal equipment, the supply situation in the mountains was serious, and the threat of a Japanese seaborne landing was real. Under these circumstances he had little time to spare to inspect his forward troops. Furthermore, to inspect Potts's brigade, forward near Isurava, would have meant an exhausting week's journey by foot. During this time the Japanese could attack either at Milne Bay, Isurava, around Wau or by sea against Moresby. If he went forward, Rowell's only communications would be by a doubtful radio or by the uncertain telephone wire that wound its way precariously along the sides of mountains and across turbulent streams.

Allen, who was concerned at the poor state of training of the militia with which he had to defend Moresby, found it equally difficult to spare the six or seven days needed to visit Potts when he was at Isurava. Later Allen wrote that:

> either the G1 [Colonel C.C.F. Spry] or myself would have gone forward before if I had not the added responsibility of the defence of Moresby area from a sea attack. The first I [Intelligence] report received was an even greater menace (if possible) than the attack over the mountains... When given the dual roles I expressed doubts to [Rowell] but at the time there appeared no alternative.

In fact Spry had volunteered to go forward to give moral support to Potts and to report to Allen. Allen doubted the value of this move and referred it to Rowell, who supported his divisional commander.[13] Later

*31* Lieutenant-General S.F. Rowell (right) with his Brigadier, General Staff, Brigadier H.G. Rourke, at HQ New Guinea Force, September 1942. (AWM Negative No. 26709)

Rowell reinstated Potts, but he was relieved again the following month after Herring had taken over from Rowell.

Rowell was now faced with his most critical and trying test. On 8 September the Japanese had attacked and isolated the headquarters of the 21st Brigade, which was fighting desperately to regroup south of Efogi, about 45 miles (72 kilometres) from Moresby. The 25th Brigade had not yet arrived in New Guinea and the 21st Brigade now had a strength of less than half a battalion. That evening Rowell wrote to Vasey: 'Today has been my blackest since we came, and none of the 28 days I've spent here has been free from worry'. Yet, depressed though he may have been, he wrote later, 'at no time did I consider that the capture of Moresby by the enemy from the north was possible'.[14]

Despite the tremendous strain, Rowell still managed to remain calm and plan for victory. MacArthur's intelligence chief, Brigadier-General Willoughby, observed at the time: 'Rowell is of dignified but amiable personality. The staff regards him with respect and affection; that feeling prevails through other command echelons.' Rowell's chief medical officer, Brigadier W.W.S. Johnston, wrote that he was 'amazed at his equanimity and unruffled demeanour in conditions which imposed terrific strain'. Rowell was still convinced that his decisions had been correct and on the evening of 8 September wrote: 'I'm confident the enemy has no hope of getting Moresby from the north. His difficulties will now start and I trust we can get him on the rebound. The AAF [US Army Air Force], gets into a tremendous panic and its on their account I'm worried. They'll probably pack up and clear out.'[15]

Lieutenant-Colonel R.R. Vial, Rowell's chief intelligence officer, identified his commander's three main preoccupations at this time in this order—the problems with Advanced Land Headquarters, keeping the Americans on side and lastly, the Japanese.[16] Rowell knew that he could hold the Japanese provided he controlled the air, and hence the sea, around Port Moresby. He had good relations with the US Air Force commander in Moresby, Brigadier-General Whitehead, but he had to reassure him constantly to prevent his moving his planes to Australia.

On the morning of 9 September, as Potts regrouped his brigade at Menari [about 35 miles (56 kilometres) by air from Moresby] before handing over to Porter, Rowell freed Allen of responsibility for the sea approaches and took over the added burden himself. That same morning the troops of the 25th Brigade arrived and the 16th Brigade was ordered to New Guinea.

Determined to instil his own confidence in his troops, Rowell ordered that demolitions were to be removed from equipment and public utilities. Until that time the policy had been the total destruction of everything liable to fall into the hands of the enemy. Rowell, who had been given authority to order demolitions as he considered necessary, pointed out that this policy was 'totally unsuitable' as it would presuppose an admission that the New Guinea Force would inevitably withdraw to the hills and, when the reserves of food were exhausted, would surrender. 'This is by no means the intention', he wrote. 'While it is not possible to guarantee that the enemy will not make an initial penetration, it is intended to drive him back by offensive action once the position has been stabilized.' The plan was for each locality to hold reserves of food and ammunition and if the Japanese attacked, the 'guns must be fought to the muzzle'.[17]

On the surface the situation looked even worse than a few days before, for by the evening of 10-11 September Porter and his scratch force was back at Ioribaiwa Ridge. However, the enemy force, now reduced by disease, had become exhausted by their hard fighting with the 21st

Brigade. They lacked food and ammunition, for General Kenney's planes raided daily their precarious supply route, and Porter was able to hold them until, on 14 September, Brigadier Eather arrived with the 25th Brigade. At last, imperceptibly at first, the initiative was returning to the Australians.

Meanwhile, Blamey had visited Rowell and Allen, and as Allen observed, 'all seemed pleased with the situation'.[18] But no sooner had Eather assumed command than he decided, in the face of Japanese pressure, to withdraw to Imita Ridge. Rowell endorsed Eather's decision but set down his views in uncompromising terms:

> Stress the fact that however many troops the enemy has, they must all have walked from Buna. We are now so far back that any further withdrawal is out of [the] question and Eather must fight it out at all costs. I am playing for time until 16 Infantry Brigade arrives.[19]

At the same time Rowell moved his last available troops, the militiamen of the 14th Brigade, to Hombrom Bluff, in the Owen Stanley foothills outside Port Moresby, as a final backstop. With professional caution, he had prepared all units in the Moresby area to fight if necessary, but his order to Allen leaves no doubt that he intended that there should be no further withdrawal.

The Japanese were at the end of their tether. They had been sent forward through the mountains with the minimum of supplies, relying on the tenuous support of a long line of porters, supplemented by what they could forage or capture. Now the Allied air forces exacted their toll, bombing the Wairopi bridge, and dispersing the carriers. By September the front-line ration was down to less than a cupful of rice per day. Then, as a result of events on Guadalcanal, orders were issued on 18 September to defend the beach-head at Buna; they therefore ceased their attacks and by 24 September the main body of the force had begun to withdraw.

From 16 to 23 September 1942 Eather's troops patrolled the area between Imita Ridge and Ioribaiwa while for the first time in the campaign Australian artillery pounded the enemy. Now the Japanese faced the same problems as those faced earlier by Potts. Their supply route was long while the Australians were one day's march from their forward depot. On 21 September the 16th Brigade arrived in Port Moresby, and two days later Eather's 25th Brigade began to edge forward. Rowell could be excused for believing that his most trying time was over, and he wrote happily to Clowes: 'I feel we are now over the worst of our troubles'.[20]

*Blamey is ordered to New Guinea*

Rowell's troubles, however, were only just beginning. Although the Allied air commander, General Kenney, had met Rowell only once (in Brisbane on 6 August), MacArthur was inclined to believe his second-hand reports that Rowell had become defeatist and that Moresby would be lost 'if something did not happen soon'. MacArthur was also deeply worried over the withdrawal of the 25th Brigade to Imita Ridge. Blamey did not share that concern, and while MacArthur paced the carpet of his office in Brisbane, he reported to the Advisory War Council on 17 September that he was 'confident of success'.[21] Nonetheless, many ministers remained fearful about the safety of Port Moresby.

It must have come as a shock to Curtin when that night MacArthur spoke to him by secraphone. He told Curtin that he was worried about the situation in New Guinea where, despite superior numbers and no reports of casualties, the Australians were still withdrawing. MacArthur had finally become convinced that 'the retrogressive nature of the tactics of Australian Ground Forces defending Port Moresby seriously threatened [the] outlying airfields'. He felt that if the Japanese advance continued, the Allies would be forced into such a defensive concentration in New Guinea as would duplicate the conditions of Malaya.[22]

MacArthur pointed out that the Australians were confident of their ability to meet the situation but that he was so far from sharing that confidence that he proposed sending American troops to stem the attack. This was at variance with the reason that Blamey had given the Advisory War Council, that MacArthur was sending a US Regiment to New Guinea so that they 'should obtain experience in operations and in the development of supply arrangements in this area'.[23]

MacArthur told Curtin that he felt that Blamey should go to Moresby to take command personally, 'not only to energize the situation, but to save himself, because, in the event of the situation in New Guinea becoming really serious, it would be difficult for General Blamey to meet his responsibility to the Australian public'. This argument may have decided Curtin, for he later confessed that, 'In my ignorance [of military matters], I thought that the Commander-in-Chief should be in New Guinea'.[24] He therefore agreed with MacArthur, and promised to speak to Blamey. As Blamey recalled: 'One night I got a ring from Mr Curtin, who said he had been talking things over with General MacArthur and they thought I should go up to New Guinea to take command, as things had not been going very well there. I raised no question.'[25]

John Hetherington, in *Blamey: Controversial Soldier*, has described the scene at Victoria Barracks, Melbourne, the following day:

> "I'm leaving for New Guinea in a few days", Blamey told Burston, his Director-General of Medical Services, on 18 September.
> "Why" Burston asked. "Are you worried about New Guinea?"
> "No", said Blamey, "but Canberra's lost it!... I think highly of Rowell and I'm satisfied he has the situation under control but I feel I must go."[26]

If Blamey felt as strongly about not going to New Guinea as indicated by Hetherington, it is strange that in his talk with Curtin he 'raised no question'. Indeed on 20 September Blamey wrote to Rowell and explained that:

> The powers that be have determined that I shall myself go to New Guinea for a while and operate from there. I do not, however, propose to transfer many of Adv HQ staff, and will arrive by aeroplane Wednesday evening... I hope you will not be upset at this decision, and will not think that it implies any lack of confidence in yourself. I think it arises out of the fact that we have very inexperienced politicians who are inclined to panic on every possible occasion,

and I think the relationship between us personally is such that we can make the arrangement work without any difficulty.[27]

This letter indicates the precarious nature of Blamey's position. After the Advisory War Council meeting on 17 September he must have been aware of the hostile atmosphere among some Cabinet ministers. MacArthur had already replaced his naval and air commanders and Blamey knew that the government respected the American commander sufficiently to follow his advice if he proposed to change the Land Force commander. Although his naval and air commanders were Americans, they were filling Allied appointments, just as Blamey was.

William Dunstan, VC, General Manager of the Melbourne *Herald*, later wrote to Rowell and explained that, at two background conferences for senior newspapermen, Curtin had discussed Blamey's position (see Chapter 11, p.000). When Blamey journeyed to New Guinea to assume command he was, in essence, fighting for his military life. A pragmatic politician, he was jealous of his own position and believed that he was the best man for the job. Indeed he had grave doubts about the suitability of some of the possible successors, like Lavarack and Bennett. A ruthless man when his own interests were at stake, it was not likely that Blamey would tolerate opposition to his plans.

On 21 September Rowell received Blamey's letter warning him of his arrival and was deeply upset. As he saw it, the issue was quite straightforward; Blamey would have to make a clear statement that he had confidence in his commander in New Guinea, and there would have to be an equally clear division of responsibilities. If this were not the case, then Rowell felt that a 'showdown' was inevitable and he would have to go. The next morning Rowell wrote to Clowes:

> The plain fact is that he [Blamey] hasn't enough moral courage to fight the Cabinet on an issue of confidence in me. Either I am fit to command the show or I am not. If the latter, then I should be pulled out. He comes here when the tide is on the turn and all is likely to be well. He cannot influence the local situation in any way, but he will get the kudos and it will be said, rather pityingly, that he came here to hold my hand and bolster me up. Shades of Greece in April 41!!

There is no doubt that Rowell was already pessimistic about the outcome of Blamey's arrival. He concluded to Clowes: 'Once I've ironed out the difficult position between TAB and myself I'll come down [to Milne Bay] if only to say goodbye'.[28] That evening Blamey arrived in Port Moresby.

*Tension in Port Moresby*

The events in Port Moresby between 23 and 28 September 1942 which resulted in Blamey sending Rowell home to Australia have been described in detail in Hetherington's *Blamey: Controversial Soldier* and in McCarthy's *South West Pacific Area: the First Year*. Their accounts of the day-by-day arguments and discussions are complementary, and are not greatly at variance with that of Rowell himself in *Full Circle*. The incident marks the climax of the ever-increasing pressure which had been applied to both Rowell and Blamey by the 'inexperienced politicians' and by MacArthur and the staff of GHQ.

On the evening of 23 September Rowell and Blamey had a long, frank, and 'at times acrimonious' discussion about a suitable working arrangement. Rowell had no wish to become merely Blamey's chief of staff, and submitted that Blamey should establish an Army Headquarters in Moresby to control all New Guinea operations, including Milne Bay and Wau, leaving him to concentrate on the operations in the Owen Stanleys and the defence of Port Moresby. Blamey would not agree. He had no intention of remaining in New Guinea and he saw no need to bring forward his staff. Furthermore, transport was in short supply and would have been further strained having to bring almost 500 personnel from Australia. Hetherington says that Blamey decided that such a headquarters 'would be redundant when the paramount need was to stop the Japanese advance on Port Moresby and turn the enemy back'.[29]

Blamey's arguments overlooked the fact that the Japanese offensive had already ceased. With the arrival of two regiments of the 32nd US Division it was obvious that unless the Japanese were greatly reinforced, the coming battles would be of an offensive nature with ten brigades, as well as independent companies and army troops, deployed in Papua. Rowell was undoubtedly right in his appreciation of the need for an Army Headquarters, and before the end of November Advanced HQ NGF was to take over the responsibility for operations around Buna, Gona and Sanananda. Rowell suggested to Blamey that the Headquarters of the First Army, under Lavarack, should be brought forward, but Blamey replied, 'to do that would be to bring in a commander I don't want'. Rowell added that if an Australian Army Headquarters was not set up in New Guinea, then he was sure that a US Headquarters would be. These were not last-minute thoughts by Rowell to save his own position. Ten days earlier he had written to Clowes and complained that his command was becoming too big: 'I can't compete with the local problem and do justice to you and Kanga [Force at Wau] as well'.[30]

Acting on a suggestion from MacArthur, on the morning of 25 September Blamey flew to Milne Bay and ordered Clowes to send a force by air to Wanigela, north-east of Milne Bay. Kenney recorded that Blamey told him that he had to '*order* General Clowes to provide the troops from Milne Bay. Clowes didn't approve of this method at all.'[31] By the time Blamey returned to Port Moresby, Rowell was already angry as he felt that Blamey had circumvented his authority. Rowell's feelings can be gauged by his letter to Clowes the next morning: 'I fairly rose. I then got off my chest what I've been storing up since April 1941. Told him he'd already dumped me twice and was in process of doing it a third time and so on. In the end he rose, as I hoped he would...'[32] Blamey replied that his words to Clowes had been a suggestion: a fine distinction when the suggestion comes from the Commander-in-Chief.

That night Blamey wrote to MacArthur, and indicated that the defensive phase was over in New Guinea. He explained that he intended

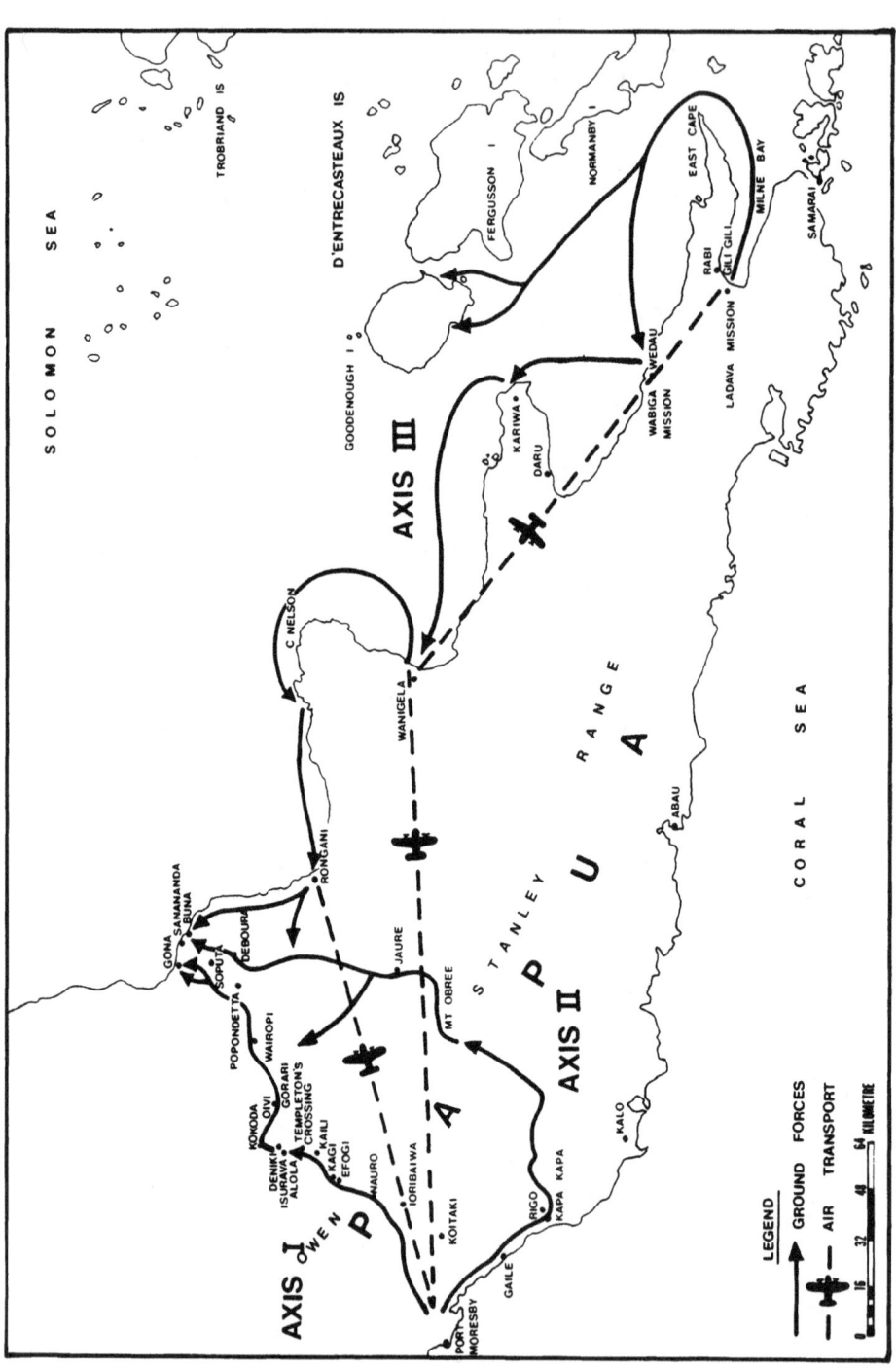

23 Axes of advance, Papuan Campaign, outlined by Blamey in September 1942

to advance on three axes. The first was the present axis of advance via Ioribaiwa to Kokoda. The second was farther east along the track through Jaure where 'a much easier route is available than the Ioribaiwa route', and the third was a sea and land route from Milne Bay. The letter indicates that Blamey was already thinking in terms of an organisation which would demand an army headquarters; that is, the 1st Australian Corps on one axis, and at least a small division on each of the other two axes.[33]

Intent on fighting the Japanese at Ioribaiwa, and protecting himself from the sniping from Australia, Rowell had not realised the change in direction that had come over GHQ between 17 September when MacArthur phoned Curtin, and 23 September when Blamey arrived in New Guinea. Rowell saw the planned movement of one regiment of the 32nd US Division over the Jaure trail as a waste of time. He was not aware of GHQ's intention that all of the division should advance over the trail, and was certain that such an advance could not be supplied. Subsequent events were to prove him correct on both counts.

From Port Moresby it must have seemed to Rowell that MacArthur's reinforcements to New Guinea were merely a stopgap. Initially, they may well have been, but MacArthur always thought in broad, sweeping terms. Blamey realised this, and in his talks with MacArthur on 19 and 22 September, before flying to Port Moresby, they no doubt discussed the plans for using these additional troops in an offensive.

Fighting a battle which changed its course day by day, Rowell never had a real opportunity to plan for an offensive, nor was he informed that he could be sure of logistic support. He has admitted that he never really made detailed plans for the reconquest of Papua,[34] and in his assessment on 20 September he envisaged a period of time when resources would be built up before the Japanese would be pushed back along the Kokoda Track.

Blamey, who was privy to MacArthur's thoughts, saw the situation in a different perspective. He recognised that the immediate task was the security of Port Moresby, but in his subsequent report he declared that:

> the main task was to develop and set in motion plans for an active offensive against the enemy which would drive him out of Papua. Up to the date of my arrival plans for this main task had not been envisaged, the attention of the Commander having been concentrated on holding the enemy in his advance on Moresby and planning to drive him back on the Moresby-Kokoda axis.[35]

What Rowell's critics have misunderstood is that this did not mean a lack of offensive spirit on his part, but a very keen sense of the possibilities in New Guinea. His new BGS, Hopkins, who had come in after his predecessor had had a nervous breakdown, shared his doubts and recognised the problems that Rowell had forecast. 'The idea of sending a US outfit over the Owen Stanley's through almost unknown country seemed crazy to me', wrote Hopkins, 'though I was BGS and had to carry it through. Of course they just disappeared and were lost to us for about 2 months. The Wanigela effort might have done better, but for the immense floods on the Musa River.' Rowell took a realistic view of the campaign. On 14 September he had told Clowes that he was convinced that Buna could not be retaken by just an advance over the Range. Buna could be retaken, however, 'from the sea, or by a move up the coast supported by sea and air'.[36]

*Dismissal*

Meanwhile, the disagreement between Blamey and Rowell had risen to a climax. Vasey, who had joined Rowell in Port Moresby, wrote to his wife on 25 September: 'Albert [Blamey] is over here again—is staying for some weeks I believe. As you can imagine Syd takes a very poor view of it and a day or so ago was threatening to get out. I told him to be cautious.'[37] Three days later Vasey again wrote that he had

> a visit from Syd this morning with dumbfounding news...
> He is to go—in fact he takes this letter with him—relieved of his command. Purely on personal grounds too—not for any failure on his part to command here. The feeling between he [sic] and Albert is really only known to themselves and this is the result. As I said in my previous letter Syd took a very poor view of Albert's arrival and conditions have got steadily worse, until this morning Syd was informed he was to go home...I find it difficult to assess Syd's actions. He may have been precipitate. I feel that without their previous association, the situation should not have been intolerable so quickly.[38]

On 1 October Blamey wrote to Curtin justifying his decision to relieve Rowell. He made various charges against Rowell which, since he had left New Guinea, he had no chance to answer. Dudley McCarthy, the official historian, answered some of these charges and Rowell in *Full Circle* answered the others. Most of them do not stand up to close examination. Rowell concluded that the charges:

> were trumped up. The comment that I was not prepared to cooperate with Blamey may have had some degree of truth initially. But when I had accepted the situation and a modus vivendi had been established it seemed that matters would settle down, and the discussion on 27 September appeared to confirm this.

He then went on to explain that the reason he was dismissed was because Blamey wanted to get rid of a troublesome subordinate and that Rowell offered a challenge to his position. As Major-General C.E.M. Lloyd said, 'This would have happened to anybody. You were getting too close to the throne!'[39]

On 29 September when William Dunstan wrote to Rowell (see p. 213) he explained that, when Curtin had discussed Blamey's position, it had been apparent that 'Canberra has the best possible estimation of you...you would almost certainly be regarded as the next man, and T [Blamey] would know it'. Dunstan continued that there had 'been a canvass of names. You, Northcott, Herring, Morshead'. Lavarack was not considered. Herring 'would rather serve than direct', and 'I can't conceive of either of the other two being considered as against you'.[40] Whether there was any truth in these statements from the manager of a paper owned by the anti-Blamey Sir Keith Murdoch, cannot be ascertained, but Blamey would have been aware of such rumours.

Rowell made much of this motive, yet he overlooked one factor—his temperament. In his book he devoted five lines to this problem, but it was the core of Blamey's complaint—'the personal animus displayed towards me was most unexpected'. Their personalities were like water and oil, and from the time when Blamey stepped onto the tarmac in Port Moresby, Rowell treated his C-in-C with less than the respect which his position demanded. The night before Blamey arrived Rowell wrote to Clowes: 'I really am sick of the whole business and hope I can keep my temper tomorrow'.[41]

In a month of deepening operational crisis Rowell had borne intense pressure from his superiors in Brisbane, and he now found himself ill-equipped to handle the clash of personalities with Blamey. A letter to his GSO1 (Int) written on 28 September revealed his feelings:

> I have had considerable heart searching during the past 48 hours as to whether I should have gone on eating dirt. But, after all, we are fighting for principles. I can hardly accept a position where my self respect is lost.
>
> The fight is by no means over and perhaps I can do something to help cut out an evil cancer in the body of the public.[42]

The same morning he explained the situation to Allen, whose troops were fighting on the Kokoda Track:

> Events moved rapidly to a crisis this morning. I had accepted the situation and was prepared to bite hard for the time being, however difficult it might have been. However, the C-in-C himself decided this morning that the position was untenable and that I was the cause of it through my temperament.
>
> I have tried not to let personal matters get on top in this difficult situation, but I am not able to go beyond a certain point in eating dirt.[43]

However much Blamey wanted to get rid of Rowell for personal reasons, and the evidence certainly suggests that he had much to gain by doing so, Rowell made it easy for him by his attitude. Indeed on 27 September Rowell told Clowes that 'We've had three first class brawls. I would never have believed a senior officer would have taken what I said to him.' Rowell despised Blamey for what he saw as a 'debauched life-style' not befitting a commander.[44] General Willoughby, in New Guinea at the time, has described the incident:

> If Blamey had arrived 5 days later the plans which Rowell and Allen had made would have been working out, and the Japs would have been pushed well back towards Kokoda. But Blamey arrived on the first day of the advance. Rowell was a proud man—I liked him—and he bridled. I saw Rowell myself, and said to him: General Blamey is the C-in-C; he can establish an Adv HQ wherever he pleases—Milne Bay, Merauke, Melbourne or Iron Range. It doesn't rob you of your command. But Rowell—well, I understand there was a background to this—something in Crete [sic]—and words were spoken. You can't call words back.[45]

Brigadier Hopkins endorsed this view. In his opinion, 'Rowell sacked himself'. Allen has also described the incident:

> It was not just that Syd Rowell objected to T.A.B. putting his HQ at Moresby, but he used R's staff, in fact expected R to be his Chief of Staff.

> I urged Rowell to grit his teeth and take it. I said: "He won't stay long". But it was too late.
>
> When Rowell went back from talking to me about it T.A.B. sacked him.[46]

It should not be forgotten that the pressure of the preceding weeks had been a terrible ordeal for Rowell during which time his nerves had been tested to the limit and not found lacking. During this time he was controlling forces in action on three fronts: around Wau, on the Kokoda Track, and at Milne Bay. Japanese bombers were overhead, and there was the constant expectation of a seaborne landing. Yet Rowell never controlled a balanced organisation, for administrative troops outnumbered fighting soldiers and the latter were of variable quality.

There were perilously few transport planes. For a while there were two, then one, and for five days there were none. Moreover, the pilots were inexperienced and the techniques of air dropping had yet to be developed properly. Thus, with the formidable nature of the terrain, Rowell found it extremely difficult to resupply and reinforce his troops near Wau and on the Kokoda Track.

Brigadier Rogers, Blamey's Director of Military Intelligence, said that Blamey told him that there was something lacking in Rowell's HQ. It was not that Rowell had 'given it away', but he could not put his finger on it.[47] Perhaps the answer was simply that Rowell lacked experience as a commander. He could not picture the problems of the men at the front and he was not assisted by his first BGS, who was inadequate for the job. Yet in the long run he never lost the confidence of his commanders and staff, and this is a good test of leadership.

Hanging over his head throughout the period had been the awful knowledge that the loss of Port Moresby would be grievous for Australia. Despite this worry, and the constant pressure from Australia, Rowell had not interfered with Clowes's handling of the battle at Milne Bay, and his troops had halted the Japanese advance at Imita, thus ensuring the security of Port Moresby. As GOC NGF Rowell had been in an invidious position. In his own words: 'Its one of the problems of a detached commander overseas that he has no one to talk to on the same plane'.[48]

Rowell did not fail as a commander in New Guinea, despite Blamey's later claims. But he did fail to establish a working relationship with a man whom he disliked intensely. Perhaps Rowell's equanimity had been lessened by the month of criticism and frustration before Blamey's arrival. Apparently Vasey believed that was the case, for he wrote: 'I rather believe the job here got on his nerves and that in turn made him act as he did in the final stages'.[49]

Rowell also failed to appreciate that the performance of his troops and his reports and requests to Brisbane were part of a delicate political balance involving MacArthur, Blamey, Curtin and indeed even the Joint Chiefs in Washington. He seemed determined to pursue the militarily correct, but nonetheless unrealistic notion, that the campaign could be conducted on a purely military basis. In reality, a commander-in-chief cannot completely isolate his subordinates from political interference, especially when national security is at risk. If Rowell had been aware of the pressure which MacArthur had been applying both to the Australian government and to the untried command framework, it would have been incumbent upon him to follow Allen's advice and 'to grit his teeth and take it'.

The trouble was that Rowell saw only personal motives in Blamey's actions. Perhaps, almost certainly, Blamey was concerned for his own position, but in his view the maintenance of his authority was important for Australia. He had no more regard for MacArthur and some members of Curtin's Cabinet than Rowell had for him, but he knew that it was important that he cooperate with them. Tactless he might have been, but he suppressed his feelings better than Rowell. A commander with more sympathy and tact than Blamey might have found his way through the shoals and rapids that became apparent once he arrived in Port Moresby. But in the circumstances Blamey had little option but to dismiss Rowell. That Rowell's career was subsequently restored indicates that the circumstances were indeed extraordinary.

# 13 Lieutenant-General the Honourable Sir Edmund Herring: Joint and Allied Commander

STUART SAYERS

On 28 September 1942, at Port Moresby, General Sir Thomas Blamey, the Commander-in-Chief Australian Military Forces and Commander, Allied Land Forces, dismissed Lieutenant-General S.F. Rowell, his senior commander in New Guinea (see chapter 12). He described what he had done in a 'most secret' signal to the Australian Prime Minister, John Curtin, and the C-in-C South West Pacific Area, General Douglas MacArthur, and recommended the immediate appointment of Lieutenant-General E.F. Herring (1892–1982) as Rowell's successor. Herring had been GOC 2nd Australian Corps at Esk in Queensland since early August, when he relinquished command of Northern Territory Force as part of a reshuffling of Army appointments, which included Rowell's posting to New Guinea as GOC 1st Australian Corps. Blamey named as his second choice for the New Guinea command Lieutenant-General Sir Iven Mackay, the GOC Second Australian Army; but he left no doubt that Herring was the man he wanted. Curtin and MacArthur accepted his advice.[1]

Herring arrived at Port Moresby by air from Townsville late in the afternoon of 1 October. He had celebrated his fiftieth birthday one month before and the appointment was his first high battle command. His posting as GOC 1st Australian Corps included command of New Guinea Force, then the junior formation. By January 1943 New Guinea Force was recognised as the senior formation, causing the renaming of the appointment to GOC NGF and 1st Corps. Except for one break while on leave he held the command until 26 August 1943, continuing as GOC 1st Corps until 10 February 1944 when he transferred to the Reserve. In the year from 1 October 1942 to 7 October 1943, when he returned to Australia after the capture of Finschhafen, Herring successfully commanded a series of operations which thrust the Japanese from Papua and the Huon Peninsula in New Guinea and demonstrated his outstanding capacity to bring together in common cause men of different services, nationalities and operational doctrines.[2]

'Am hoping I may be able to cope with the job,' Herring remarked to his wife in his first letter from New Guinea. Major-General G.A. Vasey, the commander of the 6th Division who was responsible for the defence of Port Moresby, expressed modified enthusiasm. Although convinced that Rowell possessed a greater breadth of view on Army matters, Vasey felt that he and Herring would work pleasantly together. He did not appreciate his new commander's tendency 'to solve my problems without having first solved his own' but assured his wife that 'as far as I am concerned he is amenable and I usually find that what I want is what he wants'. By contrast, Major-General A.S. Allen, commander of the 7th Division and bearing the brunt of fighting the Japanese on the Kokoda Track, was aggrieved. He had been promoted to divisional command before Herring and resented not being given the corps. He wrote to Blamey protesting that he had commanded in battle every unit from a platoon to a division. He underlined the word 'battle'.[3]

Yet the choice of Herring to succeed Rowell made sense. Blamey had named Herring in preference to Mackay as Rowell's replacement because he was younger, an important consideration in a demanding climate. But evidently it was not all important. Almost the last sentence of his signal to Curtin and MacArthur declared: 'Essential to have commander of cheerful temperament and who is prepared to co-operate to the limit'.[4] It was as telling an assessment of Herring as of Rowell.

Herring understood clearly what Blamey demanded. He described it long after to the official historian: 'I came up as GOC NGF and found that T.A.B. [Blamey] expected me to get on with the job'. He recalled talking in Brisbane with MacArthur, who sought to impress on him that the first duty of a soldier was 'to get on with the man above him'. That presented no problems: 'I was able to assure him that I would have no difficulty in getting on with T.A.B. I had already served under him in the M.E. and in Darwin and I knew he was an extremely good man to serve.'[5]

In a private letter to Herring after the war, Blamey wrote warmly of their 'partnership' in New Guinea, his friendly tone suggesting that he genuinely viewed their relations in such a light. Others have spoken differently. A close friend was still marvelling years after Herring's death that 'he never seemed to mind Blamey breathing down his neck'. D.M. Horner has argued that Blamey's insistence that Herring commanded the whole of New Guinea Force was a charade, intended to buttress his attempts to ensure that he maintained his position as Commander, Allied Land Forces. At least one contemporary observer of events and of the campaigns in New Guinea and Papua wrote that Herring 'would rather serve than direct'. Such judgments raised crucial questions about Herring as a commander.[6]

*Preparation for senior command*

Blamey and Herring knew each other well. Their acquaintance was founded in the years between the wars, when Herring was rising to prominence as an artillery officer in the militia and Blamey was commanding the 3rd Division. They were fellow members of the Naval and Military Club in Melbourne and moved in the same professional, official and, to some extent, political worlds, Herring as a successful barrister and leading figure in the Young Nationalist Organisation, Blamey as an important public servant, first when Chief Commissioner of the Victoria Police, subsequently as chairman of the Manpower Committee formed to compile a national register as a first step towards preparing Australia for war, and as Controller-General of the Recruiting Secretariat. Although utterly dissimilar in tastes and temperament, each recognised in the other intellectual and military qualities to respect and admire. Herring knew Blamey's raffish reputation but was unaffected by it. 'He never behaved that way in front of me,' he once said. Nevertheless, fellow officers were puzzled by the easy relations of the deeply religious Herring, whose taste was for innocent jokes and boyish horseplay, and the earthy, none-too-fastidious hedonist, Blamey.

References to encounters and dealings with Blamey in Palestine and Egypt in 1940 and 1941 abound in Herring's letters to his wife. The evident pleasure he found in them is of a sort more to be expected of friends than of the GOC Second AIF and a subordinate. As promotion advanced him in the army Herring's confidence in his ability to secure Blamey's support strengthened. 'T.A.B. is grand', he wrote from Darwin in 1942. 'All my requests to him & I write to him about once a fortnight, are put in hand at the toot; signals fly here & there at once & his backing of course just makes all the difference.' From Esk three months later he described resolving a difficulty with Lieutenant-General Sir John Lavarack, then GOC First Australian Army, by engineering a direct approach to Blamey at Allied Land Headquarters in Brisbane. 'T.A.B. in good form & fixed everything as wanted in a very few moments.' This close professional relationship, certainly rare, perhaps unique at such a level of command in the Second AIF, continued almost to the end of Herring's service in New Guinea. The letters they exchanged almost daily in the Papuan campaign after Herring went forward to the battlefront provide striking evidence of their confidence in each other. The consequences of the one break in their association in September 1943 would prove unfortunate. 'So long as I could make contact with Tom Blamey I could always get things organised,' Herring said years later. There is no possibility that Rowell and Blamey could have worked together in New Guinea so cordially and effectively.[7]

At the start of the Second World War Herring was Commander Royal Artillery, 3rd Division, with the temporary rank of colonel. He was the senior militia gunner, his trade learned in the First World War in the Balkans, where he commanded a battery in the Royal Field Artillery, was a major by the age of 26, was awarded the Distinguished Service Order and the Military Cross, and for a time hesitated whether to make soldiering a career or resume legal studies at Oxford. The law won, but soldiering retained its allure, the satisfaction he gained from peacetime service in the Australian Citizen Forces reinforced by the belief that another war was looming and his duty would be to fight again. In a swift pen-picture, the official war historian wrote of 'the depth of character of this small and quiet man', extolling the ideals of service 'which enabled him to give all about him something of his own quality'. Herring accepted unhesitatingly Blamey's invitation to join the Second AIF as CRA 6th Division. His appointment was gazetted on 13 October 1939, in the rank of colonel promoted to brigadier. His Army number was VX 15.

Although his experience in higher command was limited at the time of his appointment to succeed Rowell, Herring's record was excellent. He had commanded combined forces of British and Australian artillery in the Western Desert with distinction, and handled the 6th Division artillery adroitly in the mountains of Greece and on the retreat to the beaches of the Peloponnese. His appointment to command the 6th Division in August 1941 when Mackay returned to Australia to become GOC Home Forces, and the division moved into Syria on garrison duty, was well accepted, although some in the AIF professed surprise at the promotion. The cordial relations he established with the Free French, with local dignitaries in the Bekaa Valley, and in the Transjordan soon after with King Abdullah, were promising auguries of his talent for working in the Northern Territory and in New Guinea on the friendliest terms with the Americans, even General MacArthur. He looked always for the best in everybody, preferred to praise not blame, habitually labelled any man he liked a 'grand fellow'.

Yet there was an underlay of steel. The official historian described the swift and wholesale replacement in the Northern Territory of officers Herring considered incompetent as ruthless, in some cases unduly harsh, but 'electric' in effect. Men who served under him recall his startling ability to deliver paralysing rebukes without lifting his voice. One wrote of encountering at Baalbek his 'quiet, almost diffident manner, and blue-glacier stare'. He was courteous, rarely ruffled—and unusually determined. His father once declared that the boy his family and friends always called Ned had been 'the most obstinate little devil that ever was'. Yet, as a grown man, Herring could write unconcernedly of shedding tears over a novel. He reckoned himself to be incurably sentimental.[8]

Despite his mild air and unassuming manner, Herring was implacably competitive. He played cricket, tennis and golf well and with relentless determination to win. He was an energetic quoits tennis player, not often beaten. His staff remembered with despair his enthusiasm for Chinese checkers, played nightly for hours on end if the chance presented. He disliked losing.

Herring stayed cool amid the jockeying for preferment that is endemic in an hierarchical organisation such as the Army. Jealousies and professional differences seemed more than ordinarily marked in the 6th Division under Mackay, who had been a schoolmaster in civilian life, was shy, modest, relatively old and maddeningly fussy, although generally liked and admired as a capable, most courageous soldier. Herring laid much of the blame for the discord on the GSO1, the then Colonel F.H. Berryman, who became senior staff officer when Rowell joined Blamey at Corps Headquarters after the 7th Division was raised under Lavarack. Berryman, a regular soldier and Duntroon graduate, was a brilliant staff officer but a trying colleague. He was obsessively secretive, woundingly sarcastic, and scorned militia offices as mere 'week end soldiers'. Allen and S.G. Savige, the infantry brigadiers, especially resented Berryman's behaviour, their hostility barely concealed and the cause of difficulty for Mackay.[9] They were at odds also with Herring.

As veterans of the First AIF, self-made men of relatively humble beginnings, Savige and Allen regarded Herring, the former British Army officer, as an outsider, despite his Australian birth and upbringing. They were suspicious of his education and early advantages, of the Rhodes Scholarship which had taken him to Oxford in 1912, of his position in Melbourne society, secured by family connection and professional achievement. Unlike the thrustingly ambitious regular, H.C.H. Robertson, their fellow infantry brigadier, they persisted in trying to tell Herring how to do his job as a gunner. They were perplexed by and jealous of his easy relations with Blamey, who was rather feared by Allen but claimed by Savige as a close friend. They disliked his wide acquaintance with British officers in the Middle East, some fellow students at Oxford, other contemporaries from the Macedonian campaign in the First World War. They distrusted his admiration for George Vasey, then a colonel, the Assistant-Adjutant and Quartermaster-General of the 6th Division, a regular and rising fast.[10]

For his part, Herring held poor opinions of Savige and Allen. He considered they had reached the limit of their achievement as brigadiers, and would suggest after the war that their later performance proved him correct. He privately condemned their prolix reports and dramatic signals, their addiction to recording every event however trivial, as attempts to fashion history as they wished it to be. Although invariably polite, he dealt as warily with each as with Berryman. He was one of a number of strong critics of Allen's performance in Greece, perhaps, as Allen believed, the most influential. He was almost certainly responsible for Savige being recalled to Australia from the Middle East late in 1941 to be Director of Recruiting, a backwater appointment from which the entry of Japan into the war fortuitously rescued him. Allen and Savige would prove troublesome subordinates to Herring in New Guinea.[11]

*Papuan campaign*

Herring's first achievement on taking command in New Guinea was to demonstrate that a 'working arrangement' of the kind Blamey had proposed to Rowell was feasible. Blamey had suggested that he would exercise control through Rowell and his staff but did not intend to alter the method of command or do anything to derogate from the authority of the GOC NGF. It was precisely how he worked with Herring in the ensuing weeks until the Japanese were thrust out of the Owen Stanleys. Thereafter, with the command divided into a Rear and an Advanced Headquarters, operational control of the battle for the Papuan beachheads was in Herring's hands. At Port Moresby, with a staff of his own headed by Berryman, now a Major-General and Deputy Chief of the General Staff, Blamey was able to devote himself from the end of November to his responsibilities as Australian C-in-C and—rather in

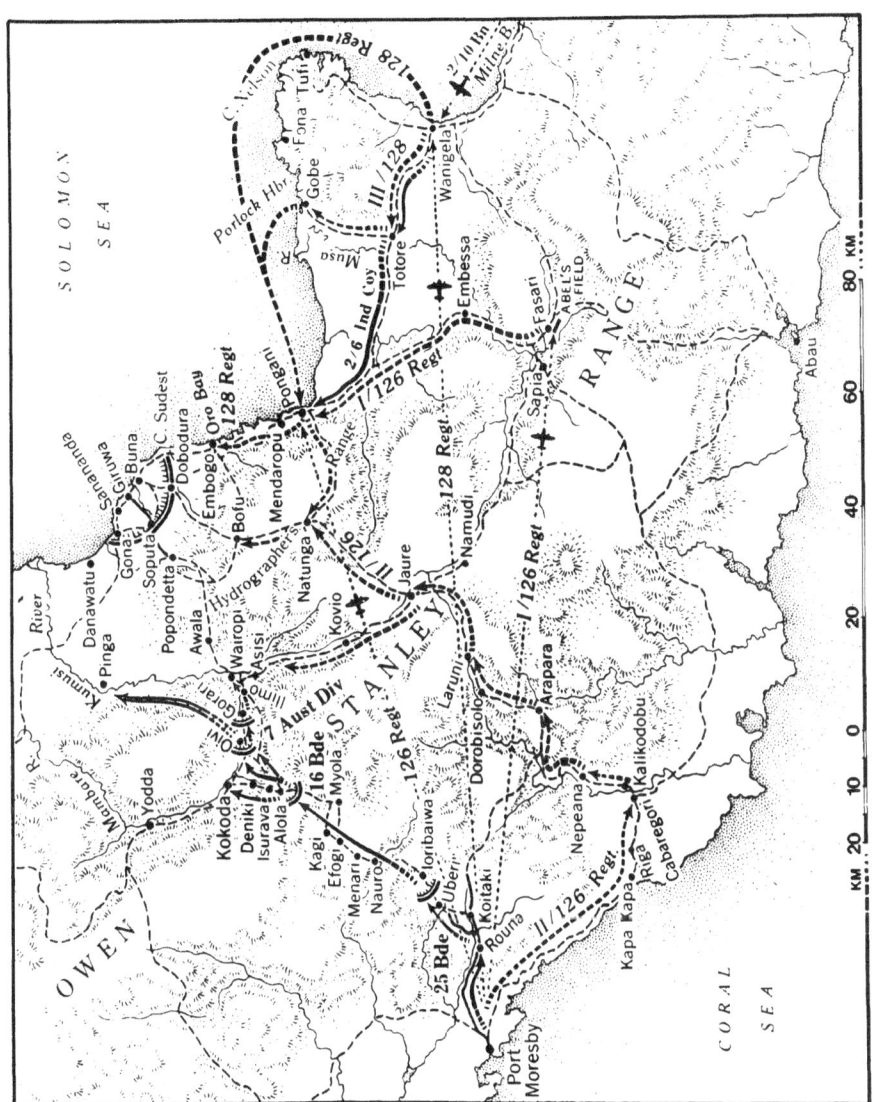

24 Allied advance to Buna, Gona and Sanananda, October-November 1942. On 28 November Herring opened his headquarters at Popondetta.

name than reality—Commander, Allied Land Forces. MacArthur, installed at Port Moresby with his staff since 6 November, treated Blamey as though he were in operational command at Gona, Buna and Sanananda, but the commander of the victorious Australian and American troops was Herring.

In a letter to the official historian in 1957, describing his early days at Port Moresby, Herring remarked of Blamey:

> He was most punctilious; he did expect to know all that was going on and I was most careful to see that he did. He had all his responsibilities as C-in-C to attend to, a full time job, and was consequently in constant touch with the C.G.S. in Melbourne and Berryman in Brisbane, until the latter came to live at N.G.F. He expected me...to get on with the campaign over the mountains...

Asides in letters to his wife in the first weeks after arriving in New Guinea are perhaps more revealing. He spoke often of his appreciation of Blamey's helpfulness, adding once at least: 'He also relieves me of much responsibility'. Long after the war he told the official historian that Blamey was meticulous not to interfere, expecting him to 'get on with the job, and really to command one's show', and his contention is supported in large measure by Major-General R.N.L. Hopkins who, as a brigadier, was Herring's chief staff officer until replaced by Berryman at the end of December 1942. Recalling the method of command at New Guinea Force headquarters, Hopkins wrote in May 1974:

> The command chain worked smoothly after Rowell left and Herring took over. Each evening I would see Herring with any papers to sign but primarily to decide the next operational move. He was a barrister and judge. His method was to work carefully round any problem, discussing factors, difficulties and courses. It took a good deal of time because he went over the same ground a good deal. But he was thorough even if I felt we had reached the ultimate conclusion twenty minutes before!
>
> Once clear on plans, we two would walk up the hill about 200 yards to Gen Blamey's hut. Gen Herring would go through our plans with the Chief and then tackle any other matters. Herring's method at least made sure that he had his brief well discussed before he saw the Chief. Blamey never gave any impression of interfering; in my view he was outstandingly helpful. Invariably, when our discussion came to an end he asked both of us if we had anything else to bring up...[12]

Nevertheless, there is ground for speculating whether Herring was at all times in full command of New Guinea Force. On his own admission he was twice obliged to protest to Blamey over what he considered unwarranted interference, first in December 1942 when he received directions about the use of tanks by Brigadier George Wootten's 18th Brigade at Cape Endaiadaire, secondly in September 1943, after the capture of Salamaua, when a peremptory signal, seemingly from Blamey but in fact from Berryman, was sent directly to his field commanders. The C-in-C settled both cases tactfully, but each is interesting as evidence of Herring's confidence in his standing as operational commander. A more conspicuous issue, although Herring wholeheartedly approved the action taken, was the replacement of Allen as GOC 7th Division by Vasey shortly before the recapture of Kokoda. The signals and letters streaming between New Guinea Force and Allen's headquarters came from Blamey, who was being goaded by MacArthur to press the attack on the Kokoda Track although neither understood the realities of the fighting in the mountains, or how stubborn and skilful an opponent the Japanese was.[13]

About the same time, the C-in-C refused to let Herring bring Brigadiers R. Sutherland and R. Bierwirth to Port Moresby to resume the positions they had held as his principal staff officers in the 6th Division and the Northern Territory Force. They were close friends of Herring, his instructors in the art of exercising command, he once said. He was anxious for them to join him in New Guinea, although telling his wife that he wanted them first to leave the Northern Territory for a period of recuperation before returning to the tropics. Bierwirth, on the other hand, never doubted that Blamey was responsible for delaying their transfer for more than six months. The C-in-C, he said, would not tolerate subordinate commanders' building teams of supporters.[14]

As D.M. Horner has argued, Herring saw his main contribution to victory in the battle for the Papuan beach-heads between 16 November 1942 and 22 January 1943 as 'ensuring the smooth operation of the Australian and American elements of his command'. He recognised that the Japanese thrust into the Owen Stanleys had spent its force by the time Rowell was dismissed, and warmly praised Vasey for his brilliant tactics in the Oivi-Gorari battle and in the dash along the Killerton track to capture Sanananda. But even as the advance towards the Papuan northern coast got under way, at a time when Australian and American confidence ran high that Buna might be captured in a matter of days, he was conceding that the Japanese was proving 'a tenacious enemy who practically has to be dug out one by one like a lot of rabbits'.[15]

A significant factor in the pattern which developed in the Papuan campaign, Herring wrote after the war, was the 'very great' capacity of the Japanese to prepare and hold defence positions, a lesson which everyone from MacArthur down had to learn 'the hard way'. In the phrase of MacArthur's Chief of Intelligence, Major-General C.A. Willoughby, the battle for Buna was 'a head-on collision of the bloody, grinding type'. The nature of the terrain dictated the tactics employed. Although the campaign gave Herring little scope to demonstrate his ability as a corps commander, it tested the leadership of his divisional, brigade and battalion commanders, as the detailed accounts of the fighting in the official and other histories make clear. The problems were immense. In the absence of a secure sea route, the supply of food and ammunition by air was an administrative nightmare. Malaria and pestilential diseases of many kinds ravaged the troops in the jungle and swamps of the coastal plain. In Port Moresby, refusing to recognise the difficulties of fighting on a battlefront he never visited, MacArthur was pressing relentlessly for quick results, once grandiosely ordering 'Gona will be captured at dawn', later describing operations on the Sanananda track as 'mopping up', and blandly announcing a fortnight before the campaign ended that the 'annihilation' of the Japanese in Papua 'can

*32* Lieutenant-General Herring accompanying General Blamey on his visit to Buna, January 1943. Leading the party is Lieutenant-General R.L. Eichelberger, Commander of the 1st US Corps and acting Commander of the 32nd US Division. (AWM Negative No. 14107)

now be regarded as accomplished'.

Herring discovered his liking for the Americans and theirs for him in the Northern Territory, where he formed a strong admiration for the US Army Air Force. In Queensland he extended his acquaintance to senior Army and Navy commanders, and came to know MacArthur sufficiently well to deal confidently with him at critical moments in New Guinea. Within a short time he was on excellent terms with Major-General George C. Kenney, commander of the Allied Air Forces and the US Fifth Air Force, and especially with Brigadier-General Ennis C. Whitehead, commander of the air forces in New Guinea and his enthusiastic collaborator in perfecting the methods first developed on the Kokoda Track of supplying forward troops by air. As he told the official historian:

> All kinds of ways and means were tried out, and the pattern was set before the end of the campaign for the campaigns of the future, Salamaua, Lae and so on... Involved in the success was the gradual build-up of goodwill and trust and confidence from the top to the bottom of the Australian Army and the 5th U.S.A.F., and if the Papuan campaign deserves to be remembered as a campaign of great importance, as I believe it does, one of the reasons is the remarkable and successful development of air supply and the close cooperation that developed between the ground forces of Australia and the air forces of the U.S.A.[16]

Likewise, the immediate and genuine trust and friendship formed between Herring and Lieutenant-General Robert L. Eichelberger, who succeeded the hapless Major-General E.F. Harding in command of the US 32nd Division at Buna, was profoundly important to the Allies' eventual victory in Papua.

The measure of Blamey's dependence on Herring's ability to gain the cooperation of the Americans was exemplified on Christmas Day 1942, when he sent him to GHQ alone to ask MacArthur to reinforce Vasey's hard-pressed 7th Division on the Sananada front by flying a regiment of the US 41st Division from Australia. A suggestion that Blamey feared Herring was flagging mentally and physically, and invited him to Port Moresby for Christmas dinner as a respite from the strain at Dobodura, where NGF Advance HQ had moved from Popondetta on 17 December, is suspect. It does not square with Herring's success in presenting his case to MacArthur nor with his calm recital of events at this time in letters home. His description to the official historian after the war of the interview with MacArthur and its outcome indicates that he returned to Port Moresby on his own initiative to ask for the reinforcements Vasey needed. The diary kept for him by his aide, and a letter to his wife on 24 December, support him.[17]

So far from feeling tired himself, Herring thought that Hopkins needed a rest and readily agreed to Berryman replacing him immediately as chief staff officer at Dobodura. He seemed also to have been concerned by signs of exhaustion in Vasey, although not so worried as to accept Allen in his place. He preferred Vasey tired to Allen fresh, he told Blamey. Long after he said that Vasey invariably held strong views, put them robustly, but 'always accepted the decision of his superior officers with complete loyalty, whether they were in accord with his view or at variance with it'. Within a fortnight of returning to Dobodura accompanied by Berryman, Herring told his wife that he felt 'very fit'. Gona had been captured after 'a pretty tough fight', and Buna also. He could not find praise enough for the Australian troops or for 'my American general friend & his men'.[18] The record of his activities at this time and after his return to Port Moresby when MacArthur and Blamey departed for Australia, gives no impression of fatigue or lapses in his control of operations. Blamey's confidence in him was confirmed when he returned to Melbourne for leave at the end of January 1943 and learned that he would continue in command of NGF and the 1st Corps for the campaigns against Lae, Salamaua, the Markham Valley and Finschhafen. He would be relieved then by his old friend, Lieutenant-General Sir Leslie Morshead, who brought the 9th Division home to Australia from the Middle East in February.

*Salamaua*

Detailed orders for the attacks on Lae and the Markham Valley were issued by Blamey on 17 May 1943 and a week later Herring resumed command of New Guinea Force in Port Moresby to begin putting them into effect. Blamey proposed a seaborne assault on Lae by a division, an

33 After the battle for Sanananda. Colonel Jens A. Doe, 163rd US Regiment, Colonel J.T. Murray, 186th US Regiment, Lieutenant-General E.F. Herring and Major-General G.A. Vasey. (AWM Negative No. 14303)

airborne landing by a second division in the Markham Valley to capture Nadzab and advance against Lae, and a preliminary landing at Nassau Bay to establish a shore base for the landing craft transporting the troops to Lae. The Nassau Bay landing was seen as a means of making junction with the Australians operating in the mountains surrounding Mubo and beyond Wau and Bulolo, where Savige, in command of the 3rd Division, had established his headquarters. Blamey hoped the Japanese would regard the landing at Nassau Bay as a threat to Salamaua, obliging them to reinforce that garrison with troops from Lae. He instructed Herring to seize the high ground around Goodview Junction and Mount Tambu and the ridges running down to the sea, but go no further. He wanted the operations in the Salamaua area to cloak the Lae assault, serving as a magnet to draw reinforcements away from Lae. Herring began to plan accordingly. He understood exactly what was intended but unhappily for him Savige did not, although Brigadier Murray Moten, commanding the 17th Brigade, was well aware that Salamaua was not to be captured by the 3rd Division.[19]

Savige appears to have misconstrued a warning order issued on 20 May by Mackay, who had temporary command of New Guinea Force while Herring was on leave. The order warned Savige to be ready by 15 June 'to threaten Salamaua by aggressive overland operation from the Wau–Bulolo valley and by threats along the coast from the Morobe area'. Savige noted the need to obtain clarification of the words 'threaten Salamaua' but seems not to have done so. An operation instruction from Herring on 27 May to the commanders of the 3rd Division, the 41st American Division, the Fifth Air Force and the American Patrol Torpedo Boats defined Savige's task as ultimately to drive the enemy north of the Francisco River and to establish a beach-head at Nassau Bay in order to open a sea line of communication into the Mubo area, enabling American and Australian forces to operate in conjunction. It gave Major-General Horace H. Fuller, commanding the 41st American Division, the tasks of ensuring the security of New Guinea from Oro Bay to Morobe, arranging the landing of a battalion group at Nassau Bay, and cooperating in driving the Japanese north of the Francisco. The instruction made plain that command of ground forces in the battle area would be exercised by Savige, who now satisfied himself that his objective was Salamaua. A conference with Herring and Fuller at Port Moresby on 31 May to settle details of the Nassau Bay landing, and another at Bulolo on 15 June with Herring, Berryman, Moten and Colonel Archibald R. MacKechnie, commander of the assault force, did nothing to alter Savige's belief that he was expected to capture Salamaua as soon as possible.[20]

As well as failing to appreciate his true role in the Salamaua operations, Savige misread the extent of his authority over the Americans at Nassau Bay and later at Tambu Bay, where Herring decided to land guns in support of the Australians on the high ground above. Herring explained at a conference in Port Moresby on 5 July that Fuller would command the move along the coast but Savige, once again misunderstanding, began soon to issue instructions which unnecessarily confused and increasingly irritated the Americans. Many signals passed between Port Moresby and the American and Australian divisional headquarters before accord was restored. After weighing the evidence, the official historian found Savige at fault, writing that 'in his earnest attempt to seek a formula satisfactory to Australian–American relations, Herring had the ill-fortune to have his orders misunderstood by the Australian divisional commander'. It was a severe test of Herring's tact and diplomacy, and an impressive demonstration of his gift for

*34* Lieutenant-General E.F. Herring, GOC New Guinea Force in his office in Port Moresby, June 1943. (AWM Negative No. 64081)

obtaining that smooth cooperation between the Americans and the Australians which was essential to success against the Japanese in New Guinea. Yet the problems remained of convincing Savige that he must not attck Salamaua, of preventing him from marching on what Herring later called 'his own sweet way', and of inducing him to move his headquarters nearer the coast, thus easing the task of supply by freeing aircraft for other pressing operational demands, and reducing the tonnage of food falling inadvertently into enemy hands, or being lost.[21]

In mid-July, by which time Sutherland and Bierwirth had at last rejoined him as his principal staff officers, Herring told Savige that while the capture of Salamaua was devoutly to be wished, 'no attempt upon it is to be allowed to interfere with the major operation being planned'. According to Savige later, it was something he had not known. Towards the end of August he handed over operations in his area to Major-General E.J. Milford and the 5th Division, and returned to Australia. Recalling the period after the war, Berryman criticised Savige's methods of command and his failure to keep Herring fully informed. He added that most of the battles in the mountains beyond Wau were brigade battles, planned and executed by the 3rd Division brigade commanders, Moten, H.H. Hammer and, later, R.F. Monaghan. It is a judgment confirmed by others since.[22]

### Lae–Finschhafen

Planning for the attack on Lae, at Land Headquarters in Brisbane and under Herring's supervision in Port Moresby, reached a peak in August 1943. The inadequacy of Nassau Bay as a staging base had been recognised and the decision made to transport the assault troops of the 9th Division, now commanded by Wootten, in the large craft of Rear-Admiral Daniel Barbey's Seventh Amphibious Force from Milne Bay and Buna. Herring described preparations to the official historian:

> The decentralisation that operated as between T.A.B. and NGF whilst he was in Australia also operated as between NGF and both 7 and 9 Divisions. Here Wootten and Vasey talked things over with me just as I talked things over with T.A.B. in the larger sphere. I left them free to work things out in the one case with Barbey and in the other with Whitehead, but I had to be kept informed as after all I was responsible for their plans, and I was in a position to help them with any particular demands that they might have to make on United States Navy or U.S. Air. During the months of preparation, Barbey, Whitehead and I were the three executive commanders in New Guinea and we were all on very good terms, so that it was only on odd occasions that I had to seek T.A.B.'s assistance to get things done...

The assault on Lae was brilliantly planned and executed. Wootten's 9th Division went ashore on 4 September. The 503rd American Parachute Regiment and a detachment of Australian gunners parachuted into Nadzab on 5 September, and Vasey's 7th Division arrived by air in the next two days. Salamaua was captured on 11 September, and Lae on 16 September. Herring immediately set in train the seaborne attack on Finschhafen.[23]

Blamey had arrived at Port Moresby on 15 August, ten days before MacArthur. He assumed command of the Lae operations on 20 August, as Commander, Allied Land Forces, and formally separated New Guinea Force Headquarters and 1st Corps Headquarters a week later. Herring went forward to Dobodura as GOC 1st Corps on 29 August, unaware that Blamey had decided to name Mackay immediately as GOC New Guinea Force, although the appointment did not become effective for three weeks. The consequences were unfortunate.

Herring's plan for the attack on Finschhafen was taking shape well before Lae fell, and was finalised there on 18 September at a hastily assembled conference with Wootten, Brigadier W.J.V. Windeyer, commander of the assault force, and Rear-Admiral Barbey's representative, Lieutenant-Commander C.C. Adair. The landing was timed for just before first light on 22 September, but was advanced by half an hour the next day after much wrangling in a meeting at Buna with Brigadier-General S.J. Chamberlin, MacArthur's senior operations officer, and Barbey, who was alarmed by the possibility of daylight air attacks on his destroyers and landing craft, and wanted to put the troops ashore at 2 am. Barbey thought Herring inflexible, arguing confidently but mistakenly as events would prove, that his radar would pinpoint the assault beach. Berryman attended the conference and irked Herring by his failure to support him. His dealings with the American admiral in the past had been scored on his own admission by constant 'clashing': the argument over the Finschhafen landing became unusually acrimonious.

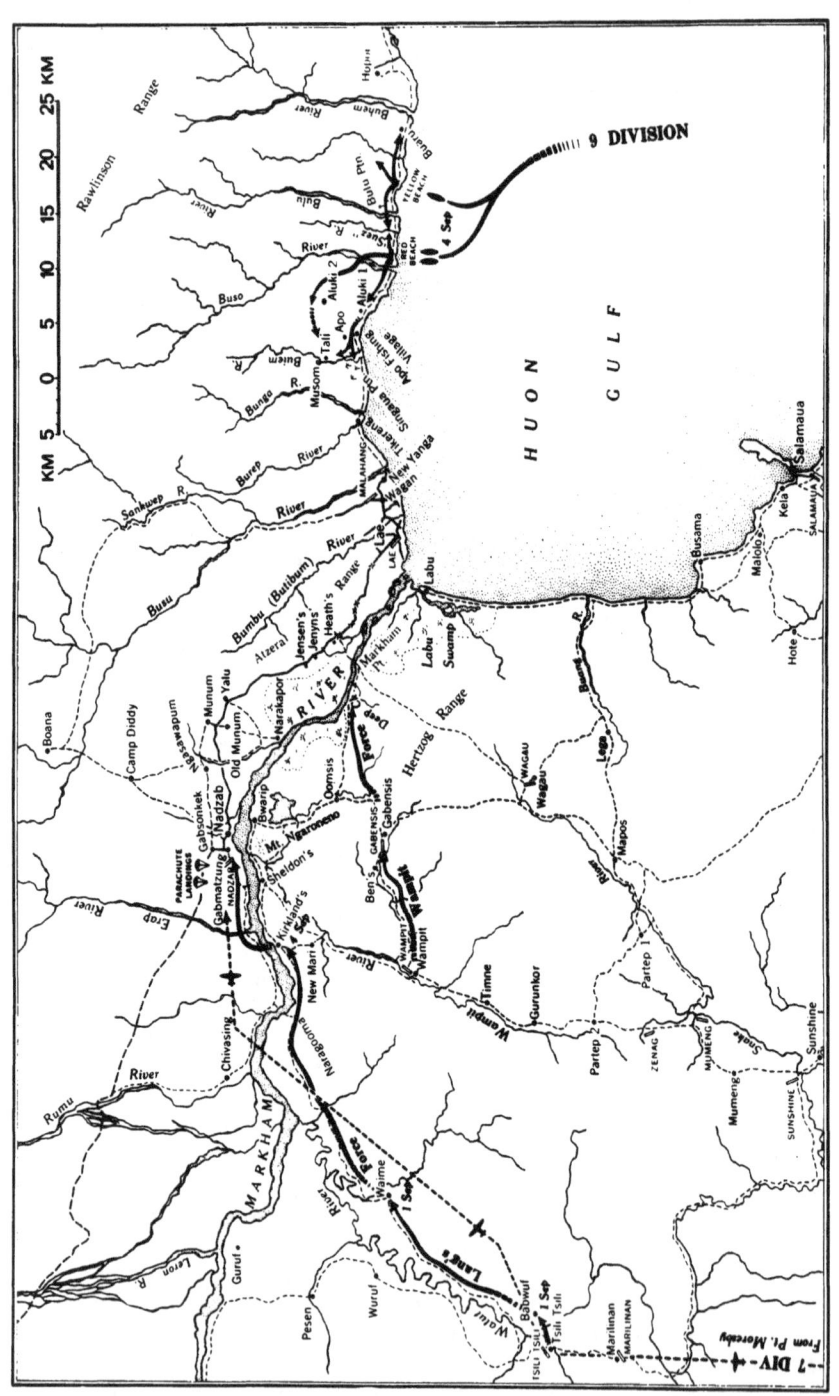

25 The advance to Lae-Nadzab, 1–5 September 1943

He had been almost without sleep for two nights on end, was conscious, moreover, that he was no longer GOC New Guinea Force and therefore less well placed to assert authority, and felt that Berryman was siding with Barbey against him. In the event he was obliged to give way, although not so much as Barbey.[24]

Three days later, Blamey returned to Australia, leaving Mackay in command of New Guinea Force with Berryman as his chief staff officer. Until Mackay's arrival at Port Moresby on 21 September, the day before the assault on Finschhafen, there had been no inkling that he would be given the command. Herring's staff thought the appointment should have reverted to him, but he recognised and later argued with characteristic logic that his imminent, long-intended relief by Morshead to command the forthcoming operations against Madang made the appointment of a temporary commander of New Guinea Force necessary. The disadvantage for him was that his direct access to Blamey, enjoyed for so long and, he suspected, resented by Berryman, was now blocked.

Only hours after Windeyer's 20th Brigade went ashore five miles (eight kilometres) north of the Finschhafen area, and shortly before leaving New Guinea for Australia, Blamey ordered Herring to send in Wootten with his 9th Division Headquarters and another brigade to support the attack. Herring saw Barbey at Buna the next day to make the transport arrangements but found the American unwilling to cooperate. Although MacArthur had agreed on 17 September that Blamey could send a second brigade to Finschhafen if needed, Barbey had wind of a new operations instruction from MacArthur's headquarters which he believed would oblige him not to do what the Australians wanted. He seems to have grown resentful of being regarded as provider of sea transport for another country's army, almost certainly was aware of the strengthening determination at MacArthur's headquarters to relegate the Australians to a secondary role in the Pacific war, and was not sure what his superiors expected him to do. He resolved his dilemma, terminating a fruitless discussion which had lasted seven hours, by sailing abruptly for Milne Bay.

Herring immediately rang Mackay to report his inability to reinforce the Finschhafen front and was dismayed by what appeared to him the lukewarm response at New Guinea Force Headquarters. Once again he blamed Berryman, insisting later that 'both Mackay and Berryman must face the fact that it was NGF's job to see that T.A.B.'s orders were carried out, and that they failed to get this done, and what is perhaps more difficult to understand, passively resisted all my attempts to restore the position'. He flew to Port Moresby on 24 September after spending some hours with Vasey at Nadzab discussing the operations of the 7th Division in the Ramu Valley. He talked briefly with Mackay and Berryman before dinner and at greater length after. Next morning, at his insistence, Mackay and Berryman accompanied him to GHQ to meet Kenney and Vice-Admiral A.S. Carpender, commander of the Seventh Fleet, who agreed eventually that the Navy's obligation to lift a second brigade to Finschhafen must be met. As Herring recalled the occasion for the official historian, he carried the burden of arguing the Australian case, although convinced that the responsibility properly belonged to New Guinea Force. Yet Mackay and Berryman 'seemed to adopt a detached attitude, almost as if the matter had nothing to do with them'.[25]

Armed with Carpender's assurance, Herring returned to Dobodura and warned Wootten to prepare for the move to Finschhafen with his 9th Division Headquarters and staff. That evening Berryman rang from Port Moresby to say that Carpender had changed his mind. 'He didn't seem disturbed about it and apparently NGF didn't propose to do anything about the matter,' Herring recalled. As he could no longer speak directly to Blamey he began a series of increasingly vitriolic telephone calls to Mackay, demanding urgent action by New Guinea Force. His calls and a sharply worded letter on 26 September undoubtedly offended Mackay and led Berryman to confide to his private diary doubts about Herring's mental state. 'I suppose I was pretty well worked up over the whole affair', Herring wrote later. Since then, some play has been made with extracts from Berryman's diary which show Herring in an unflattering light. The extracts do not say much for Berryman. So far from trying to understand what had moved this normally calm, self-controlled man to such uncharacteristic behaviour, he seemed intent on maligning him. The official historian read Berryman's diary but does not refer to or make use of it. Herring, in his official report on the Lae, Salamaua and Finschhafen operations as GOC 1st Corps, confined himself to a straightforward recital of facts. He ignored the personal differences which, in his belief, contributed significantly to the delay in sending reinforcements and supplies to Windeyer and 'very nearly, as it turned out, led to a most serious disaster'.[26]

Late on 26 September, with increasingly disturbing reports arriving from Finschhafen, Mackay at last signalled Blamey and MacArthur, seeking 'early action' to reinforce and supply Windeyer. It was a milder request than Herring thought necessary, but it alerted Blamey and MacArthur and resulted in the despatch of a battalion, not a brigade, to join in the assault and the eventual capture of Finschhafen on 2 October. By then Herring's service in New Guinea was almost over. He had narrowly escaped death on 28 September in an aircraft accident in which his close friend and chief staff officer, Roy Sutherland, was killed. He was shaken and deeply grieved, and greeted the arrival of Morshead at Dobodura with relief. The exchange of command was completed on 7 October. At the end of January 1944 Herring was appointed Chief Justice of Victoria, his active service with the Army terminating on 10 February, when he was transferred to the Reserve.[27]

Herring's relief by Morshead after the capture of Finschhafen had been foreshadowed by Blamey as early as the end of January 1943. The plan then, as explained by Blamey, was that Herring and the 1st Corps would control the operations in New Guinea to clear the Japanese from the Huon Peninsula and from the Ramu and Makham Valleys and that Morshead would take over with the headquarters of the 2nd Corps for the attack along the coast to Madang. Herring would resume the New Guinea command for operations after that. By late 1943, as Herring wrote later, the question of what lay ahead for the AIF was obscure: 'As far as we could see then, with the American army taking up the running, it rather looked as if the main work of the AIF for World War II was over'. It was an important consideration in his decision to nominate for the post of Chief Justice of Victoria, an appointment Blamey was anxious he should accept not only because of the high professional achievement represented, but also because of the credit reflected on the Australian Army by it and by Herring's eventual succession as Lieutenant-Governor of the State. Blamey, as Herring would recall, was 'all for my accepting the office'. He believed it would be 'good for the Army' for one of its generals to be appointed to so high and honourable position.[28]

*Allied commander*

In the year he commanded New Guinea Force and 1st Australian Corps, Herring displayed an administrative ability of the highest order. He was responsible for several of the most important campaigns fought by the Australians in the Second World War, although the fact must be recognised that his command during the battles at Buna, Gona and Sanananda, at Salamaua, and at Lae and Finschhafen did not oblige him to demonstrate strategic flair or to react rapidly to fast-changing situations on the battlefield. Strategic decisions were made by MacArthur and relayed through Blamey, ostensibly the Commander, Allied Land Forces, and essentially the battles were fought by brigade and battalion commanders. But preparation and control of the battles demanded rare organisational and planning skills to ensure that supplies and reinforcements were brought forward through difficult terrain in unpredictable weather, and that medical services were maintained in appalling conditions not encountered before. Herring was responsible also for ministering to the needs of the native populations of Papua and New Guinea, for establishing law and order between the different, often antagonistic tribes, and ensuring that justice of some sort prevailed. Above all he had to orchestrate the land, sea and air resources of two nations, and bring Allied power to bear smoothly and successfully against a determined, resourceful enemy. Undoubtedly, Allied flexibility in the use of air and sea power, supported by the daring and skill of the Australians in jungle fighting, drove the Japanese from the key central area of New Guinea. For that achievement, Herring deserves great credit.

Certainly, his contemporaries thought so. In *Jungle Road to Tokyo* Eichelberger wrote of the lifelong friendship he formed in New Guinea with Herring, his immediate superior, declaring that 'from Australian commanders in the field I received co-operation, much sound advice, and the fraternal understanding which arises from what St Paul describes as the "fellowship of suffering"'. Kenney, in his memoirs, declared Herring to be 'exceedingly capable'; MacArthur welcomed his appointment to succeed Rowell in New Guinea with enthusiasm. Lieutenant-

General Sir Vernon Sturdee, Chief of the General Staff from 1940 to 1942 and again from 1946 to 1950, considered Herring to be one of the top commanders in the Second AIF with an unmatched talent for integrating Allied forces. Curtin, speaking of Herring's appointment as Chief Justice of Victoria, praised his 'great and distinguished services'. Mackay, according to his biographer, regarded Herring as 'a relentless thruster in war', one of the finest field commanders produced by Australia in the two world wars.[29]

Soon after the start of the Korean war in June 1950, Herring returned to duty as Director-General of Recruiting. He was the logical choice after Morshead, who had been on the Retired List for six months and declined an offer of the post. Herring—to give him full title, Lieutenant-General the Honourable Sir Edmund Herring—was by then Lieutenant-Governor as well as Chief Justice of Victoria, and still two years short of the statutory retiring age for an officer of his rank. Sir Frederick Shedden, nearing the end of his distinguished career as Secretary of Defence, and known to have been influential in Blamey's appointment to the comparable position of Controller-General of the Recruiting Secretariat more than twenty years before, nominated Herring and argued persuasively in his favour. Whether history would have been repeated and Herring become C-in-C of the Australian Military Forces in the event of another world war can only be speculation, for undoubtedly several regular officers with wartime experience were eligible for such an appointment. But it is a measure of Herring's standing that he was considered by many who had served under him to be the likeliest candidate.

# Major-General George Alan Vasey: Commander, 7th Australian Division

D.M. HORNER

No Australian divisional commander during the Second World War faced a more difficult situation on assuming command than Major-General George Alan Vasey (1895–1945) when he took over the 7th Australian Division on 28 October 1942. His predecessor, Major-General A.S. Allen, had commanded the division since June 1941 and had led it both in the bitter fighting in Syria and in the early battles on the Kokoda Track. His men had retreated back along the Track, had held the Japanese at Imita Ridge, and had then begun the slow advance forward along the steep, slippery Track towards Kokoda. Short, heavily built, 'Tubby' Allen was an able commander; indeed in two world wars he had commanded infantry units at every level from platoon to division in action. He was loved by his soldiers and was fiercely loyal to his officers and men.

It was a cruel blow to the troops of the division when General Blamey, Commander-in-Chief of the Australian Military Forces, and Lieutenant-General Herring, GOC of New Guinea Force, relieved Allen for not pressing the advance rapidly enough towards Kokoda. It mattered little that, in the words of Allen's GSO1, Colonel Spry, 'We were living, fighting and dying on two and a half pounds per man per day, for food, ammunition, everything'. The troops of the 16th Brigade who, in chilling rain were fighting the stubborn Japanese at Eora Creek, were not advancing quickly enough for either Blamey in Port Moresby or MacArthur, the Allied Commander-in-Chief, in Brisbane. Few divisional commanders in history have been replaced by their army commander without having been visited by either the army or corps commander, or indeed by anyone above the rank of lieutenant-colonel.[1]

It needed an extraordinary commander to step into Allen's shoes. George Vasey was such a man. A professional soldier, he graduated from Duntroon in 1915 at the age of twenty and served with the Australian artillery during the battles on the Somme in 1916. By August 1917 he was brigade major of the 11th Australian Infantry Brigade. It is a sad commentary on the development of the Australian Army between the wars that he was not promoted to substantive lieutenant-colonel until after the outbreak of the Second World War. The confident, outgoing Vasey, so evident during the fighting on the Western Front, gradually withdrew into himself during these heart-breaking years. He contemplated leaving the Army and studied accountancy. But a number of postings to India relieved the monotony and added to his military knowledge and experience. He attended the Staff College at Quetta in 1928 and 1929. For eighteen months (1934-36) he was brigade major of the Bareilly Infantry Brigade, and he was GSO2 Rawalpindi District for a further year until he returned to Australia in March 1937 to become GSO2 and then GSO1 (Training) at Army Headquarters.

At the outbreak of the Second World War Vasey was appointed chief administrative staff officer (AA & QMG) of the 6th Australian Division with the rank of colonel, and he departed for the Middle East with the advance party in December 1939. He retained the position for the next year, including organising the administrative support for the capture of Bardia in January 1941. Then he was senior operational staff officer (GSO1) of the division during the advance to Benghazi before assuming command of the 19th Brigade in March 1941.

In grim fighting in Greece Vasey showed outstanding ability both as a determined commander and leader of men. His 19th Brigade took the first shock of the German assault and he fought a determined rearguard action at Veve on the Greek-Yugoslav border. Later he held the vital Brallos Pass. His tall, gaunt frame, with his head of wiry black hair parted in the middle, could always be found in the forward areas. He talked to his soldiers in picturesque language which soon became lengendary, but never seemed to offend. When called to the phone by a senior officer he would commence the conversation with 'Bloody old George here', and he would greet an old friend with, 'You dear old bastard'. Despite his brusque exterior he was sensitive and intense. Sir Kingsley Norris wrote: 'Polite or profane as was suitable for the company or the occasion, beloved by all ranks under his command, including the padres, George was inspiring by his example, but was humanely harsh in the fairness of his abrupt condemnation, which in some way did not engender resentment'.[2]

As commander of the Australian forces on Crete he was faced with a series of desperate situations. Taking control of the rearguard as the Allies withdrew over the island's mountainous spine he held the Germans long enough for a large number of the fighting troops to be evacuated. He came out of the campaign with an enhanced reputation for coolness and determination and with a sharpened awareness of the requirements of modern warfare, which he applied later in the war.

*Kokoda Track*

On the outbreak of war in the Far East in December 1941 Vasey returned to Australia as a major-general, initially as chief of staff to Mackay's Home Forces, and then, after Blamey became Commander-in-Chief, as Deputy CGS. In effect he became Blamey's chief operational staff officer at Advanced Land Headquarters in Brisbane. Finally, in September 1942 he was sent to New Guinea to command the 6th Australian Division at Port Moresby. He was there a little over one month before he was sent forward to the 7th Division, then fighting its way tortuously over the Kokoda Track—that narrow footpath which wound its way precariously over the towering Owen Stanley Range.

As the new divisional commander Vasey was fortunate in reaping the rewards of Allen's careful preparations. One week after assuming command he wrote to his wife:

> Poor Tubby, it was just beyond him, he simply has no idea what a div staff is for or can do. The following day [after he took over] I moved HQ forward and again the next day so as to get on the tail of the brigadiers and get them forward. One of the grouses against Tubby was that he let the situation control him instead of controlling it. I never believed in allowing that to occur and decided to take control quickly.

At times Vasey tended to play up his own performance in his letters to his wife (not a unique trait), and the above account is less than charitable to Allen. Indeed Colonel Norris, the division's chief medical officer, has recorded that as he and Vasey walked back after saying farewell to Allen, 'the silence was broken after a while. "Poor bloody old Tubby. He really has done a grand job" [said Vasey]. George could be very gentle and generous.' Furthermore, it was all very well for Vasey to speak of moving his headquarters forward, but it was only possible to do so because, in the face of determined attacks, the Japanese abandoned the whole Eora Creek position on the night of 28–29 October after holding up the Australians for some seven days. Vasey assumed command on the morning of 29 October.[3]

Vasey immediately impressed his personality upon the division. Brigadier Eather, commanding the 25th Brigade, recalled that when Vasey took over the troops were a bit shocked at Allen's dismissal, but it was not long before they realised that although they had lost a good commander they had gained one equally as good. The Japanese had developed a habit of sniping at those whom they recognised as officers, and to overcome this practice officers had tried to make themselves look as much like private soldiers as possible. Divisional headquarters officers had followed suit. Vasey believed that confidence between officers and soldiers was not gained by officers becoming 'one of the boys' but by knowing and practising their duties and obligations at all times. Shortly after he arrived he boomed, 'Put back those bloody badges of rank—I don't know whether I'm talking to the cook or my G1'. And he insisted on wearing his scarlet-banded general's cap at all times. Officers were at pains to point out the dangers of this rash conduct, but he would not be dissuaded. It was probably his only known act of shownmanship. 'I want them to know me', he said, 'Good for morale.'[4]

Vasey also had an immediate impact on operations, urging his brigades forward with characteristic enthusiasm: 'The enemy is beaten. Give him no rest and we will annihilate him. It is only a matter of a day or two. Tighten your belts and push on.'[5] On 2 November 1942 the 25th Brigade entered Kokoda while the 16th Brigade bypassed it to the south and set out in pursuit of the enemy. On the afternoon of 3 November they struck the Japanese before Oivi, and by 5 November they were up against the main defences. For the next day the 16th Brigade, short of ammunition, sought to break the enemy's line, and one battalion was sent south in a broad sweeping movement in an attempt to cut the Japanese lines of communication to the Kumusi River. Meanwhile the 25th Brigade was being replenished from supplies flown in to the newly opened Kokoda airstrip.

On the evening of 6 November Vasey decided 'to risk having no backstop on this front', and verbally ordered Brigadier Eather to take his brigade, swing south of the enemy's position, assume command of the battalion of the 16th Brigade, and cut off the enemy from the Kumusi. This bold move was the first of its type made by Allied troops in brigade strength in the jungle in the Second World War. In Burma the British were still on the defensive. Although Allen had foreshadowed similar tactics, the battle illustrates the rapidity with which Vasey had seized control and had implanted his personality on the division. Colonel Spry met Vasey on the Track soon after he had ordered Eather to begin his advance. 'Well,' he said to Spry, 'I've bloody well fixed that.' It was not quite that simple; but it did not mean that Vasey was impetuous. As he had shown in Greece and Crete, and was to show again, he was the master of the quick appreciation which he followed with immediate orders.[6]

The 25th Brigade contacted the enemy on 8 November and a heavy fight began which lasted for three days, but with the help of pressure from the 16th Brigade the Japanese regiment, forming the rearguard of the force preparing to hold the beach-head, was crushed. Possibly 600 Japanese were killed in the area; others were lost in the rugged country or, like their commander Major-General Horii, were drowned trying to cross the Kumusi. By the afternoon of 13 November the first troops of the 25th Brigade were on the Kumusi. The next day Blamey summed up the operation: 'The greatest factor in pressing the continuous advance has been General Vasey's drive and personality. He pushed right forward himself the whole time and so kept his brigadiers pressing on.' In later years Vasey liked to recount that after the battle his forces captured the Japanese commander's plans and found that he had a similar plan for enveloping and destroying the Australians. Vasey had beaten him to it by half an hour.[7]

*Gona and Sanananda*

The battles of Buna, Gona and Sanananda, which eventually drove the Japanese out of Papua, were some of the hardest fought of the war. Two Allied divisions, the 32nd US Division and the 7th Australian Division, were employed to destroy approximately one Japanese division fighting stubbornly, indeed fanatically, from prepared defences. The terrain was a mixture of swamp and jungle with the Japanese occupying the firmer ground.

While the Americans attacked Buna, Vasey's 7th Division had the tasks of capturing Gona and Sanananda, but his two brigades were exhausted and understrength after their gruelling march over the

35 Headquarters 7th Australian Division, Soputa, during the Buna, Gona, Sanananda battles, December 1942. (AWM Negative No. 30209/5)

mountains. Nevertheless, with the help of troops from the 126th US Regiment, they pushed on until held by the Japanese defences. In the last week of November a further brigade, the understrength 21st, arrived to relieve the 25th Brigade, and on 9 December, after amost 40 per cent casualties, they took Gona.

Meanwhile the battle on the Sanananda Track wore on. Early in December the 30th Brigade joined the fight, but little progress was made. It was not until after the fall of Buna to Eichelberger's Americans and the 18th Australian Brigade in early January 1943 that the full Allied strength could be brought to bear on Sanananda. Wootten's 18th Brigade, supported by tanks, was brought across from Buna, and on 22 January 1943 the shattered village of Sanananda was captured.

During the fighting eight brigades had been under Vasey's command, but at no stage did his forces equal the strength of a full division. It was a terrible trial for a commander since there was no alternative to attacking each defensive position individually. Each time they attempted to outflank the Japanese positions, the Australians found themselves stuck against a further fanatically defended strong-point. Vasey's artillery consisted of only four guns, the meagre supply of ammunition had to be flown over the mountains in uncertain weather, and the troops were sick with malaria.

It was Vasey's task to maintain the morale of both his soldiers and his commanders. He made it a rule that he or one of his senior staff officers should visit one of the units at Sanananda each day. Usually this involved slogging through mud up to the ankles or often higher, even up to the knees. Each officer was timed and there was rivalry among Vasey and his staff to see who could complete these visits in the fastest time. Lieutenant-General Sir Reginald Pollard, then GSO1 of the division, recalled that on one occasion he accompanied an American ration party which was to take supplies to an American road-block. The route led through jungle possibly alive with Japanese. The party returned without delivering the supplies and Pollard came back to Vasey's headquarters fuming. 'You were with them. Why didn't you order them to go forward', commented Vasey. 'I couldn't', replied Pollard, 'I was only accompanying them, and their commanding officer wouldn't listen to me.' 'All right then', rejoined Vasey, 'you will lead them in tomorrow.'[8]

Despite his efforts to maintain his soldiers' spirits Vasey himself began to feel the strain of the constant close-quarter fighting. In mid-December Blamey in Port Moresby suggested to Herring, the corps commander at the battlefront, that Vasey might be 'over-wearied; he has had enough to cause this'. He said that he was prepared to replace him with Allen who was 'quite fresh and fit. It would be a mistake to let Vasey wear himself out, there is still much ahead of us.' Herring replied that he 'preferred Vasey tired to Allen fresh'.[9]

Vasey's problems were compounded by MacArthur's anxiety to conclude the campaign successfully before the Marines on Guadalcanal had achieved a similar success. At one stage he issued an order to 'Take Buna at dawn', and he told Eichelberger 'to take Buna, or not come back alive'. Yet MacArthur never understood the situation at the front. While his headquarters thought that the enemy strength was no more than 4000, in fact the Japanese force was close to 9000 strong. In Port Moresby MacArthur told General Herring: 'This situation is becoming very serious. If we can't clear this up quickly I'll be finished and so will your General Blamey.' Already Vasey had written to his wife that he had no idea that the Japanese 'could be as obstinate and stubborn as he has proved to be... the Jap is not playing to our rules'. Some idea of Vasey's dilemma can be gauged from his reply to a battalion commander at Gona who complained of the heavy and unnecessary casualties. Vasey sent him a short note: 'Canberra must have news of a clean up and have it quick or we will both go by the boot'.[10]

Caught between the vice of pressure from the Japanese, apalling terrain, lack of supplies and MacArthur's anxiety, on 2 January 1943 Vasey wrote to his wife: 'For weeks and weeks now I have been trying to make bricks without straw, which in itself is bad enough, but which is made much worse when others believe you have the straw'. By mid-January he was at the point of desperation, but the Japanese situation

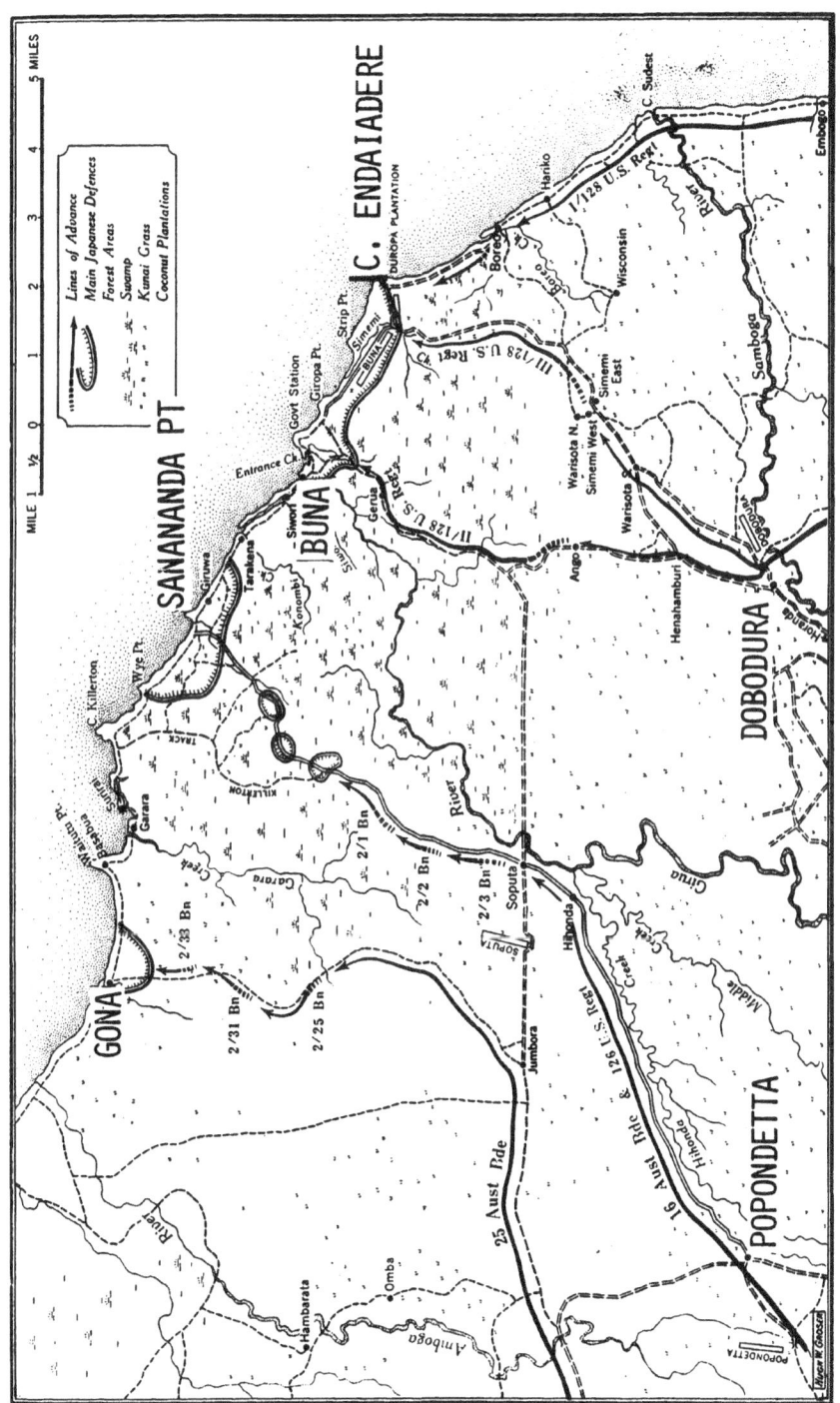

26 Allied advance on Buna, Sanananda and Gona, 16–21 November 1942

was even more desperate. They broke first, and the following morning they began to withdraw. Vasey's depression vanished and he phoned the corps commander, Eichelberger, who had relieved Herring, with the good news—in his own words, 'the bugger's gone'. With characteristic dash Vasey gave them no quarter and quickly captured Sanananda.[11]

The battles at Buna, Gona and Sanananda proved to be a test of command and leadership in every respect and at all levels. General Eichelberger wrote of Vasey that, 'Even after many weeks in the jungle, he looked like a commander'.[12] He had won the respect and confidence of his soldiers and his superiors in a campaign which, after the rapid advance from Kokoda to the beach-heads, offered few possibilities for tactical manoeuvre or for the deployment of superior firepower. It was a grim, slogging match with a premium on individual initiative and fortitude, and on mutual confidence between officers and men. It had been an exacting introduction to jungle warfare on which Vasey was to build as he planned the next round.

*Nadzab and the Ramu Valley*

The months following the gruelling battles at Gona and Sanananda showed Vasey's full value. On 18 April 1943, after leave and sickness, he returned to command the division on the Atherton Tableland in northern Queensland to find his formations decimated by malaria and battle casualties. Brigadier Chilton recalls taking command of the 18th Brigade and finding between 200 and 250 men on parade. Yet by August Vasey had the division ready for a new offensive. Reinforcements were absorbed, units were trained and commanders studied the lessons from the Papuan campaign. Vasey himself prepared detailed papers on the lessons, stressing that the Australian soldier was superior to the Japanese. He emphasised physical fitness, determination and a high standard of individual training, but he resisted what he described as the 'tendency of clouding "Jungle Warfare" with too much "Hoodoo"'. He thought that 'the first essential was thorough training in more open warfare followed by such specialist training as was necessary for the jungle'.[13] He took particular interest in training his officers in tactics.

While the 3rd Australian Division carried the fight to the Japanese around Salamaua, plans were prepared for the capture of Lae. It was decided that the 9th Australian Division would land by sea north-east of Lae and advance west along the coast, while the 7th Australian Division would either advance overland or air-land in the Markham Valley and converge on Lae. On 25 July Vasey and his advanced headquarters flew to Port Moresby and that day he attended a planning conference at HQ New Guinea Force. His chief administrative staff officer, Lieutenant-Colonel (later Major-General) L.G. Canet, recalled that after the conference Vasey returned to his headquarters and disgustedly exclaimed, 'They've got a dog's breakfast of a plan to capture Lae'.[14] The plan relied on using the Bulldog–Wau road, not yet completed, to move troops and supplies into the Bulolo Valley, and then on using

36 Major-General George Vasey, Commander of the 7th Australian Division, 1943. (AWM Negative No. 52620)

another road to advance to the Markham Valley. Although an American parachute battalion was to capture Nadzab it was uncertain whether the following Australian brigade would reach Nadzab by land or air.

That same day Vasey returned to New Guinea Force Headquarters and proposed a plan to fly the bulk of his force into Nadzab. Eventually it was agreed that he should use the whole of the US 503rd Parachute Infantry Regiment to seize the Nadzab airfield. Both the 25th and 21st Brigades would fly into the airfield while the 2/2nd Pioneer Battalion crossed the Markham River from the south. Although Vasey and his staff had some experience of the problem of air supply during the Papuan Campaign, Nadzab was the first Allied operation of its kind and the staff had little knowledge on which to base their plans. Vasey worked closely with Major-General Ennis Whitehead of the US 5th Air Force and plans were developed which went down to small but vital details such as the weight of loads for each aircraft.

On 4 September 1943 the 9th Division began its seaborne landing and the following day Australian artillery and American paratroopers under Vasey's command seized the Nadzab airfield. The 25th Brigade flew in soon after, and Vasey urged it on to capture Lae. Blamey had intended the 7th Division to play second fiddle to the 9th Division, holding off Japanese reinforcements while the 9th Division captured the town. Vasey had grander plans and after a number of sharp engagements, on 16 September troops of the 25th Brigade entered the damaged township, two hours ahead of the first troops of the 9th Division.

Now Vasey reversed his axis of advance by 180 degrees, and set out up the Markham Valley towards the head waters of the Ramu Valley. The 2/6th Independent Company (Commando) was flown to a pre-war airstrip some 40 miles (65 kilometres) north-west of Nadzab and on 17 September, with great audacity captured Kaiapit from a Japanese detachment. A proud Vasey called it an 'extraordinarily fine effort'.[15] Immediately he began to fly in troops of the 21st Brigade, but the operations of the 9th Division in the Finschhafen area were absorbing most of the available air transport. Strength could be built up only gradually. Nevertheless, resupplied by air, Vasey pushed his troops forward a further 40 miles and occupied Dumpu on 4 October, enabling MacArthur to move his advanced air bases forward. These operations were outstanding examples of the use of air power at a time when it had not been exploited by the Allies in such a fashion. Slim's forces in Burma were to follow a similar pattern months later. Vasey's quick thinking and aggression had thrown the Japanese off balance and his troops had gained important strategic positions.

While the fighting at Finschhafen continued, Vasey was forced to creep forward slowly into the rugged Finisterre Range toward Bogadjim. On 7 October the commander of New Guinea Force, Lieutenant-General Sir Iven Mackay, wrote to Blamey: 'There never was such a mobile force as Vasey's and the difficulty of 1 Corps and NGF will be to restrain him from going too far—he still has his eyes on Bogadjim and even beyond'. On 18 October, Vasey's corps commander, Lieutenant-General Sir Leslie Morshead, wrote to him: 'I appreciate the urge to go forward and to do things, but now it is a case of holding your horses'.[16]

By late December Vasey was once more able to advance. On 27 December 1943 the 21st Brigade stormed Shaggy Ridge, and during the following month, when given orders to 'contain hostile forces' while the Americans advanced along the coast, he organised a number of small offensives which by the end of the month found his troops over the crest of the Range. It was in accordance with his principle of doing what the enemy lets you do until you are stopped. He wrote later: 'I have little doubt that, had we been told to do so, we could have got to Bogadjim shortly after we arrived...It will always seem to me a pity that we did not take advantage of that favourable situation.' Although Vasey gave

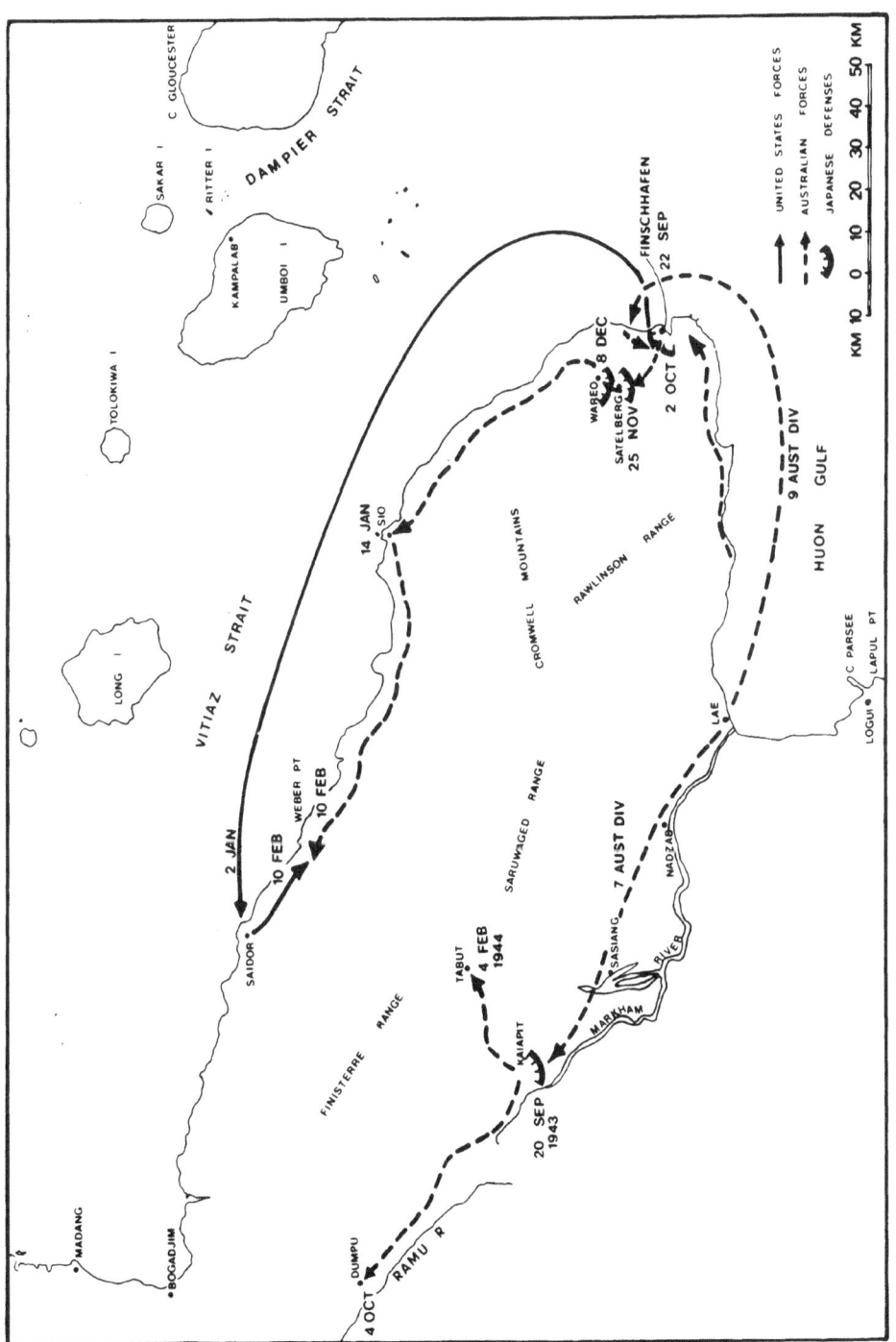

27 The envelopment of the Huon Peninsula, September 1943–February 1944

37 Major-General George Vasey (GOC 7th Division) points out positions in the Finisterre Ranges to General Sir Thomas Blamey and his Chief-of-Staff, Major-General Frank Berryman, after their arival at Dumpu, November 1943. (AWM Negative No. 62284)

credit to the brigadier who commanded the final operation, he himself had been responsible for the plan; it was a measure of the man that he always gave his subordinates full credit for successful operations.[17]

Shortly afterwards the 7th Division was relieved, but by this time Vasey had become ill and had been evacuated. He had spent himself completely. During the early months of 1944 he again began to retrain his division, but in June 1944 he became seriously ill with polyneuritis, a viral infection of the nervous system, and was forced to relinquish his command. While he was in hospital in Brisbane, close to death, soldiers from the division would stop nurses in the street to ask after their former commander. He fought back with typical courage and it took many months of convalescing before he began to regain his strength, but although not completely well he was determined to command a division in battle once more. In early 1945 he was given command of the 6th Division and almost immediately flew north to take over the division, fighting around Wewak on the north coast of New Guinea. On 5 March 1945, just out of Cairns, the plane crashed into the sea and everybody on board, including Vasey, were killed. He was not quite 50 years of age.

*An Australian style of leadership*

In over four years of hard campaigning Vasey was faced with a succession of difficult assignments. As AA & QMG of the 6th Division he had had to organise the reception and training camps in Palestine and then to overcome the exceedingly difficult supply problems during the Cyrenaica campaign. Then followed command in Greece and Crete in which his men were outnumbered and forced to give ground, eventually being evacuated in most trying circumstances. It is easier to appear in a good light when your troops have the initiative; it is a greater test of leadership when faced by a retreat. The next tough assignment was to take over from Allen and to lead his division in a campaign which tested the morale and stamina of commanders and soldiers alike. Finally, the Nadzab operation demanded most intricate planning in the first Allied operation of its kind in the war. This and the subsequent maintenance of the 7th Division in the Markham and Ramu Valleys pioneered the supply entirely by air of large isolated forces.

John Hetherington observed that: 'New Guinea revealed Vasey's wealth of knowledge of every aspect of military tactics, organization and administration, his inventiveness and adaptability, his willingness to experiment, his vast powers of administration.' General Canet recalled one occasion when Vasey explained a proposed course of action: 'You wouldn't get full marks at a staff college for that solution', said Canet. 'Of course I wouldn't', replied Vasey, 'but the Japs will be expecting me to follow the conventional staff solution. We need to do something different.' Yet Vasey never hesitated to point out that 'it is only complete familiarity with your text books and manuals which makes constructive thought possible'. Vasey's GSO1, Colonel W.T. Robertson, recalled that the GOC often produced his plan before he, as GSO1, could finish his appreciation. On the other hand Robertson thought that there was a close working relationship on the headquarters with little interference from the GOC.[18]

Vasey believed that a commander's first task was to build up *esprit de corps*, and that this was based on discipline and confidence. These views were similar to those expressed by General André Beaufre over 20 years later: 'An army is an organized mob and the cement which bonds it together is discipline and mutual confidence. More therefore than all plans and schemes based on material factors, the art of battle consists in maintaining and strengthening the psychological cohesion of one's own troops while at the same time disrupting that of the enemy's.'[19] Vasey linked discipline to the personal qualities of the commander. In his view the commander had to know his job and his own mind, as none was quicker at finding out a commander's weakness in these matters than the soldier. His training methods centred around developing a spirit of confidence. First, the soldier needed to have confidence in himself; that is, that he was well trained and physically fit. Second, he needed confidence in his weapons: that he was well equipped and possessed a mastery of his weapons. Third, the soldier needed confidence in his leaders, and it was in this respect that Vasey concentrated most of his energy.

Building on an essential basis of knowledge and experience he set out deliberately to implant his personality on his division. He explained later:

> The day of a commander in the field is one of constant vigilance. He must continually be getting about to see what is actually happening and be sure that his subordinates are in fact doing what is required of them in the best possible

way. This getting about not only keeps the commander informed by first hand knowledge, but the confidence of the troops is raised when they constantly see him amongst themselves. No walk I ever did in New Guinea failed to pay a dividend, either by seeing or improving something or by maintaining the morale of the soldiers by having a chat with those whom I met.[20]

Vasey, of course, had the happy facility of claiming men's affection as well as their respect. Partly it was his picturesque language and his sense of humour, certainly it was his lack of vanity and self-importance, but most important, it was his deep humanity—his concern for his soldiers as individuals—which won their affection. In the middle of the Sanananda campaign he signalled HQ NGF asking for rations to have the greatest possible variety, whether air-dropped or otherwise, and he demanded 'smokes on a liberal scale', 'authentic world news', and 'canteen stores'. Chaplain D.L. Redding, who served under him, said that Vasey:

> knew the very thoughts of the men he commanded. He knew of their work, their difficulties, their trials and disappointments because, as far as possible, he lived amongst them. He talked with them not only on ceremonial occasions—not only in settled camps—but on the jungle tracks and mountain steeps—where the going was hard, and tucker poor. He talked with them as a man talks with his friends. Many of us learnt the inspiration of his presence—many of us took fresh heart because he was on the trail. He never lost the common touch, and that too is, I am sure, a mark of real greatness.[21]

This close relationship with soldiers was vital in instilling confidence and *esprit de corps*, but it provided Vasey with something more. It gave him a 'feel' for the battle. As Brigadier Shelford Bidwell put it, 'Good generals have what superficially appears to be a sixth sense, a sort of intuition or insight—what the Germans call "fingertip-feeling"'. Instinctively they place themselves at a point of crisis and arrive quickly at correct decisions without a long preliminary appreciation. The higher the command position, the more difficult it is for a commander to maintain the feel of the battle. Field Marshal Montgomery relied heavily on liaison officers, but as an army group commander his ability to see all of the battlefield was limited. On the narrow fronts in New Guinea Vasey could and did move close to the front line. One day near Lae he came across some men resting beside a jungle track. On learning that they were short of tobacco he handed out a packet of cigarettes. As he moved away one of the soldiers looked up at his red-banded cap and said to his mate: 'You can't beat these Salvation Army bastards. They're with you all the way.'[22] Vasey's personal style of command was so much part of his nature that it is doubtful whether he would have been happy at a higher level of command where there is less opportunity for direct contact with fighting troops. Certainly he was not anxious for a corps command provided those who were promoted were not greatly junior to him or lacking in ability.

Vasey epitomised the heroic style Australian commander and leader who, although more common at the battalion level, has been found less

38 Major-General George Vasey chatting with a soldier in the foothills of the Finisterre Ranges, November or December 1943 (Vasey Collection NLA)

frequently at the divisional level. One Australian general said that it was impossible to fault Vasey's leadership. 'The only doubt remaining in my mind of his ability to be a great commander', said the general, 'was whether he would have been ruthless enough. In practice, he might have been, but his capacity for ruthlessness—that is, ruthlessness in the military sense, which means that one must sometimes sacrifice lives now to save many more lives later—was never proved.'[23]

In his profile of Vasey, Hetherington claims that the attack of polyneuritis, with symptoms of pain and numbness to the extremities of the body, had caused permanent injury to Vasey's health and that the strain of commanding the 6th Division in 1945 'would almost certainly have left him a chronic invalid'. Further research indicates that this was indeed the case. Vasey lobbied strongly to return to action, and against his better judgment Blamey bowed to the wishes of the government and appointed him to command the 6th Division. But the clearance by his medical officer was more a tribute to his persuasiveness and determination, than to his restored health. There is much truth in the view of some of his staunchest friends that his death was 'a benign act of Providence'. As Brigadier Clive Disher, senior medical officer of the 6th Division in the Middle East, and then of New Guinea Force explained, he might have made a superb CGS, but if he had failed through poor health, it would have been an even greater tragedy.[24] With competition from outstanding staff officers like Berryman and Rowell it is unlikely that Vasey would have become CGS, and after the exhilaration of war it is doubtful whether the man of action could have settled easily to a senior staff appointment.

The loss of Vasey was deeply mourned in the Australian Army. To Blamey he was the ideal commander—brave, resourceful, 'a cheerful and inspiring leader under all conditions... destined by training and capacity to rise to the very top of his profession'. General Kenney, commander of the 5th US Air Force, wrote that 'Vasey was one of my favourite generals of anyone's army and one of the ablest leaders I have ever come in contact with'. General MacArthur, who took a keen interest in the Nadzab landing, personally seeking Vasey's views in visits to his headquarters, has the final word on George Alan Vasey, CB, CBE, DSO and bar, Greek MC, American DSC: 'I regarded him as a superior division commander. He was an excellent soldier in every sense of the term and I held him in deep admiration and esteem. He was typical of the Australian fighting man. None better have ever been produced...'[25]

**Part Three**
Post-Second World War

# Lieutenant-General Sir Horace Robertson: Commander-in-Chief British Commonwealth Occupation Force

RONALD HOPKINS

Lieutenant-General Sir Horace Clement Hugh Robertson (1894–1960) provided, in his colourful career, some of the great paradoxes of Australian military history. As a graduate of the Royal Military College, Duntroon, in an era when these officers were expected to form the expert staff of units and formations commanded by citizen soldiers, Robertson made his mark as a dashing commander in the First World War and continued his reputation into the Second World War. Yet partly because of his own self-confidence and ego he was to be denied the opportunity to win further laurels as a commander in battle. His finest achievement was command in peace—admittedly a very special command—that of Commander-in-Chief of the British Commonwealth Occupation Force (BCOF) in Japan, which required wide-ranging skill and experience. His success has left military historians pondering how he would have performed if given greater opportunity during the Second World War.

Robertson was born at Warrnambool, Victoria on 29 October 1894. His parents sent him to Geelong College where one of his contemporaries was John Rogers who became Director of Military Intelligence during the Second World War. Rogers remembers that Robertson was particularly mild-mannered, never displaying any of the ego which later became the hallmark of his personality. From Geelong College he was successful in the competitive examination for entry to the Royal Military College in March 1912. The College had opened in 1911 so that Robertson was one of the 41 cadets of the second intake.

In his training and studies as a staff cadet, Robertson was above average; he had a low placing on entry but graduated eighth in a class of 30. The course included considerable academic study as well as practical and theoretical military training. In his third year, when most of the appointments to cadet rank were filled by the senior class, he was one of the first group of staff cadets in his own class to be promoted. This marked him, in the eyes of the College, as possessing above average leadership qualities. The 1913 entry to the RMC brought a staff cadet named Robinson with hair as black as Robertson's was red. From then on, throughout the rest of their lives, they were known as 'Red Robbie' and 'Black Robbie'.

## First World War

War with Germany broke out in August 1914. As soon as recruiting began the two senior classes at Duntroon were specially graduated for active service, commissioned in the Permanent Military Forces and posted to units in the Australian Imperial Force. Lieutenant H.C.H. Robertson joined the 10th Light Horse Regiment in Western Australia. Together with the 8th Light Horse (Victoria) and 9th Light Horse (South Australia and Tasmania) it formed the 3rd Light Horse Brigade. They were all in Egypt about Christmas 1914 and commenced training in their mounted role from their camp at Maadi on the outskirts of Cairo.

It had not originally been intended that the Light Horse would be required on Gallipoli, but within a month of the landings on 25 April 1915 reinforcements were needed desperately. On 16 May the 3rd Light Horse Brigade, without their horses, sailed for their grim introduction to battle. Early in August they were ordered to capture enemy positions at the Nek and Baby 700, but as C.E.W. Bean observed, their 'unaided' effort was 'almost hopeless'.[1] The Light Horse were faced with a narrow frontage confined by declivities on either side. Dominating hostile machine guns had not been silenced as ordered, nor was artillery fire available to force the enemy to take cover at the moment of assault. The attack failed with disastrous casualties.

Robertson was one who came through unharmed although losses were such that his promotion to captain dated from that day. Not long afterwards, the 10th Light Horse seized and consolidated enemy trenches on part of Hill 60 in a difficult and confused situation. The strength of the regiment had been reduced by casualties to 180; they had been organised into two squadrons for the attack and one of these was commanded by Captain H.C.H. Robertson (aged twenty). His reputation for courage of a high order and as a leader of promise had quickly become established. He remained with his unit on Gallipoli until the evacuation. He would not have found harder fighting nor more difficult and unpleasant conditions in all the rest of his service. As a result, his confidence was established on a firm basis and his knowledge of men was greatly enhanced. Important tactical lessons would have been ingrained; in particular, the relationship between fire and movement and the value of careful reconnaissance.

Back in Egypt the lighthorsemen joyfully regained their horses and concentrated on making themselves fit for mounted operations in the Sinai Desert, where the Turkish forces were threatening the destruction of the Suez Canal. Early in April 1916, Major-General Chauvel moved his ANZAC Mounted Division, including the 3rd Light Horse Brigade, about 16 miles (25 kilometres) into the desert east of the Canal.

It was not until 5 August 1916 that the Turkish force was defeated at Romani, about 25 miles (40 kilometres) east of the Suez Canal. The 10th Light Horse played little part in the battle until late in the day when the 3rd Light Horse Brigade began to threaten the enemy's southern flank.

During their slow and strongly contested withdrawal, the Turks endeavoured to hold onto the scattered oases where water was available and the Australian mounted troops were forced to move wide into the desert to outflank their positions. Robertson gained plenty of experience as he had been promoted to major in May and commanded a squadron for the remainder of the year.

The ANZAC Mounted Division then advanced across the Sinai Desert, and on 23 December attacked Magdhaba. The 10th Light Horse played an important part in the battle, and Major Robertson, acting in command of the regiment, was awarded the DSO for his skilful and determined leadership. The Official history carries a full description of the action at Magdhaba.[2] At the outset, Robertson was directed to block a Turkish escape route. After a fast movement under fire he ordered a charge which captured a number of enemy. He then directed the regiment against hostile redoubts, pinning down the garrisons and over-running them after heavy fighting. The 10th Light Horse took 772 prisoners in all.

The ANZAC Mounted division then moved to Rafah on 9 January 1917 where the enemy forces were overcome after a full day of heavy fighting, in which the 10th Light Horse participated. As at Magdhaba, victory came only at the last moment. A day earlier, the commanding officer of the regiment recorded his 'high appreciation of the manner in which the duties of senior major have been carried out by Major Robertson who has acted in that capacity since 18.11.1916'.

Shortly afterwards Robertson left the regiment for staff training which, in due course, brought his appointment as General Staff Officer, Grade 3 (GSO3), at Headquarters, Yeomanry Mounted Division. But an accident during the training period, resulting in a broken leg, caused a set-back to his career. He was two months in hospital and returned to active operations only to find that his leg continued to give trouble. Robertson stuck it out during the long, rapid advance up the Palestine plain but was obviously not fit for such demanding work. Early in 1918 he was transferred to a less onerous job at HQ Delta Force in Cairo. It was characteristic of the man that he never admitted physical incapacity. It seemed as if he feared the consequences of an adverse medical report appearing on his official record.

The year 1916 had been a formative one of continuous mounted operations. Robertson gained great experience in the use of ground, the value of mobility and the overriding influence of high morale and the other soldierly qualities. Time and again he was involved in outflanking movements designed to force the enemy from their positions by manoeuvre. When they stood fast, and this was not infrequent, the skilful use of ground or the ability to call up the led horses and make a rapid shift of position could give the attacker an advantage. He realised that his own determination and will to win were leading to success. Constant decision-making in contact with the enemy undoubtedly produced a sense of timing and an ability to calculate the risks as well as the prizes resulting from bold, resolute and rapid action. Lessons such as these, and the judgment they developed, remained with Horace Robertson. Like most battle experiences, they were unforgettable. Strongly evident in his maturity during the campaigns of the Second World War, these attributes so strengthened his grasp of leadership and tactical skill that he appeared head and shoulders above his contemporaries during the early stages of that war.

He seemed to have regained his fitness by the end of 1918, although still in Egypt and not directly involved in Allenby's final offensive. Nevertheless, that great victory, and its lessons, reinforced his conclusions regarding the importance of seeking to destroy the enemy's determination and confidence. At war's end Major Robertson was awarded the Order of the Nile (4th Class); he had already been thrice mentioned in despatches.

*Between the wars*

The writer served on Robertson's staff at AIF HQ in Cairo in mid-1919 where, as Assistant Adjutant-General, he had charge of all Australian administration and repatriation. He impressed me with his efficiency and invariably appeared to be in complete control of any situation. This characteristic stayed with him throughout his career. He was friendly with a warm personality and a good-humoured approach to official relationships. The autocratic manner of later days was not in evidence nor had he developed the trait of displaying his knowledge and judgment in many fields to the discomfort of his immediate disputants.

Return to Australia and peacetime soldiering brought appointments concerned with Citizen Force training. He threw himself wholeheartedly into this. A novelty of approach and an insistence on realism brought sustained interest and enthusiasm. He was an early selection for the Staff College and attended Camberley (UK) in 1923–24. Afterwards he specialised in small arms training before returning to become Chief Instructor at the Small Arms School, Randwick, NSW. He modernised and developed this training throughout the Australian Army and also became responsible for four Vickers medium tanks which arrived in 1929.

Robertson's next appointment, on the staff of the Inspector-General (IG) of the AMF, was short-lived. After only seven months with the IG's Department he somewhat suddenly arrived in Brisbane where I was then Brigade Major of the 1st Cavalry Brigade. He was considerably my senior with a distinguished war record yet was posted as Brigade Major of the 7th Infantry Brigade. There was no doubt that he had fallen foul of authority. The details were not known to me although he made no effort to conceal his resentment at the treatment he had received. My own position might have been invidious but Robertson could not have been friendlier.

There was a sequel. Some time later, he came to Melbourne on leave and called to see me at Victoria Barracks during office hours. His only purpose, he stressed, was to arrange golf; otherwise he would not have come near Army Headquarters. He was still antagonistic about his treatment in 1930. My reaction was to tell him that his proper course would be to make quite sure that the Army authorities understood that he bore no ill-will for his previous treatment, that he was keenly interested in his infantry post and not in the least put out by being denied a cavalry one. To this end, if he would give me his card (visiting cards were *de rigeur* in those days), I would take it into the CGS. I remember adding that his visit most certainly had been remarked upon already by one or another of the inmates of our large headquarters; if he did not recognise the fact, it also would be remembered against him. He saw the point, was ushered into the CGS and apparently found that a placatory attitude was appreciated. In consequence he appeared to forget his rancour and became a great success in his next major posting to the Royal Military College.

Robertson's appointments as Instructor in Tactics and later Director of Military Art between 1934 and 1939 were formative periods enabling him to consolidate his thinking about higher defence matters. He had an important essay on 'The Defence of Australia' published in the UK *Army Quarterly*, but such was the attitude in the Australian Army that this contribution to the defence debate was not praised, but criticised for its impertinence. Also during this period he began to display attitudes which earlier had not been quite so noticeable. Brigadier Geoffrey Solomon, then a cadet at Duntroon, recalled of Robertson:

> Vain, self-centred and arrogant, he looked and acted the part which he had written for himself. Fortunately he had the other qualities which were necessary if he were to bring it off. Altogether he was an infuriating man of great abilities, as many had found already and many more...were to find during his remarkably successful career.[3]

Robertson's appointment in March 1939 as Commandant 7th Military District, which included control of the newly raised Darwin Mobile Force, seemed a certain indication that his high professional qualities were receiving recognition at Military Board level. At the time, shortly before the outbreak of the Second World War, the Darwin Mobile Force was the only regular infantry unit in the Australian Army.

*Second World War*

After the war began in 1939 and the Australian government decided to raise the 6th Australian Division for service abroad, the Prime Minister also announced that the brigade and unit commanders would be selected from experienced Citizen Force officers rather than from the regulars of the Staff Corps. There was naturally much unhappiness among members of the Corps who had been debarred from unit and higher command in the war of 1914–18. The official attitude was softened before long so that regulars as well as citizen soldiers became available for command. But the many Staff Corps officers selected to join the 6th Division were all employed in staff appointments. When the 7th Division was raised its commander was Major-General J.D. Lavarack, a regular, and Brigadier

H.C.H. Robertson became one of the three brigade commanders. Although he and Milford, who became CRA, were the first RMC graduates to gain selection for senior AIF command, this was, to some extent, fortuitous; the 6th Division already contained RMC graduates who probably would have been considered equally suitable.

At this stage in Robertson's career it is of interest to review his attributes and limitations. The official historian, Gavin Long observed:

> Robertson...was considered an outstanding leader and a particularly able trainer. He had served in the light horse on Gallipoli and in Palestine, and between the two wars had left his mark particularly as Chief Instructor of the Small Arms School, as Director of Military Art (senior member of the teaching staff) at the Royal Military College and later in command of the garrison at Darwin. He was a confident commander, sure of himself and of any troops he had trained. He believed that physical hardness was one of the first necessities of military efficiency, he took special pains to insist on smartness of dress and deportment. Some successful commanders have been unassuming and inconspicuous, others have striven to present their men with a picturesque figure to look up to and talk about, perhaps to emulate. Robertson was of the latter type, just as definitely as Mackay, for example, was of the former. He was ambitious and was criticized for making no secret of it, but this ambition was with him more than a personal affair, and embraced the men he commanded'.[4]

He was lean in appearance with ruddy complexion, flaming red hair and moustache. One's impression on meeting him was of piercing blue eyes and a strong personality. It was said of him, noted the *Bulletin* in 1960, that he enjoyed being a soldier and considered the army to be a splendid machine provided its various parts functioned correctly. If a cog seemed rusty it was cleaned and oiled; if it failed then it was replaced—with no beg pardons.[5] Robertson had proved his courage in the First World War, and this, coupled with wide knowledge, physical endurance and a strong and unyeilding spirit made him sure of himself.

These qualities also brought the admiration and respect of his men. In retrospect it became close to affection. Like Sir Arthur Wellesley leading a British army in Portugal ten years before Waterloo, his men believed that his personal leadership, and presence, brought success. In the midst of the battle for Tobruk, two sergeants of the 19th Brigade met briefly. 'Where's Red Robbie' said one. 'I haven't seen him about.' Said his mate, 'By now he'll be swimming out to the *San Giorgio* with a cutlass between his teeth!'[6] One of his soldiers wrote, long after the capture of Tobruk:

> Robbie had a brief command as Brigadier of the 19th Brigade. Its brevity did not prevent it from becoming historic, neither did it prevent the name of 'Red Robbie' becoming a legend...he became the youngest brigadier...and he received the toughest nut to crack. The 'nut' was the 19th which comprised 2,000 troops from three different States. They had been cast out from their original formations and did not like it. Robbie did not even think about the difficulties or the lack of *esprit de corps*. In four months he led one of the finest brigades to see action in the Middle East.[7]

It is an astonishing thing that, except for a fairly quiet period at the end of the war, Robertson's actual operational contact with the enemy in the Second World War lasted no more than 55 days. This period began on 1 January 1941, when the 19th Infantry Brigade came into divisional reserve before the battle for Bardia. It ended with the withdrawal of the 19th Brigade to Gazala on 24 February after the capture of Benghazi on 7 February.

In the 250 miles (400 kilometres) from Bardia to Benghazi the 19th Brigade had endured extremes of cold, rain, dust and gale-force winds. For much of the time they existed on hard rations. Often their captures of vehicles and petrol were the only means of maintaining the momentum of the advance. Fitness, high morale and determination enabled the brigade to gain an outstanding success. A great deal of the credit for the conduct of operations, as well as the standard of fitness and training of all ranks, clearly belonged to Brigadier Robertson. He showed marked tactical skill, high personal qualities and a capacity for command unmatched in his division. His energy and determination were outstanding. Witness the situation near Martuba, some nine miles (fourteen kilometres) east of Derna, early on 25 January 1941. Sporadic firing could be heard ahead as the long column of motor transport carrying the 19th Brigade halted and the troops scrambled out of their vehicles. Only a few light tanks of the divisional cavalry could be ahead, yet, standing on the roadside was their brigade commander, waiting to give his orders for the advance on Derna. This was his kind of leadership and all ranks gained inspiration from it.

At Bardia, Robertson arranged for the barrage to be maintained well beyond the enemy defensive line although this entailed commencing the attack without it, owing to a temporary shortage of ammunition. Later he reported with satisfaction that his judgment had proved correct as the enemy morale did not break until the barrage had almost reached them. Failure in the attack could have resulted if the barrage had stopped at that point. Capture of the enemy redoubts, which the 17th Brigade had failed to take on the previous day, ensured the collapse of hostile resistance. Robertson's brigade took a large number of prisoners, including two Italian divisional commanders. Being well forward, he could take steps to complete the defeat of the enemy.

Somewhat similarly at Tobruk, he had joined the leading elements of the attack. Realising that enemy morale was cracking, he quickly ordered two light tanks forward under a white flag to threaten the enemy with total disaster unless they surrendered immediately. When this was signalled, he drove in himself and secured the surrender of the Fortress commander. His actions were criticised on each occasion as self-advertisement. But the opportunities had been created by Robertson's brigade and had not been recognised by Major-General Mackay and his staff.

Robertson had the principal task in the Tobruk battle. He selected the objectives cleverly and controlled the operation personally. His aim was the destruction of the enemy centres of command and communication. In the process, he accurately assessed the surprise effect of a fast advance and the enemy's inevitable loss of confidence. After observing Robertson's performance at Tobruk the war correspondent, Chester Wilmot, thought that he possessed 'undoubted brillance'.[8]

But at Tobruk Robertson began to sow the seeds of his own fall from grace. When Major K.A. Wills, an intelligence officer of the 1st Australian Corps, asked him to take the surrender of the colonel commanding the Italian artillery, Robertson replied: 'Throw him back. Nothing below admirals and general for me today.'[9] At a dinner for Prime Minister Menzies in Benghazi he incurred the displeasure of General Blamey by talking about his own exploits and by displaying a captured flag.

I received a letter from Robertson about this time, dated 28 January 1941. His brigade then was close to Derna. If boastfulness was evident, it was mostly about his troops: 'My people got the comd 2 Corps' and 'the following morning they got comd 61 Div.' But certainly he wrote: 'I dashed in with a couple of carriers to take the surrender of a Rear Admiral in the town'. I had sent him news of the start of armoured training in Australia. His letter concluded typically, 'I think I should have an armoured division!'

Chafing at the delays after Tobruk and frustrated at the way the Italian Army was escaping, Robertson sought out his divisional commander. In a brief interview on a hillside beyond Derna he persuaded General Mackay that the correct way to conduct a pursuit was to place all divisional resources behind one of the brigades—and that *his*, Robertson's 19th Brigade, should be selected for the task of rounding up the fleeing Italians. Mackay agreed. The next three days saw the Italian delaying position bypassed by a brilliant manoeuvre and the end of resistance with the surrender of Benghazi.

After the Cyrenaican campaign on 24 February 1941, the 19th Brigade was moved back to Gazala in order to relieve administrative pressures. By 1 March, the 6th Australian Division had been warned to make ready for the campaign looming in Greece. At that moment, General Blamey transferred Robertson to be commander of the AIF Reinforcement Depot in Southern Palestine. He was replaced in the 19th Brigade by Colonel G.A. Vasey who had been senior administrative staff officer at HQ 6th Australian Division during the recent operations. Robertson is said to have protested, without success,[10] but he had unwittingly confused the situation. His varicose veins were becoming troublesome and the withdrawal of his brigade to Gazala carried the promise of a period of rest after a very strenuous campaign. It is believed that he took advantage of this interlude to enter hospital for treatment.

John Hetherington has referred to this incident in his biography of General Blamey,[11] stating that Robertson began to complain about his legs and went into hospital for treatment *after* the 6th Division had been told it was to be part of the expeditionary force for Greece. Hetherington concluded that although Robertson did not lack personal courage he was 'unwilling to face an enemy whom he knew he could not defeat'. Robertson's categorical statements that he did not intend to go to Greece were recorded by visiting officers but these statements have been countered by others.[12] Hetherington's very serious charge appears to have had many doubtful features.

Leaving aside the reported statements and counter-statements, there can be no doubt that if General Blamey had had the slightest suspicion that Robertson was planning to escape going to Greece, he would never have employed Robertson again. But Blamey also seems to have had other plans for Robertson before the Greek campaign was mooted. In late 1940, Blamey had written to the CGS, Sturdee, that he needed an officer of 'very firm character and great administrative ability' to take charge of the AIF Reinforcement Depot in Palestine; he rejected a number of contenders available in Australia.[13] In selecting Robertson for the appointment Blamey showed that he had every confidence in his ability. Perhaps the final judgment rested with his men. They would have known, as soldiers always do, that he had never evaded his duty; this was evident from the warmth of their greeting when he met his old 19th Brigade in Palestine after their return from Greece. As they swarmed out of the train, he was recognised standing alone on the platform. As the story goes, there were delighted shouts of 'Robbie, you old bastard' as they crowded round him.

Reflecting back on this period, Gavin Long wrote in his notebook: 'It is doubtful whether there was another leader in the Australian Army with greater devotion to his job of training and commanding soldiers, or with more military learning or a more complete and carefully worked out doctrine of war'.[14] Robertson's success in training the AIF reinforcements was remarkable and the Commander-in-Chief, General Auchinleck, ordered all training organisations in the Middle East to follow his methods.

In April 1942, Robertson was promoted major-general and returned to Australia to succeed Northcott in command of the 1st Australian Armoured Division. This appointment was ideal for a cavalryman with wide experience of mobile operations. He set to work with tremendous enthusiasm to fit the division to join the Allied forces in the Middle East where its deployment had been planned. But the threat of a Japanese invasion of the mainland caused the Armoured Division to be retained in Australia and organised into three smaller armoured formations.

Towards the end of 1942 Robertson's 1st Armoured Division moved to Western Australia, where he joined that other outcast from the Australian Army establishment, Lieutenant-General H. Gordon Bennett. Robertson's exile has been widely blamed on his earlier disagreements with Blamey and was exacerbated by the so called 'Revolt of the Generals' in March 1942 when Generals Herring, Steele and Vasey tried to persuade the Minister for the Army, F.M. Forde, that Robertson should become the new Commander-in-Chief of the Australian Army. Robertson himself had advised Forde that he was prepared to accept any post offered to him, even 'that of Commander-in-Chief'.[15] It is little wonder that Blamey kept Robertson at arm's length in Western Australia. However, the latter was severely ill during his time in Western Australia and would not have been fit to command troops in action in

*39* A review of Robertson's 1st Armoured Division, Puckapunyal, Victoria, June 1942. (AWM negative No. 25467)

any event. Blamey probably knew this fact, and if so was wise not to select him.[16]

After the Armoured Division was disbanded Robertson held several temporary posts before being given command of the 6th Australian Division at the end of the Wewak campaign. When the war ended he was temporarily in command of the First Australian Army in New Guinea in the rank of lieutenant-general. In March 1946 he returned to Australia as GOC of Southern Command.

*British Commonwealth Occupation Force*

Within a month, and quite unexpectedly, Robertson was offered the appointment of Commander-in-Chief of the British Commonwealth Occupation Force in Japan, in succession to Northcott who had been appointed Governor of New South Wales. In selecting Robertson the Acting Commander-in-Chief, General Sturdee, described him as 'an outstanding commander of troops, possessed [of] a strong and magnetic personality,... our best field commander'.[17] This was the second time that he had followed Northcott, and in each case the result was electrifying. While there was more scope for dramatic action when the Armoured Division changed hands, Robertson in Japan immediately lifted morale with his personality and drive.

Some of us who had been in the BCOF before Robertson's arrival had become aware that the administration, particularly in the Base organisation in Kure, was showing signs of deterioration. Lowered morale in the Base area carried the implication of supply difficulties for the whole force. In addition everyone was hearing of a multiplicity of luxury leave centres available to US troops while only a single establishment at Ebisu,

40 General Blamey, Brigadier M.A. Ferguson and Major General H.C.H. Robertson at the saluting base, 1st Armoured Division review, June 1942. As C-in-C Blamey faced an immense and complicated task. Robertson had offered himself as C-in-C. (AWM Negative No. 25454)

a suburb of Tokyo, was available for the 40 000 personnel of the BCOF.

General Robertson brought an immediate improvement because he visited every part of his command, talked to all ranks, inspected conditions and generated hopes for rapid improvement. These were not misplaced. Poor accommodation was rapidly improved or better buildings acquired very quickly; this had particular application to Kure where conditions had been very poor. The new C-in-C also negotiated the transfer from the US Army authorities of an important group of leave centres which were exclusive to BCOF personnel; these were located in a number of historic and picturesque regions beyond the confines of southern Honshu, which was devoid of attractive tourist resorts.

Before he reached Japan, Robertson studied the basic organisation which had already been adopted. It provided for a central, integrated headquarters through which all matters affecting every portion of the force was to be handled. Such a policy took little account of the wide variations between national components as well as between the sea, land and air forces which composed them. Robertson believed that this policy overloaded Headquarters BCOF with a great number of matters which were nothing to do with it. He wished to allow each national component the freedom to train and administer its own forces in direct communication with its home government and defence organisation. HQ BCOF, in his view, should be concerned only with purely Occupation Force matters: tasks and areas, strategic and tactical control, major administrative policy concerning the whole force. These ideas were accepted by the Joint Chiefs of Staff in Australia (JCOSA) and marked Robertson's first reform.

The new C-in-C established satisfactory relationships and excellent cooperation both with General MacArthur, the Supreme Commander for the Allied Powers (SCAP), and his staff. Robertson proved to be a friendly and sensible negotiator in the many matters of joint US/BCOF interest and the United States authorities were equally helpful. In this, both parties belied the reputation they had acquired of being difficult and overbearing.

Some of Robertson's early problems concerned complaints from the British Component, the most important of which was their claim that they were not receiving their full ration entitlement. His handling of the ration problem should be taken as a model for students of administration anywhere. The fault finally was sheeted home to the British/Indian Division itself, which had received full rations but had failed to pass them on to the troops.

Difficulties were also posed by the reluctance of the British and New Zealand commanders to surrender control of some matters which should have been administered centrally. Such things may have been the result of misunderstandings but needed tactful handling by the C-in-C.

Robertson, from the outset, insisted that he was the sole representative of each of the four British Commonwealth governments participating in the Occupation of Japan. But the British government already had two other representatives: one a civilian as Head of the British Mission; the other a lieutenant-general who was the personal representative of the Prime Minister. These sniped at each other and at Robertson while the British government generally failed to support their Commander-in-Chief. The worst example of this two-faced approach concerned the withdrawal of the British Component from the BCOF, which was negotiated by the Prime Minister's representative direct with General MacArthur, without the knowledge of the C-in-C under whose command they had been placed.

The British Army had a well-known and long-established practice that Commanders-in-Chief of British forces had the honour of knighthood bestowed on them. Undeniably, Robertson was entitled to expect that he too, as C-in-C of a British Commonwealth Force, would be treated no less generously than an equivalent British service officer. But he was ignored. Angry at the unfairness of the British attitude and very conscious of the position in which future Commanders-in-Chief of British Commonwealth Forces might find themselves, Robertson decided that the precedent would have to be established at the outset, and he personally took the matter up with the appropriate authorities on a visit to London. As described to me after we had both retired, there were several Gilbertian efforts at evasion on the part of the War Office but to no avail. He was knighted in 1950. Sheer force of character and determination appeared to have played their part in establishing the rights of the appointment whatever the nationality of the occupant.

Robertson did all in his power to maintain the prestige and influence of his position, and no better man could have been selected, but a major area of difficulty lay in his relationship with the JCOSA. This body included representatives of the nations which made up the BCOF. Its members directed the policy that Robertson was to follow but seemed unable to give him clear-cut decisions or even helpful advice and guidance. Unfortunately it was not a simple matter; the source of the trouble was to be found in the differing political and diplomatic aims of the governments concerned.

The United States forces in Japan were relatively low in strength and not particularly well equipped. Japanese labour was used on a considerable scale to offset the reduced numbers of servicemen. The defence of Japan against a foreign power was therefore a serious problem and General MacArthur was known to be relying on the seven squadrons of aircraft which were part of the BCOF command, if such an eventuality arose. The MacArthur/Northcott Agreement which governed the BCOF operations seemed to commit the BCOF to this defensive role but Robertson thought differently. He 'insisted that no matter what powers had been signed away to General MacArthur...the sovereignty of the British Commonwealth Governments did not allow them to hand over the unrestricted use of their forces to any Allied commander no matter how close relations might be'.[18] The position of the BCOF in the event of an attack by another power was of great urgency and needed clear and realistic decisions, but it was not settled until a year before the Korean war when Robertson was informed 'that the defence of Japan should be regarded as included in occupation duties'.

*Korea*

When North Korea invaded South Korea in mid-1950 the Commonwealth countries rallied to the support of the United Nations and the United States forces assisting the South Koreans. Robertson played an important role in maintaining close and friendly relations with General MacArthur and his staff and in reporting the complex and rapidly

*41* Lieutenant-General Sir Horace Robertson, C-in-C British Commonwealth Forces, Korea, Mr Thomas MacDonald, the NZ Minister for Defence, and Brigadier R.S. Park, Commander of the NZ 'K' Force, after a ceremony in Japan marking the formation of the 1st Commonwealth Division in Korea. (AWM Negative No. DUNK 4615)

changing situation to the Australian Chiefs of Staff.[19] At the time of the invasion, the BCOF had been reduced to a more-or-less token force of two small RAN vessels, the 3rd Battalion, Royal Australian Regiment, and the 77th Squadron RAAF. The Communist aggressors gained a startling initial success and the South Koreans, with considerable assistance from US troops, had the greatest difficulty in preventing total defeat. In the emergency General MacArthur became the commander of the United Nations forces. He was hard put to find enough troops immediately available to halt the North Korean thrust.

The complex problems which faced Robertson in these circumstances gave him new experience of the higher direction of war. He became deeply involved in political and diplomatic matters as the changed circumstances influenced the role of Australia and the other British Commonwealth nations. The calls for support being made by the hard-pressed Americans were growing louder almost daily.

At the outset Robertson rightly understood and anticipated his government's reluctance to allow Australian units to become involved in the

Korean operations. The units were under strength and poorly equipped for war and it was thought that their small contribution would have little influence on operations. But the underlying reason was the need for regular troops to act as instructors in the National Service training which was soon to be launched. As it was, the Australian government felt compelled to offer 77 Squadron RAAF and the small naval component of BCOF to augment the United Nations forces within a few days of the outbreak of hostilities.

Robertson, on his own initiative, staved off early requests for the 3rd Battalion by accepting additional Occupation tasks in order to release further US troops for operational duty. But the pressure mounted. The Australian government became torn between the political choices of providing land forces for Korea at a cost of slowing down the plans for National Service, or denying the United States appeal and forfeiting possible opportunities for securing a treaty of alliance with them.

Forthright as ever, Robertson told the Australian Chiefs of Staff that MacArthur would like very much to be offered Australian troops. Not only did he admire their fighting qualities but their presence would immediately create an Allied command for other nations to join. Robertson warned, however, that any Australian commitment, to be of value, would need to be at brigade level since any future British Commonwealth contribution would scarcely be less than a division. This realistic report actually forecast the scale of support which would finally be adopted.

As well as commanding the BCOF, Robertson was also C-in-C of the British Commonwealth Forces in Korea. In fact he had no operational role but was responsible only for the administration of the force. Nevertheless he took every opportunity to visit the forward units and, with the glint of battle in his eye, was present in Korea in April 1951 when most of the 29th Brigade—including the 1st Battalion, the Gloucester Regiment—after a heroic defence, was overrun by superior Chinese forces. The brigade suffered severe losses with many men captured. Undaunted by the confusion of the withdrawing US units he insisted on trying to visit one of the forward British battalions in the midst of battle because he had, on the previous day, promised to see them on their regimental birthday. He was finally stopped only by the road-blocks created by the counterattacking battalion. He returned that day to Japan and peremptorily ordered his staff to round up all available reinforcements in Japan. Through his drive and organisation the 29th British Brigade was up to strength in the forward areas by noon the next day. Robertson's wisdom and experience provided great strength and support to the young Commonwealth brigade and battalion commanders in Korea and he safeguarded admirably the interests of the Commonwealth governments and their Forces. The Commonwealth Division, universally regarded as the best in the theatre, owed much to the personal efforts of the Commander-in-Chief both on its formation and when it operated in that most difficult campaign.

*A colourful and dominating commander*

In June 1951, the Australian government needed a replacement for Sir Edmund Herring, Director-General of Recruiting, who wished to retire. Robertson was thought to have suitable qualities. The requirement was for a prominent leader with strong public appeal. His experience equalled that of any other senior officer and he had other attributes, as noted succinctly in the Official History of Australia in the Korean War: 'He was undoubtedly charismatic, exerting a strong individual style of leadership... At times he aroused an antagonism, particularly among his superiors and contemporaries, for his undisguised egoism, self-assertion and occasional vulgarity.' He had been a contender for the position of Chief of the General Staff, which went to General Rowell, but as O'Neill observed, 'the authority and independence of his command in Japan had given him both compensation and insulation from the factionalism and power politics which existed within the small coterie of senior officers in Australia'.[20]

The background to Robertson's egoism and self-assertion sprang from an intense will to dominate all with whom he came in touch. This was the basis of his tactical skill and was part of all his activities. He only used aggression when he felt it was needed. He could dominate an argument by sheer logic and superior knowledge; even in sport he strove to dominate. I have known him work out a way of lowering his opponents' self-confidence in the course of a golf match; it was as natural to him as his plan to gain victory at Tobruk. In Japan, we met casually at a cocktail party. He told me he would be needing a new Australian ADC as the present one was moving back to Australia. An excellent young officer came quickly to mind and I said so. My nominee, I remembered was an old Geelong Collegian but I had no time to mention that. The C-in-C said, 'I should like to see him in my office at 10.30 tomorrow morning'. Our relations were easy and I replied, 'I'd be glad of a little more time, Sir, he is not stationed locally'. In fact, he was at Fukuyama about 60 miles (100 kilometres) away with rail or sea the only means of travel. The C-in-C's eyes went cold and he repeated, 'in my office at 10.30 tomorrow'. Back at my headquarters the phone ran hot, staff captains were alerted and, fortunately all went well. Late next morning my phone rang and I heard Robbie's voice. 'Thank you, Ron,' he said, 'So-and-so will do very well—but you needn't have thrown in the old school tie!' Had he thought I was bluffing, or becoming complacent and needed a jolt? I never discovered. I believe he was probing for a weak spot. It was his nature.

Robertson began his new appointment on 16 November 1951. Subsequently he was GOC, Southern Command before retiring in October 1954. But the period of over five years as Commander-in-Chief BCOF was the high point of his career. He had to face substantial problems of administration and diplomacy during the first year of his command and towards the end of it he had the additional burden of administrative responsibility for the British Commonwealth Forces in Korea. Lieutenant-General Sir Sydney Rowell, CGS from 1950 to 1954, observed that he:

> was one of the most colourful figures in Australian military history... However successful he may have been in battle, I believe his greatest work from a national point of view was the command of the Occupation Forces in Japan from 1946 onwards. Although he was not an easy subordinate and at times I differed from him widely on methods, it should not be forgotten that he possessed, to an outstanding degree, the capacity to inspire the loyalty and affection of those who served under him. No military commander can ask for more.[21]

Robertson was the first Duntroon graduate to be knighted and the first to become a commander-in-chief. He led the first mounted charge by Australian horsemen in the First World War and was selected to command the first regular infantry unit in the Australian Army. At his life's end in April 1960, the Sydney *Bulletin* published this striking obituary:

> Robertson was a strong personality, sure of himself and sure there was only one way to go about anything—the right way. He was forthright in speech and act, and contemptuous of half-measures and compromise. As a result, he made a few bad friends both above and below him but... any reasonably good soldier of any rank admired him as a soldier and aspired to be like him. Perhaps Horace Robertson knew this and set himself such lofty standards of military skill and behaviour for these very reasons. In off-parade moments he was a courteous, kindly, amusing and charming man.[22]

An estimation of Robertson's standing as a commander of Australian troops must be limited because of the brevity of his operational experience in the Second World War. As a junior officer in Sinai he performed very well. Had there been no artificial restriction on promotion, he might have risen quickly to higher rank in the mobile, open warfare of Palestine. In the Second World War, in command of a brigade, he seemed so thoroughly competent, so entirely swift in thought and action, and so confident in the quality and loyalty of his men that his, and their, success seemed preordained. But there was no opportunity to judge him against a German or Japanese enemy, nor in adversity and, perhaps, defeat. It was only in the occupation of Japan that he could display his mature qualities in high command. His work as Commander-in-Chief of BCOF was outstanding, and he played an important part, offering sound judgment and advice to the Commonwealth governments and force commanders, in the difficult days of the Korean war.

In both these roles he was completely in control of any situation which arose. This, perhaps, is the guarantee that his performance, in the worst possible circumstances or up against the toughest of enemies, would not deteriorate. Whether he missed his opportunity through ill-health, bad luck, his own abrasiveness, or through the jealousy of others, cannot be determined. One thing remains clear. His strength of purpose and determination were such that it would have been entirely in character for him to reach new heights of brilliance under the spur of adversity or in the exercise of high command in war.

# 16 Air Chief Marshal Sir Frederick Scherger: Chairman, Chiefs of Staff Committee

HARRY RAYNER

The RAAF has always been a much more individualistic service than the Army or Navy, but to some extent it was far too small a Service for anyone to become a 'great' commander such as Sir Keith Park in the Battle of Britain, where the fate of the whole military machine depended on the skill and way in which he led his forces. Furthermore, during the Second World War many thousands of RAAF aircrew were dispersed into the Royal Air Force without being appointed to senior command posts. Yet in the period of limited opportunities for a regular airman during the Second World War and the post-war conflicts, the RAAF found, in Sir Frederick Scherger (*b*. 1904), a commander of outstanding ability. He became the first Australian to reach the rank of Air Chief Marshal and he made a lasting impact on the Air Force and the Australian Defence Force.

During the early years of the Service the top echelons of the RAAF knew each other only too well; indeed, strengths and weaknesses were common to most of them; it was a commonality which the conditions of service almost dictated, and the personal judgments of men or the decisions on promotions and/or appointments were invariably within the parameters of this conformity. Nevertheless, on entering the Army and then joining the Air Force Scherger could not be described as the emerging stereotype of the Australian serviceman. His career as cadet, as junior officer and later as commander within a small, specialised force cannot be satisfactorily assessed, therefore, without something more than a passing nod to his antecedents, to his personality, and the events and environment which moulded that character.

Of German migrant grandparents, Frederick Rudolph William Scherger was born in the small Victorian country town of Ararat in 1904. Ten years later, at the outbreak of the First World War, the slightly built schoolboy had to endure a rabid xenophobia, especially against anything remotely associated with Germany. Coming from such a background it is not surprising that he was subjected to unusual pressures when he entered Duntroon in 1921, and as he remembered later, 'The RSM was one Chumleigh, with a voice like a soprano bull, and no great liking for me with my German sounding name. They were not easy days.'[1] It is also significant that despite his winning the King's Medal for graduating first

42 Air Chief Marshal Sir Frederick Scherger. Chairman, Chiefs of Staff Committee 1961-1966. (AWM Negative No. CNB 66/51/1)

in overall academic results in 1924 at the Royal Military College, he was the only cadet of his intake not to be granted one of the honorific cadet ranks. Similar pressures were to continue. When soon after graduation he transferred to the RAAF, he observed that he 'was intensely unpopular because I was a graduate of the RMC Duntroon';[2] at the flying school at Point Cook there were many wartime officers who feared for their promotion prospects against such young newcomers fresh from a military college. In 1941 Prime Minister Menzies received an anonymous letter, postmarked Wagga, claiming that the writer's 'blood ran cold' at the thought of an 'officer named Scherger' being allowed to command young Australians at the Wagga flying training school.

These episodes probably combined to develop within Scherger the seemingly careful appraisal of people among whom he found himself, along with an accompanying cynical disdain of affectation in language or attitude, especially among senior officers. His ready wit probably had with it a certain defensiveness but there was also a particularly sharp tongue which he used on occasions with devastating effect. He was nevertheless a handsome and appealing officer-type, popular with both men and women, and he knew how to carry that popularity, invariably avoiding face-to-face confrontation among his peers either in administration or in command. Seldom in direct conflict with anyone Scherger was at his best in the 'making peace' role, getting the very best out of everyone. Nevertheless he was 'not a man with whom you take liberties'.[3]

Like his contemporaries, Scherger had to endure the heart-break of service life between the wars. Indeed the RAAF had a grim struggle to survive as an independent Service. But as a determined regular airman Scherger threw himself into his work, as Wing Commander F. McNamara, VC, recorded in 1930: 'A keen, determined officer; he takes a strong interest in the comfort and well-being of his men'.[4] In 1938 he became Director of Training and saw at first hand the impact of the visit of Air Chief Marshal Sir Edward Ellington with the subsequent criticism and removal of the Australian CAS, Air Vice-Marshal (AVM) Williams. Later he saw the establishment of the Empire Air Training Scheme and the resignation of the acting CAS, Goble. The new CAS was Air Chief Marshal Sir Charles Burnett, RAF, who was brought out of retirement; in his view the RAAF was merely a training school for the RAF. No doubt Scherger was pleased to leave Melbourne to command the Wagga flying training school in 1940. The following year his ability was recognised by the Air Member for Personnel, AVM Anderson, who noted that he '... fights for his unit and his officers; has strong views and extremely loyal; is inclined to be over-critical'.[5]

On occasions Scherger's critics have claimed to see anti-British sentiments in some of his decisions, especially during the years in which he headed the Air Force and, later, the Chiefs of Staff Committee. In the early years of the RAAF certainly there was resentment among officers at the patronage and seeming arrogance of those Royal Air Force members who were brought to Australia to report upon or to command the infant 'colonial' air service, and there are well-recorded instances of his critical comments on aspects of Australian–British defence relations. On becoming Chairman, Chiefs of Staff Committee he was quick to oppose the existing practice of senior Royal Australian Navy officers being included on the Royal Navy List, and also was instrumental in the elimination of the notation, '... and Chief of the Australian Section of the Imperial General Staff' which up to that time had been included alongside the title of the Chief of the General Staff printed in the Australian Army List. In retirement Scherger claimed: 'Mountbatten was one who always had the idea that Commonwealth forces were colonial forces, and were in fact extensions of the British Services. This was firmly in the minds of fellows like Ellington and Burnett.'[6]

*Darwin, 1942*

It was that same British officer, Burnett, who relieved Scherger of his post at Darwin after the Japanese raids which began on 19 February 1942. On that fateful day the AOC North Western Area, Air Commodore D.E.L. Wilson, was absent on an official visit to the ABDACOM Headquarters on Java, and Group Captain Scherger, as the Senior Air Staff Officer, was the next senior officer in the area. After the first air onslaught, and a sequence of garbled orders by RAAF officers for their troops to move to safety away from the actual environs of the base, a lot of airmen took part in the general stampede southwards. At a parade the next day over 270 airmen were absent from the RAAF Base. Scherger remembers that, 'As a result of that raid the three Service commanders, Wilson and myself, and the two senior Army people, and the two senior Navy personnel, were removed and I was unemployed for one month, which isn't very encouraging in the middle of a war'.[7]

The outcry which followed the raid was such that the government was forced immediately to set up a Royal Commission, with Mr Justice Lowe as commissioner, and hearings commenced in the devastated northern township. It was left for Scherger, appearing before the inquiry virtually only hours after the raids, now to demonstrate his moral courage in challenging the efficiency of the RAAF system, in effect challenging his own hierarchy. Scherger put his career on the line when, in answering Lowe's query whether he considered that some officers at the RAAF Base had demonstrated a lack of leadership, he said, 'I put the blame on the Air Force system, because in the Air Force very few officers know anything about leading men. It is not part of their training.' By comparison he praised the Army's promotion of NCOs, who were chosen as leaders and controllers of men. Midway through Scherger's evidence Mr Justice Lowe commented: 'You have with complete frankness, courage and a high sense of public duty, expressed what is at fault in the system of air training, for which the Government may be exceedingly grateful'.[8]

It was not a situation which appealed to the Air Board and finally the Chief of the Air staff, Burnett, had Scherger called back to Melbourne, where he was virtually ostracised by an Air Board which refused to recognise him or to provide any post, or work, for such an experienced officer. Recognition and relief for Scherger only eventuated after the government received the Lowe Report, in which the commissioner stated that, 'Scherger on the day of the raid acted with great courage and energy. I desire to record the view that, on all the evidence before me, his conduct in connection with the raid was deserving of the highest praise.' The Air Minister agreed with Mr Justice Lowe and directed the Air Member for Personnel to appoint Scherger to a suitable and honourable post. After a period commanding the RAAF station at Richmond, he was appointed Director of Defence on the headquarters of the newly established Allied Air Forces, South West Pacific Area.

*South West Pacific Area*

Scherger is unique in the RAAF in having commanded international forces on joint operations in different theatres of action, first under General MacArthur and later as Air Officer Commanding (AOC), RAF Command, Malaya. In late 1943 when the Allies were beginning to make headway in New Guinea the Australian government was keen to ensure that a national air component was seen to be playing its part alongside the Americans. No. 10 Operational Group RAAF was formed, made up of a wing of three squadrons of Kittyhawk fighters and a second wing of three squadrons of Vultee Vengeance dive-bombers, along with ancillary support units, to operate with the US 5th Air Force. Scherger's ease with the Americans had earlier been noted and it was no surprise that the Air Board nominated him as Air Officer Commanding. By the beginning of 1944 he had his group in action in New Guinea. Scherger's task was not easy. No. 10 Group was to become a central issue in the bitter feud between the RAAF Chief of the Air Staff, Air Marshal Jones, and the Australian who was heading the RAAF Command in the South West Pacific Area, Air Vice-Marshal Bostock.

Within a short while the RAAF was made aware that the Americans would not countenance some of the luxuries of administration to which the RAAF had become accustomed, and when General Ennis Whitehead, heading the US 5th Air Force forward headquarters, threatened to split the RAAF fighter wing by dispersing the three squadrons to individual airfields in rear areas it seemed clear that the Australian air effort must lose its national recognition. Scherger was finally able to argue Whitehead out of the decision, and then set about building up the operational hours of all three fighter squadrons so that they really warranted their space on the limited airfields available in the forward area.

Meanwhile all this time he was being forced to tread a difficult diplomatic service tightrope in reporting both to Jones and Bostock. His ability and drive, however, was readily recognised and in April 1944 MacArthur called on Scherger to command the American and Australian Air units for use in 'Operation RECKLESS' (the assault on Aitape), and then in the following June he was again given control of the Allied air task force for the assault of Noemfoor. The citation of a Distinguished Service Order, awarded to Scherger a short time afterwards, succinctly outlined his part in the operations:

> In the landings at both Aitape and Noemfoor Island Air Commodore Scherger operated his air forces with great skill and success. He was himself in the forefront of the landing of the ground troops, and by his personal courage and leadership proved an inspiration to all personnel under his command.

In both operations he had demanded that support personnel and material for the Australian task force be kept to a minimum, so that rapid deployment was not hindered. He observed the American leave system for squadrons, which gave its members regular rest periods before returning to the same unit for service, and later argued that it was preferable to the RAAF custom of progressively 'turning over' the entire manpower of a squadron, which tended to produce 'highs and lows' of capability in the Australian squadrons rather than a progressively strengthening morale and experience which could be seen with American units.

By August 1944, therefore, No. 10 Operational Group was in extremely stout heart, with Scherger well pleased with the recognition his group had won, when Fate struck a cruel blow. In a collision between an American amphibious vehicle and Scherger's jeep he received extensive injuries, including a fractured pelvis, which resulted in his being invalided from Noemfoor Island to Australia.

Two months later No. 10 Operation Group became the 1st Tactical Air Force (TAF), RAAF, under the command of Air Commodore A.H. Cobby. But as a subsequent investigation observed, 'From the beginning of January 1945 there was a widespread condition of discontent and dissatisfaction within 1st TAF at Morotai'.[9] The discontent had many causes, but a catalyst was a request to Cobby by a number of highly decorated pilots to 'resign' their commissions, maintaining that they were being wasted on operations of no real consequence while the Americans were advancing northwards. Allegations of illicit trade in liquor involving some very senior officers also clouded the issue, although liquor sales by Australians, especially to Americans, was clearly accepted by one American general, who wrote: 'According to current rumours, two bottles of almost any liquor was good for a load of lumber, and for a case a Quonset hut might be unloaded by mistake'.[10] Superimposed over these problems, along with the national jealousies, there continued the sniping match between Jones and Bostock.

Finally the Australian government decided to appoint Mr J.V. Barry, KC, as a commissioner to investigate, and in his findings later he included the accusation: 'As that widespread condition developed and existed without his being aware of it the AOC 1st TAF failed to maintain proper control over his command'.[11] It was a bitter Cobby, a heroic air figure from the First World War, who was relieved of command and a still convalescent, eager Scherger who arrived in May to take over. It was somewhat of a reversal of roles, for it was Cobby who had taken over from Scherger at Noemfoor after the jeep crash. The appointment was clearly recognition of Scherger's command ability as well as his compatability with the Americans, and especially significant in that while Bostock had originally called for Scherger's appointment as a replacement for Cobby, Jones also had been prompt to approve the request.

Scherger arrived at Morotai to find a force divided within itself and strained by conflicting chains of commands. The 1st Tactical Air Force essentially had emerged from No. 10 Operational Group but it was nothing like the lean, flexible force that Scherger had left behind. It had grown to over 22 000 men, a swollen force whose own requirements militated against its effective use in the assaults upon Borneo in Operation OBOE. Its situation within the RAAF could of course be seen as symptomatic of the general Australian desire to be involved in the final stages of the war, although underneath was the haunting query as to whether the whole operation would be a waste of lives in achieving results which would eventually have little bearing alongside the weight of American arms forcing Japan to its knees. Scherger quickly set about reducing the 'fat' from 1st TAF while at the same time winning cooperative agreement and support from the Americans, and in particular, the Australian Army. While the history books outline the Australian successes in Borneo, and the part played by Scherger's force, there is still argument as to whether the lives lost in those operations were in vain.

Scherger's style of operational leadership appears essentially as an amalgam of the mateship of Australian bush literature and of the ANZAC tradition. During the New Guinea campaigns, where camp sites were rough and facilities invariably rudimentary, he moved among his troops with an easy camaraderie, often working stripped to the waist or wearing a bush shirt which carried no rank badges. Concern for the welfare of his men was always paramount, but while he was particularly mindful of the physical and psychological problems of flying personnel he was a martinet when it came to pilots, trainee or otherwise, who needlessly stressed aircraft in unnecessary flight manoeuvres. As officer commanding an Empire Air Training School in 1941 he had been known for the extremely tolerant attitude he took concerning young trainees' pranks on the base or in the township of Wagga, but he meted out heavy penalties on young pilots who allowed exuberance to make them forget flight regulations or the limitations of individual aircraft types. A savage dressing-down by Scherger, brutally stressing that trainees were available in endless numbers while machines were in short supply, remains today a vivid memory for some very senior Australian airmen. (It needs to be recorded, however, that Pilot Officer Scherger, in the late 1920s also stood to attention before his flight commander to be castigated for incidents connected with inverted and very low flying.)

Throughout the island campaigns there never was any doubt over the easy acceptance of Scherger among senior American staff officers. Lasting friendships were developed between the Australian and a number of key figures in the American defence and aviation world which later were to stand him, and Australia, in good stead either as Head of the Australian Joint Services Staff in Washington or as the leader of various specialist teams visiting American bases and factories investigating aircraft types and equipment. He was always able to call upon these reserves of friendship to open doors and to secure concessions. Even before the New Guinea campaign US General Lewis Brereton, deputy commander of the Allied air forces in the ill-fated ABDACOM, remarked on the energy, efficiency and impatience of Scherger to 'get on with the job' in Darwin prior to the Japanese attacks of February 1942.[12] This American acceptance of Scherger paid ample dividends to the RAAF squadrons, invariably short of supplies in New Guinea. The commanding general of the Allied Air Forces, Pacific, General George

C. Kenney, wrote: 'I felt that he, Scherger, instinctively knew what to do in almost any situation and I should help him whenever he needed anything...I gave General Whitehead orders to dip into the American aviation field supplies and to take care of Scherger's needs until his own service met his requirements to keep No 10 Operational Group going.'[13] Alex Stewart, a senior Australian journalist in the theatre of operations at the time, summed up this recognition of Scherger's ease of contact with the Americans with the terse, 'He talked their lingo'.[14]

*AOC RAF Malaya*

After the war Scherger became Air Member for Personnel, then Head of the Australian Joint Services Staff, Washington, and from 1947, Deputy Chief of the Air Staff. Two years later the CAS, Air Marshal Jones, noted that Scherger was 'most impressive. Stands out among his fellows.'[15] His appointment in 1953 as Air Officer Commanding, Royal Air Force Command Malaya, was applauded by ex-service organisations as forecasting his ultimate appointment as CAS.[16] There were, however, some cynical airmen who believed that it was both a British political ploy to keep him away from a top RAAF post at a time when British aircraft salesmen were hopeful of achieving major military sales in Australia, as well as a sop to colonial resentment at the earlier appointment of a Royal Air Force officer to command the RAAF.

By 1951 the command structure of the RAAF had become ossified; its leadership was static. Air Marshal George Jones had headed the RAAF for nearly a decade; a steady, sincere but colourless leader, his retirement had been mooted for some time. Jones was then aged 56 and while Scherger, then aged 47, was almost universally seen as a future chief, there were a number of older two-star airmen (air vice-marshals) who were known to be competitors for their place in the sun. They included Air Vice-Marshals J.P.J. McCauley (then AOC Eastern Area) and F.M. Bladin, both aged 53, and J.E. Hewitt and E.C. Wackett, both 51. The entire Air Force was to be shocked, however, when Prime Minister Menzies announced that yet again a Royal Air Force officer, Air Marshal Sir Donald Hardman, was to head the RAAF, from January 1952. Menzies' announcement was accompanied by the stinging comment that, 'There is no RAAF officer of sufficient age, or sufficient operational experience, to take the post of Chief of Air Staff'.

Scherger's arrival in Malaya at the beginning of 1953 was watched with keen interest by all British forces, as he was the first Australian seconded to head the RAF Command Malaya. His experience in the New Guinea campaigns made him quickly aware of an inherent problem of air operations then in Malaya. While the air commander possessed a multiplicity both of international squadrons and of aircraft types, spread around a number of airfields, with operational requirements ranging from simple supply drops, to locating targets deep within almost impenetrable jungle, a major difficulty was the siting of the air headquarters. Previous AOCs had preferred to base themselves in Singapore, even though 'The Supremo', the redoubtable General Sir Gerald

Templer, and his staff, as well as the senior Army command, were based at Kuala Lumpur. Scherger acted in character, brooked no opposition, and within the year had moved his staff from the fleshpot of Singapore to a site adjacent to Templer's headquarters.

Other changes soon became noticeable. Wasteful saturation bombing of general jungle areas merely suspected of providing terrorists' hideouts was superseded by pinpoint bombing of especially important targets decided upon from grid-markings, made after close cooperation had been built up between Army ground patrols and light aircraft observations. Bombers were 'led-in' by light aircraft marking the target by a smoke 'puff' in advance of the bomb run. The jungle canopy often caused premature explosions of bombs, so barometric fuses were used to bring about explosions closer to ground level. Scherger was responsible for supervising the introduction of 'voice' aircraft as a psychological weapon of war, in which loud-speakers set underneath low-flying, slow aircraft called upon terrorists to surrender. The voice aircraft, the coordination of light aircraft and helicopters in conjunction with the bombers, the denying of jungle 'gardens' by defoliant sprays, all can be seen as forerunners of operations in Vietnam a decade later. As AOC in Malaya Scherger pushed for a major expansion in the use of helicopters, particularly in jungle air supply for the evacuation of casualties, as distinct from the surprise delivery of troops into combat areas. This support for the helicopter continued. On returning to Australia he was a strong advocate, and as CAS had oversight, of the introduction into the RAAF of the Iroquois. He always stressed, however, that the real military potential of the helicopter could only be fully used when total air control of the battlefield was available.

His personal style among his peers along with his ease with rankers which had been so noticeable with the Americans did not alter in his associations with the British in Malaya. He moved around the countryside with regularity, spending a great deal of time which both Air Force and Army units in the field. He appeared especially to enjoy visits to Gurkha troops. Their fighting qualities, linked with a deep sense of humour, appealed to him greatly, prompting Major-General Lance Perowne, GOC 17th Gurkha Division, to comment that relations between Scherger's Air Command and the Gurkha units had developed into 'an exceptionally cordial cooperation, which was really efficient'.[17]

Any comparison of American and British reactions to Scherger's leadership must reflect differing national characteristics: the accepted idea of the brisk, outgoing Americans compared with the cool, reserved English. The wary British watchfulness of the colonial who arrived as their Air Officer Commanding in January 1953 changed somewhat before his departure two years later, when a number of his British peers within the Malayan Command were generous in their assessment of his ability and his development of strong operational cooperation between the Army and the Air Force components. The opinion of one RAF operations room officer at Kuala Lumpur, then a squadron leader, is somewhat more illuminating: 'On the first visit to Kuala Lumpur by

AVM Scherger we gained the impression of a brash and rough Aussie, who had a poor opinion of the British and particularly the British Army'. This impression, however, changed over the following two years and, giving praise for Scherger's work during the Emergency, he added: 'I respected him for his great drive and determination to get things going his way, and single-minded attention to the task, but did not take to him as an individual. I suspect this view was shared by many of my contemporaries'.[18] Perhaps a final word on this aspect of his personality should come from the chief of staff to General Templer, Major-General Sir William Oliver (later High Commissioner to Australia) who said: 'You always got a straight answer from Scherger, even when you didn't like it. He was very pro-Australian, and why not?'[19]

*Chief of the Air Staff*

On return from Malaya Scherger became Air Member for Personnel, and two years later, in 1957 was appointed Chief of the Air Staff. He assumed command of the Air Force at a time when both the RAF and RAAF appeared to be on a downward slide. An editorial in *Aircraft* in February 1957 claimed that 'Strong political pressure by the Army through the Returned Soldiers' organisation is credited with playing a major part in the abandonment of the 1952-1954 decision for a strong RAAF', and in the September issue the editorial claimed that 'The RAAF has failed to make good its case'.

While Scherger would be the last to claim that his occupancy of the offices of CAS and Chairman of the Chiefs of Staff Committee provided the catalyst, there is clear evidence that in less than a decade the RAAF had assumed a foremost stance in political and public acceptance. Nor would he claim the modern Air Force as his child alone, acknowledging always that so many defence projects begin from group discussions and pass through many changes and refinings before emerging as practical and effective action. Nevertheless, his early decisive reversals of some Air Board attitudes and his positive support for other ventures are evident in much of today's Air Force structure as well as in its armoury.

First was his success in overturning a decision to extend the RAAF Base runway at Laverton, near Melbourne, and diverting the planning—and the funds—into airfields and establishments further north. This moving of the Air Force 'centre of gravity' northwards was later followed by the Army which had considered basing a battle group on Mornington Peninsula. It set in train long-term planning for the upgrading of Townsville and Darwin airfields, which today are capable of accommodating the world's largest and heaviest aircraft.

His second coup was in smothering off public acceptance of a plan to buy the Lockheed F104 Starfighter, and then winning approval for acquisition of the French Mirage. The original Air Force approval for the F104 was symptomatic of the psychological urge in every Service chief to have his term of office crowned by the acquisition of some exotica of defence equipment. History has shown that the F104 would have been disastrously incapable of meeting Australia's highly specialist criteria for fighter aircraft.

A few years earlier the Air Force had changed from the cumbersome geographical command system to a functional command system. As Deputy Chief of the Air Staff (DCAS) Scherger had argued, unsuccessfully, with Air Marshal Jones for such a change. Now, as a very visible, articulate and popular CAS it was natural that the main body of the force would be lifted in morale and gratefully link Scherger with the increased Budgets which followed and the arrival of new equipment and weapons. He was seen at hand to greet the landing of new Hercules and Mirage aircraft; he publicly fostered a 'state of the art' introduction of missilry with the Sidewinder and Bloodhound missiles, while his continuous advocacy of the helicopter culminated in the purchase of the UH1B Iroquois. Of great importance also was the acquisition of support items such as simulators for both Hercules and Mirage, the updating of radar installations within the air defence system, and equipment to broadcast facsimile weather charts. The RAAF administrative system was first to adopt electronic data processing, as also with the introduction of science degrees for Point Cook cadets (by cooperation with Melbourne University). Always mindful of maintaining good internal public relations he approved and supported inauguration of the *RAAF News* paper, but there were occasions on which Service eyebrows were raised at some promotions. Scherger's only-slightly-humorous adage that equipment needs could be described as, 'One on, one off, and one in the wash' became a byword, just as his 'Julius Caesar's centurions had three pairs of shinguards, and the British still issue three sets of underwear as Army personal issue' apparently has been adopted by following planners, even in the acquisition of ships.

During the 1950s the obvious keenness of competition between the Services, rather than concentration upon any national enemy, became a dominant feature of the Australian defence scene. Each Service was highly attuned to the arguments against the other Services, in order to protect its own little corner of the yard. While appointments to the top posts are often dictated by retiring ages and rotational averages as between competing individuals and Services, Scherger's postings from 1957, always accompanied by successive extensions, were evidence of government approval; his personality and nationalism won support from both sides of the political spectrum.

As a peacetime commander Scherger's supervision and control of staff was deceptively easy, consistently delegating authority, a practice which promoted some comment from visitors that he appeared lazy in administration. But officers under his immediate eye were quickly made aware that he demanded purposeful research and attention to detail. He encouraged staff members to submit their ideas to him, but was ruthless in commenting upon position papers which he considered not concise enough or insufficient for executive discussion, and 'a flea in the ear' was very often the return ticket for sloppy staff work. Nevertheless, he made visitors to his office, whether civilian or Service, comfortable and at ease although he rarely allowed his real feelings to show, or to display temper—that remained for personal staff to experience, and when intensely irritated his flow and force of language was something to be remembered.

Scherger's style was his own, deliberate and planned. Prior to assuming the office of Chief of the Air Staff he had signalled the style of control he was going to exercise: 'I'm not going to allow myself to be bogged down with minor matters of detail. We have the command system. Broad policy comes from the top. These decisions have to be implemented in the commands—and that's the way its going to be.'[20]

Relations between RAAF members and civilians working on remote bases or in small units, where both are fairly closely integrated in their work, almost invariably have been harmonious and effective, and this is especially true still in those areas where flying training or operations are being carried out. It has not always been the case between uniformed personnel of the Services and the Commonwealth Public Service administrative staff in capital city concentrations, where the smooth maintenance of working relationships often can pose a thorny path for departmental civilian heads along with Service chiefs. While this tension was not unduly a problem in Melbourne, with its diversification of housing locations and distractions, when the transfer commenced in the late 1950s of the Defence group of departments from Melbourne to the seeming isolation and insularity of the national capital, some disparities in the provision of housing and rates of pay between the 'frocks and smocks' could and occasionally did develop into a highly charged acrimony, with the very real possibility of a breakdown of cooperation.

The Department of Air was first in the major transfer and while a great deal of psychological and social environmental research was instituted beforehand, there were qualms in high places about the success of the movement. Scherger took the new location and changed atmosphere in his stride, quite clearly setting out to become known and popularly accepted within all branches of the Defence Public Service. He stressed to Air Force staff working within the department that while they invariably would be transferred to other duties within a few years, the civilians could spend their entire working lives in that area; it was essential that these 'civilians' understood and were sympathetic towards the demands of the RAAF. The proximity of Parliament House, the Canberra 'cocktail circuit' and Royal Canberra Golf Club provided an additional milieu in which it was possible to influence important figures in political and key departmental areas. He tackled the Canberra transfer with gusto. He was fortunate also in that throughout his term as Chief of the Air Staff both Defence and Air Departments were headed by ex-Air Force officers with whom he shared an excellent rapport. He got on particularly well with Mr (later Sir Edwin) Hicks, who headed Defence, and there had been wartime experiences shared with the Secretary of Air, A.B. McFarlane, a former pilot and decorated group captain.

He knew the value of good personal public relations, accepting also that political liaison and public acceptance was an area in which he had to be active, and in this regard has been critical of two very able contemporaries, Air Marshal Hancock and General Wilton, who appeared to have had difficulties in getting their views across because 'they would not accept the view that you can't sell your ideas unless you can sell yourself, and if you can sell yourself you're half way to selling the ideas that you've got'.[21]

His general practice of appearing casually as 'one of the boys' when off duty or in the officers' mess might have given an impression that the authority of the most senior position in the Air Force was being downgraded, but woe betide the officer or airman who presumed in too familiar a fashion. As one senior public servant observed, 'There could be an impression that Scherger was a terrible larrikin, but he had a backbone of steel, but he never tried to show it; he'd rather do things the gentle way'.[22]

Scherger was the supreme pragmatist. At a time when some Australian Service leaders were under a delusion that they alone were responsible for maintaining the defence forces and, where necessary, dictating any changes in military strengths, equipment or dogma, Scherger was quick to acknowledge the overriding position of the government and Parliament; he quite shamelessly played the military political game.[23] Along with practising the 'art of the possible', he drew upon his affable personality and gregarious nature in courting holders of the Defence group of portfolios along with business and community leaders. While he always argued the Air Force case with the utmost vigour, he also knew when to temper the argument, to draw back, or to delay and assemble the argument, although when he settled for less he was often to see his original proposals ultimately come into being. He was not a Service leader who disagreed publicly with the government or who threatened to resign in protest. If advice given to a minister was not taken he believed that:

> the minister should, if he can, explain why he hasn't taken the advice. There are occasions where he cannot explain why he hasn't taken the advice, but he is the man who has to be sure that political considerations are not going to be forgotten. If they are forgotten he can't blame the Service officer. On the other hand the Service officer must not try to anticipate political requirements and spend most of his time giving the minister the advice he thinks the minister would like to have; if you do it that way the whole system collapses on itself.[24]

That he was successful in securing political support during his term as Chief of the Air Staff can be assessed from the steady expansion in the allocation of funds for the RAAF in successive Budgets; when he moved from the CAS position to become Chairman, Chiefs of Staff Committee in 1961 the RAAF, with a total strength of only 16 263, had a Budget allocation the same as that of the Army (strength 47 417).[25] He passed on to the Air Force an armoury, in being or under order, within a force much as he had predicated in 1957: 'Our task is to keep our Air Force strong in quality within the limitations imposed, pending the day when expanding population allows us to develop a really strong air force as Australia's first line of defence.'[26] This ability to push the RAAF argument in competition with the Navy and Army later drew rueful and cynical comments from Service and political contemporaries. In retirement Scherger was occasionally moved to express regret at his seeming insularity in pressing the RAAF case, sometimes against the wider aspects of national defence, but there is no doubt that he enjoyed his role as head of the Air Force, especially when the Service chiefs met each year to learn how the 'defence cake' was to be apportioned.[27]

*Chairman, Chiefs of Staff Committee*

Scherger's appointment on 28 May 1961 as Chairman, Chiefs of Staff Committee (CCSC) marked an important development in the organisation of defence in Australia. Early in 1958 a committee chaired by Lieutenant-General Sir Leslie Morshead had recommended many changes to the Defence group of departments, one of which was the amalgamation of the three Service departments and the Department of Defence. Prime Minister Menzies did not accept all of Morshead's recommendations but he gave the Minister for Defence complete authority in the field of policy, and he decided to appoint a separate Chairman of the Chiefs of Staff Committee to ensure that military advice reaching the government represented more than a form of compromise between able men holding strong Service views—as could be the case with a committee comprising, up to that point, only the three Chiefs of Staff. As Bruce White, formerly Secretary of the Department of the Army, has observed: 'The formal provision of a separate and continuing office of Chairman, Chiefs of Staff Committee did something to remedy the lack of cohesiveness and objectivity in the presentation of military advice to the Government'.[28]

For a while Lieutenant-General Sir Henry Wells was recalled from a short retirement to become the first separate Chairman, but it was only after Scherger's appointment that the Chiefs of Staff Committee began to fulfil its prescribed functions. Meanwhile, the Defence Committee, with the Service Chiefs and Secretaries of Defence, Treasury, External Affairs and Prime Minister and Cabinet, was developing as a forum for the exchange of ideas and as an important body to advise the government on defence policy. The Chairman, Chiefs of Staff Committee was an important addition to the Defence Committee. And there were flow-ons within the Department with the establishment of joint committees such as the Joint Warfare Committee under the Director, Joint Service Plans, who was responsible to the Chiefs of Staff Committee for the formulation and review of joint Service policy, doctrine and the standardisation of techniques for joint warfare operations and training. Nonetheless, Scherger had only a minimal staff and was thus prevented from exercising a joint role with any effect.

As CAS Scherger had been a strong protagonist of the 'my service first' attitude, but on becoming Chairman, Chiefs of Staff Committee he appeared to change and to respect studiously the susceptibilities of the three Chiefs of Staff. He was promoted to Air Chief Marshal and thus became the first Duntroon graduate to reach four-star rank. But he did not pay frequent visits in a 'commander-in-chief' role to any of the Service units, a practice which could have been resented by the base commanders as well as by their chiefs. Of course, he had no authority to exercise command over the Services; he was only an adviser, responsible for putting the joint Service view to the minister. As time progressed, however, he gained a certain amount more freedom of activity. Moreover, the Defence Committee and other interlocked groupings began to accept and look more like an overall defence system.

Not everything he espoused or demanded was achieved. Indeed, as CCSC he was not in a position to demand anything. While he did not favour total integration of the Services he consistently argued for 'one Australian Defence Force', under one Minister for Defence, comprising three fighting arms (not 'services') with two supporting arms—made up largely of civilians—for logistic and technical backup. Challenged years later as to why he had not achieved something towards this objective during his term as CCSC he replied: 'Vietnam was no time for changing horses in mid-stream, or even changing the colour of the horses'; he added an assurance that 'with the end of Vietnam I'm reasonably confident that the Government will do something about getting on with integration', claiming there would be a pruning of a defence structure which was overloaded with five ministers and five departments.[29] He was soon to be proved right.

It was fortunate that Australia's Chiefs of Staff Committee had been improved because soon after Scherger's appointment the government was faced with the crisis of the Indonesian Confrontation, with the requirement to deploy forces to Borneo. Also during this period there was the slow build-up of advisers in Vietnam, culminating in the decision of 29 April 1965 to send a battalion of troops to fight in Vietnam.

As Chairman of the Chiefs of Staff Committee and as a member of the Defence Committee Scherger played an important role in the decision. At the Guam Conference in August 1961 he had taken particular care to leave no implication with the Americans that Australia would be unwilling to support intervention in Laos if it became necessary. This was in keeping with the Australian populist theory that it was far preferable 'to fight in somebody else's backyard' rather than see 'the Red Menace from the North' creep too far south. The Australian Services in general did not disagree with this concept, and while in the years following the actual Vietnam operations some senior personnel joined in the fashionable lamenting of the ultimate Australian involvement, there can be no doubt that at the time there was an acceptance almost akin to eagerness, to commit forces. The Australians saw it both as a vindication of their existence and an opportunity to hone the skills of the fighting arms. Following the Honolulu Conference of April 1965, involving Scherger with US Admiral Sharp and New Zealand's Rear-Admiral Sir Peter Phipps, he returned to tell the Defence Committee, 'I believe that we should contribute...and that a battalion should be offered'. The Committee agreed that it was 'essential for Australia to show a willingness to assist'.[30] There appeared very little public or media opposition at the time, and as for Scherger: 'It never was conceivable by us that America could lose—no way. But lose she did.'[31]

Praise or blame in military successes or disasters can be a fickle mistress. In a democracy it is not the Service chiefs who finally decide to go to war; in theory that is an option retained by the politicans—and public opinion—but defence opinion also is one which invariably gets quoted. Unlike Korea where Australia's involvement followed almost smoothly after the Second World War, public opinion was, and still is, strongly divided about Australia's entry into the Vietnam conflict and its aftermath of an extensive use of National Servicemen. Scherger did not shirk moral responsibility in voicing an opinion. 'If you want allies, you've got to support allies' and 'National Service was the only way' was his reply on all the implications of Vietnam participation, the ANZUS Treaty, and the buildup of the Australian Army.

In many respects Scherger was well equipped to advise the government on defence during this period. To his extensive combat experience was added years of cooperation with the British and American Allies and first hand knowledge of counter-insurgency warfare. He could not be described as an intellectual but he had learned to operate effectively at the important political-military interface for many years. He was forthright on a number of politically contentious issues. A strong advocate for the acquisition of nuclear weapons, he also supported subsidies for Australian defence industry. When he handed over his position to General Sir John Wilton in March 1966 the Defence organisation was a much more cohesive organisation and the Chairman, Chiefs of Staff Committee had gained authority and prestige among the Services as well as credibility with the government.

## 'A bit of class'

Between 1957 and 1966 Scherger became the most quoted and best known of any contemporary military leader. In the Canberra of that period he became an identity, and whether in mufti or in uniform was recognised by civilians and greeted with respect, often with warmth, in the near-parochial atmosphere of those growing years of the nation's capital, equally as within the corridors of the Defence complex of buildings. Not only was he popular with civilians he was, as Sir Edwin Hicks observed, 'a most attractive figure to the airmen. He had come out of Duntroon with a chip on his shoulder, and the way he was going to overcome that was by being a devil-may-care airman'.[32]

Scherger's personality, his style, his appearance, had a particular appeal within a Service which during the early days of flying in Australia had produced some national 'characters' along with quite a few of the heroic figures of two wars. The Service as a whole generally warmed to a leader who retained something of the cloth-helmet-and-goggles, open-cockpit flying charisma about him, yet patently was a decisive leader of the Royal Australian Air Force in a world in which defence was coming under closer public and political scrutiny.

43 Air Chief Marshal Sir Frederick Scherger (right) hands over to General Sir John Wilton on his retirement from Chairman, Chiefs of Staff Committee, March 1966. Between them is Sir Edwin Hicks, Secetary of the Department of Defence, and Sir Allen Fairhall, Minister for Defence. (AWM Negative No. CNB 66/25/3)

Scherger was always leadership material, and his peers readily acknowledged his capabilities. AVM E.C. Wackett (along with Scherger under consideration for the CAS appointment) claimed, 'Scherger was the first to go to the CAS post with a bit of class; the others approached it in a registered, a pedestrian fashion. There was a strain of ruthlessness in Scherger; he would not spare you if his interests and yours conflicted.'[33] Scherger's antecedents, early environment, and then the military machine which progressively provided appointments with, increasingly, both challenge and opportunity, produced a shrewd, forthright, man-management-wise leader; a courageous man, demonstrated more in his flying than on operations, where Fate on two occasions cut short command opportunities. Although a complex character, he certainly was not the aloof commander on Olympian heights, but rather will be remembered as the quintessential Australian, an earthy commander understanding his men and himself and capable of getting the most from both for the task in hand.

# 17 General Sir John Wilton: A Commander for his Time

IAN McNEILL

It was July 1953. The weather in Korea had changed from the severe cold of winter to the debilitating heat and humidity of summer. Brigadier J.G.N. Wilton (1910–1981) had taken over command of the 28th Commonwealth Infantry Brigade only three months before from his friend and colleague from Duntroon days, Brigadier T.J. Daly. Patrol reports and intelligence information had led the troops to expect a massive Chinese night attack. Mortar bombs and shells had already begun to fall.

Wilton was in the command post. Quiet, somewhat reserved, even shy, Wilton was given rather to introspection than to those more obvious characteristics which draw men behind a leader. He was lean, almost frail-looking in his open shirt, jungle green trousers, heavy boots and gaiters. One who did not know him might ask how he came to be there on that summer night, commanding a brigade in what was perhaps the best division in the line. He ordered that the brigade front be illuminated by star shells.

They saw the Chinese advancing shoulder to shoulder, in human waves. When the leading elements reached the prearranged target areas for the artillery, one hundred 105 mm howitzers lay down a barrage. Receiving horrendous losses, the Chinese pressed forward. Those who survived the artillery were caught in the meticulously directed crossfire of machine-guns along predetermined lines. Riflemen added their fire. 'After dawn next day', recorded Wilton, 'we saw that the approaches to our battalion's position and the valley below (no man's land) were literally carpeted with dead bodies. They were lying almost three deep in the area about thirty metres in front of our bunkers. It was a terrible and gruesome sight.'[1] Only a few days later, on 27 July 1953, the Armistice and cease-fire took effect. It was the last time that Wilton, or any of his close contemporaries for that matter, commanded troops in battle.

Wilton seems not to have fitted the traditional image of the military leader. A former British brigade commander with Wilton in Korea commented:

> As a Brigadier commanding troops he was most thorough. He stood no nonsense and planned his operations with great care and completeness. He was very quiet and in no way flamboyant and carried out his tasks to ensure that no detail was left untidy. I think he was tough and fit. He was not the natural

44 General Sir John Wilton, Chairman, Chiefs of Staff Committee, 1966-70. (AWM Negative No. MAY/68/439/HQ)

leader in the true military sense—more the staff officer, but he would not have been selected for this special and unique role if he had not possessed the capabilities required.[2]

Was Wilton a commander, or was his zenith as a staff officer? Lieutenant-General Sir Sydney Rowell, Chief of the General Staff from 1950 to 1954, wrote concerning the selection of brigade commanders for Korea: 'With Australia in a position to claim a brigade command, we were able to give two of our best officers this high post. They were Tom Daly and John Wilton, both of whom had been marked out immediately after World War II for advancement to senior posts.'[3] What then can we understand of Wilton, the officer whom few outside his family circle found easy to approach? What was his background; what were his experiences? What were the attributes which contributed towards his success? Was his style of leadership out of step with others, or was it a sign of society's response to changing times?

*Early regimental experience*

John Gordon Noel Wilton was born of English parents on 22 November 1910 in Sydney. After attending Grafton High School he entered the Royal Military College, Duntroon, in 1927. During his high school days John was beginning to display some of the characteristics by which he became known. His stepmother recalls that he was always a very clear thinker and a young man of integrity, taking after his father in both these respects. His brother, Maurice, saw John as a loner—studious, non-aggressive, developing an interest in sport only in his later years at Grafton High. He was always, observed Maurice, the quiet, determined, achiever.[4]

Wilton entered the Royal Military College as a skinny sixteen-year-old. Life there was monastic, made harder by the constant tribulations of the junior class meted out by the senior classes. Lacking in robustness and not yet mature, Wilton admitted it was a tremendous task to keep up with those who were a year, sometimes two, older than he. Gradually he got on top, playing breakaway for the First XV, hockey, and winning prizes in swimming and diving. Nevertheless Wilton always kept slightly apart from the other cadets. Sir Thomas Daly, who entered Duntroon in 1930 and sat on Wilton's table, well recalls him then as a courteous, gentle person, good-looking although somewhat withdrawn. He never took part in the harassment of his juniors. Wilton's final report from Duntroon was brief. In all categories he was marked 'good' or 'very good' with a one-sentence summary: 'Will make a good regimental officer, reliable and hard-working'.[5]

Nineteen-thirty was a sombre year for the defence forces. The onset of the Depression had resulted in the abandonment of universal military training, thus reducing the need for staff corps officers. Many cadets from Duntroon, threatened with retrenchment, transferred into the Public Service. The College took no new intake in 1931 and, to achieve economy, moved to Victoria Barracks, Paddington, until 1937. The British government offered commissions in the British Army to six graduates. Wilton, facing an uncertain future in the Australian Army which was almost bereft of regular units, and liable to be retrenched at any time, was one of four who accepted. Only twelve cadets graduated on 9 December 1930 from an original class of 21 in 1927. Four went to the Australian Army, four to the Air Force, and four joined the British Army.

Wilton remained in the Royal Artillery until 1939, serving all except the first twelve months with British regiments in India and Burma. His unit always at full complement and with the exciting task of helping to watch over the jewel of the British Empire, he received the best regimental experience available when it was most advantageous to his career. While his contemporaries in Australia were languishing behind desks or training half-strength units at weekends, Wilton was practising man-management and leadership in foreign jungles. He even saw minor action against dissident tribes in the mountains of Burma. In 1938 on a return visit to Australia to marry he was asked if he would rejoin the Australian Army because of the impending hostilities. He accepted the offer and after a short spell back in India, reported for duty in the Royal Australian Artillery as a captain, his first commission in the Australian Army, in Sydney on 26 May 1939.

His first year was spent in coastal artillery, a branch in the doldrums because of lack of men and equipment. 'The outstanding impression I received of Australia's military preparedness', Wilton confided in his diary, 'was that it was tragically inadequate. Australian Defence is the plaything of Australian politics, and the general public is apathetic and sees in Defence matters nothing to concern them!'[6] Years later, as top serviceman in the country and close to the centre of political power, Wilton's determination to change the unwieldy defence structure and his bipartisan political approach mirrored closely the sentiments recorded here. The remarks also hint at what was to become a widely recognised ability of Wilton's: his skill at perceiving and isolating the cause of both staff and command problems.

From coastal artillery Wilton was posted to command a field battery in the 7th Division. His commander, was, said Wilton, 'a most ineffective little man, he can't make decisions by himself'.[7] Wilton found himself virtually running the regiment. The Division sailed for the Middle East in September 1940.

*Second World War*

During the Second World War Wilton served in the Middle East, New Guinea, the Pacific islands, and in Washington. He saw service directly under some of Australia's ablest and best known leaders—Blamey, Berryman, Sturdee, Lavarack, Milford and Savige. After initial duty as a battery commander, he, like most of his Duntroon contemporaries, was allocated to staff appointments. He became Brigade Major Royal Artillery of the 7th Division and the principal general staff officer on the artillery headquarters of the 1st Australian Corps.

On return to Australia he was posted in June 1942 as General Staff Officer Grade 1 to the 3rd Division, commanded by Major-General Savige, then preparing for operations in New Guinea. In December of that year, Wilton was warned of promotion from lieutenant-colonel to brigadier, as artillery commander of the 7th Division which was engaged in heavy fighting in New Guinea. His reaction, as noted in his diary, was typical: 'I am in a cleft stick,' he wrote; 'I don't want to desert the general nor the division but am afraid I won't have the option—I can hardly refuse the appointment which is to a fighting job—the promotion aspect does not interest me.'[8] Wilton went straight to Savige who had not been informed of the impending move. Savige was reluctant to lose

Wilton. Not only was he of high calibre, but Savige had suffered five changes of his principal staff officer in five months. Nor was it a question of stopping Wilton's advancement, as Savige himself intended to arrange his promotion to brigadier at a later stage. Savige therefore wrote to Army Headquarters requesting that Wilton be retained, concluding: 'Wilton is an extraordinarily able man, and one of great loyalty too. He hits it remarkably well with Commanders and produces excellent results. I would be happy to have him as a Brigadier commanding one of my brigades later on.'[9] Wilton stayed with the 3rd Division when it went to New Guinea in March 1943 and provided invaluable assistance to Savige in what has been described as a 'very messy and difficult' campaign to capture Salamaua.[10] As Wilton and Savige were in a forward command post to watch the last of the attacks around Salamaua, Berryman flew in to the area from Port Moresby. With him he had Wilton's posting order to the Australian Military Mission in Washington.

Wilton arrived in Washington in September 1943 and remained until March 1945. He was the senior general staff officer under, first, Lieutenant-General Sir Vernon Sturdee then Lieutenant-General Sir John Lavarack. Here he saw at close hand the planning of military strategy such as the allocation of resources between theatres of war. Wilton also visited American military establishments where he lectured on New Guinea and exchanged information. One trip was of particular importance. On Blamey's instructions, he was sent to Europe in November 1944 for two months to learn what he could of European-type warfare and the army structure there. Wilton visited bases from Eisenhower's to Montgomery's and finally down to battalion and company level. It was an unparalleled opportunity, and one which Wilton regarded as being of utmost value in shaping his ideas of military organisation on the grand scale.

Wilton was posted to Advanced Headquarters Australian Military Forces at Morotai in March 1945 and was sent to the forward detachment with MacArthur's headquarters at Manila. Blamey, the Commander-in-Chief, and Berryman, his Chief of Staff, were there, and the discussion centered around Australia's role in the Pacific—whether to participate in the invasion of Japan or in mopping-up operations on islands bypassed by the main forces. Wilton returned to Morotai after three weeks and was promoted to colonel, becoming the senior general staff officer. Operations were being launched on Tarakan, Brunei and Balikpapan and it was a hectic period. Close coordination was required with the Air Force and Navy.

With the unexpected surrender of the Japanese on 15 August 1945, Wilton's career entered a new phase of frenetic but interesting activity. He was working direct to Blamey, the latter making plans to reconstitute forces from Australia, if necessary, to put on a show of strength when taking Japanese surrenders in the scattered islands. No one was certain whether the Japanese, in view of their strong cultural tradition, would accept orders to surrender. There were many who supported Blamey's inclination that a division was needed, but there was none available. Wilton, however, noted that one infantry company on a destroyer had taken without trouble the surrender of a whole brigade on a nearby island. He became convinced that no strong force was needed. Wilton put this view to Blamey, who, after railing over the effrontery of his staff, agreed the next day to try the idea. It worked. Small detachments in naval ships accepted surrenders of the Japanese throughout the islands. Plans to stop the demobilisation of troops were shelved.

*Post-war planner*

The period from March 1946 to November 1951 formed a plateau in Wilton's career. It was the longest time until then that he had stayed in any one place. The urgency of the war period had passed and defence matters took a back seat to national reconstruction. Demobilisation was a higher priority than training. Wilton spent this time at Army Headquarters, one year as Deputy Director of Operations and five years as Director of Military Operations and Plans. It was a consolidation for Wilton in the rank of colonel, and an opportunity, as the right-hand man and confidant of the Chief of the General Staff, to observe and participate in the running of the Army and the formulation of the strategic basis of Australia's defence policy. As Deputy Director of Military Operations he was part of the Joint Planning Staff, and as Director of Military Operations and Plans he became a member of the Joint Planning Committee. He accompanied the Australian Military Mission to Malaya in July and August 1950, and led his own mission to Malaya in February 1951, both delegations concerned to appraise the Communist threat and discuss Australia's response.

The general period was one of flux, when former alliances were being overturned and new power blocs emerging. It was enlivened towards the end by the outbreak of the Korean war and the introduction of National Service. The post of Director of Military Operations and Plans was probably the most important to Wilton in his rise to the position of Chief of the General Staff. He learned not only Army management and how Australia's strategic basis was formulated, but was exposed for the first time to the realities of domestic political and budgetary restraints. When Wilton left this appointment in November 1951 it was to attend the Imperial Defence College, London. The direction was set for his ascent to star rank.

In reviewing Wilton's career to this stage, two trends are clear: the first is his retention exclusively in regimental and general staff appointments; the second is that his staff appointments were close to the highest levels in the theatre in which he was serving. His regimental experience was extensive, in India and in the early part of the Middle East campaigns. His staff experience was in a variety of key operational posts up to corps and theatre level. He had become well versed in joint operations and had witnessed decision-making between allies on questions of grand strategy. In Australia, he had worked closely with the Chief of the General Staff, becoming involved in national defence policy and strategy which took into account all the aspects of his country's strength.

Reports on Wilton present a picture of a remarkable officer, equally suited to staff or command. Consistently praised were his intellectual capacity and his ability to get on with commanders: from his tour in Washington, 'Colonel Wilton has greatly impressed all with his sincerity and intelligence'; from Lieutenant-General F.H. Berryman,

> An outstanding officer. Excellent in a crisis...tact and temper admirable. Has served me as BMRA and BM to a force in Syrian Campaign and Col GS Adv HQ MOROTAI with great distinction and ability. Initiative, ability to work a staff and get results very good. I should be glad to have him as a commander or staff officer;

from Major-General Rowell, when Vice Chief of the General Staff, 'an officer of great ability...high personal qualities...one of the best logical brains in the services...equally well for command and staff work...particularly tenacious in conference...he will go far in the service.'

Rowell also touched on other traits which were linked to characteristics Wilton had displayed since childhood—his quiet reserve, his seeming inability to come close to people. 'He can be quite ruthless in the performance of his duties,' wrote Rowell, 'and in some moods he may, quite unwittingly, give the impression that he lacks warmth of nature.' It was apparent, however, that responses to Wilton were mixed. As Rowell continued: 'But those who know him well and have his respect have great affection and loyalty towards him'. What this might simply have meant is that Wilton did not suffer fools gladly; he certainly did not in later years. When Wilton became Director of Military Operations and Plans, Major-General R.N.L. Hopkins, Deputy Chief of the General Staff, reported: 'A keen, quick thinking and highly efficient staff officer. Very well balanced and possessing a cool logical judgement...' To this Rowell, having risen to Chief of the General Staff, added: 'An officer of the very highest professional attainments whose reputation and stature increase year by year'.[11] Rowell's faith in Wilton was to be vindicated.

### Shaping Australia's military strategy

After attending the Imperial Defence College in 1952 Wilton was promoted to brigadier to command the 28th Commonwealth Infantry Brigade in Korea. On returning to Australia he became Brigadier in Charge of Administration, Eastern Command. This position in the Army's largest command rectified a deficiency in Wilton's career, his lack of an administrative appointment. It was an enjoyable posting, providing a house at Victoria Barracks, Paddington, and an opportunity for his family, which now included two sons and a daughter, to be together. Accustomed to procedures designed to achieve quick action and precise results, Wilton was dismayed by what he found: a cumbersome staff system, red tape, and lack of delegation especially in financial matters. He was able to affect some improvements, but many changes required policy decisions from Canberra. He did not forget this when he became Army chief.

Wilton moved to more familiar ground in November 1955 as Brigadier General Staff Army Headquarters. He controlled Army intelligence, operations and plans under the Chief of the General Staff, Lieutenant-General Sir Henry Wells. But for the addition of intelligence, it was similar to his old appointment. New concerns, however, were emerging, in particular the participation of Australia in the recently formed South East Asia Treaty Organization (SEATO). As the senior Australian planner, Wilton attended successive meetings throughout South-East Asia, and accompanied the Defence Minister and Australian military adviser to the annual SEATO council meetings. Moreover in the first year of his appointment Australia committed a battalion with ground support to the British Commonwealth Far East Strategic Reserve (BCFESR) in Malaya—an addition to the two RAAF squadrons stationed there since 1950. Australia's defence strategy was now being considered against the background of the Cold War—the hegemony of China in South-East Asia, the possibility of Communist domination of Malaysia, and the future intentions of Indonesia. Wilton helped shape Australia's reaction to these events which were to dominate the remainder of his military career.

On 24 March 1957, at the age of 47, Wilton was promoted to major-general and became Commandant, Royal Military College. Since his own cadet days, Wilton had been imbued with what College graduates called the 'Duntroon Spirit', a code of honour, loyalty, and duty, together with a sense of common identity amongst those trained at Duntroon. His acceptance of the military virtues had been reinforced during the Second World War, a point commonly noted by reporting officers. As Commandant, Wilton was responsible for inculcating the Duntroon Spirit in cadets, as well as, in accordance with the charter of the College, 'a correct understanding of the place of the Armed Services in the Australian Nation'.[12] Wilton himself referred to these values as the 'military ethos' which he found difficult to define, seeing it as a stream of ideas. 'To me,' he said, 'it means the Army's way of life, its way of thinking, its standards, its loyalties, its dedication to its work—all those things are wrapped up, in my mind, in the word "ethos".'[13]

Wilton also believed that officers needed a tertiary education, culminating in a degree, to equip them for the technological age and the modern Army. The two convictions, the need to imbue cadets with the military ethos, and the necessity for tertiary training, led Wilton to attempt to have Duntroon given the status of a degree-granting institution. As cadets would need to undergo their higher education in a military environment, he was opposed to the general proposition that universities outside the military institution should be responsible for the academic teaching process and conferring of degrees. A devotee of the integration of the Services, he supported the idea of a tri-service academy where cadets from all three Services came together for their tertiary training and returned to their own Service for specialist training. While Commandant he attempted to gain approval for Air Force cadets to attend a common academic course at Duntroon but did not succeed. Wilton attributed his failure to the strength of Service rivalries and jealousies which then existed. The idea of a tri-service academy was eventually accepted, largely as a result of Wilton's later influence when Chairman, Chiefs of Staff Committee.

From Commandant, Royal Military College Wilton went to Bangkok in June 1960 as Chief of the Military Planning Office, SEATO. The United States had expressed a keen interest in having an Australian officer in the position and stressed that professional competency was paramount.[14] The Defence Committee had no hesitation in endorsing Wilton as the nominee, an officer whom they considered to be eminently qualified by both experience and ability.[15] The tour of duty was a challenging and interesting one in which Wilton, acting for all members of SEATO, supervised the production of joint SEATO plans by representatives of the member nations. Wilton travelled widely throughout the SEATO countries and other parts of Asia. One particularly nostalgic visit was to India where he had served in the British Army. Wilton also visited Pakistan several times where he was highly respected. As a former part of India, Wilton knew its territory and many of its Army officers from his Indian days. He also spoke fluent Urdu. Later, when he was Chief of the General Staff, the Pakistani government made several requests for Wilton to visit. That he did so in 1964 was not so much for military reasons but because the Department of External Affairs, knowing the esteem in which he was held, asked if Wilton would represent pressing Australian diplomatic views to the President, Field Marshal Ayub Khan.[16] Such was his experience as Chief of the Military Planning Office, his tours, his contacts, his discussions, that, added to his responsibilities in former appointments, Wilton had now become an expert not only on SEATO matters but on South-East Asia as well.

While Wilton was in Bangkok the government considered the appointment of the next Chief of the General Staff to replace Lieutenant-General Sir Reginald Pollard, due to retire in January 1963. Pollard recommended Wilton as his successor ahead of others of higher seniority on the grounds of his exceptional ability, and his highly trained and logical mind which was capable of brilliance. Pollard felt compelled to add that although Wilton 'was not a colourful personality' he was nevertheless highly regarded by the Army because of his competence, strength of purpose, and integrity. The recommendation was strongly endorsed by the Minister for the Army who added that Wilton had performed 'outstanding service in command, planning and administrative appointments'.[17] On 21 January 1963 Wilton was promoted to lieutenant-general to become Chief of the General Staff, 1st Member of the Military Board, and Chief of the Australian Section of the Imperial General Staff.

*Preparing for the Vietnam war*

In 1963 the Regular Army was in a state of flux, ill-prepared to meet the competing and increasing demands which were being placed upon it. It was a time of growing international tension. Australian security was based on a strategy of forward defence—the stationing of troops in South-East Asia to hold a line against Communism while at the same time providing depth to the defence of Australia. The strategy depended largely on the maintenance of bases in South-East Asia by first Britain, then America, to which Australia could contribute. Membership of reciprocal alliances such as SEATO and ANZUS bolstered the forward defence stance. The Regular Army then had only four battalions, two large battalions on the 'pentropic' establishment, and two on a 'restricted' establishment which was compatible with that of British (and American) battalions. One of the restricted battalions was stationed in Malaysia as part of BCFESR and the other was held in Australia to relieve it. A small group of 30 officers and senior non-commissioned officers had been sent to South Vietnam as advisers in 1962 and were being increased in number. Subalterns, warrant officers and sergeants were in short supply and were required in battalions, in the Citizen Military Forces, in training establishments, and in Vietnam. With responsibilities in Malaysia and to the commitment of troops to SEATO plans, and in the face of increasing instability in Asia, the Regular Army ceiling was not nearly enough. Even if a strength of 33 000 calculated by the Army as the minimum necessary to meet the growing demands were agreed by the government, there were insufficient recruits offering.

The first major task to which Wilton applied himself on becoming Chief of the General Staff was a re-examination of the pentropic organisation. The Pentropic Division had been introduced after much fanfare in 1960 as the organisation best suited to limited war in South-East Asia at a time of restricted manpower in the Army. Battle groups were formed with a large battalion of five rifle companies as the nucleus—with artillery and other support being integral to the organisation. The group was commanded by a colonel. As each battle group was an independent, relatively self-contained force, the intention was that they could be deployed individually overseas rather than brigades which were far bigger and more costly in manpower. But the battle group was not compatible with any of the organisations of Australia's Allies—the United States had experimented with a similar 'pentomic' organisation and had abandoned it just as Australia embraced the pentropic organisation.

Wilton had perceived flaws in the pentropic division from the start. In 1959, while Commandant of the Royal Military College, he had voiced his disquiet to the Chief of the General Staff, Sir Ragnar Garrett, who was the driving force behind the new organisation. By the time Wilton had returned from Bangkok to take over from Garrett's successor, Pollard, the reorganisation was a *fait accompli* and he found himself in the position of having to consolidate it. Wilton's early objections stemmed from several grounds: it was too unwieldy for the jungle; it meant keeping two kinds of battalion organisations in the one small Army so that the battalion which went to Malaya would fit in with the British; and it was too difficult to command. At Exercise 'Nutcracker' in October 1962 when the battle group was tried out, Wilton's fears were confirmed. Scattered widely through the close country, with poor radio communications, difficult to control, the new, proudly acclaimed organisation was, Wilton thought, 'a dog's breakfast'.[18]

By 1964 some doubts were beginning to be felt in political circles about the efficacy of the pentropic division. When these reached Wilton he knew he had to make up his mind. The radical change to the infantry division by the introduction of the pentropic organisation had resulted in reverberating changes throughout the Army. New tactical methods had been adopted, new supply and transport systems evolved, new combat support organisations created, new text books and training pamphlets printed, and even new accommodation planned for the soldiers. Changing the structure of the Army again, especially amid a period of worsening international relations, would be a weighty responsibility. Wilton was nevertheless undeterred, and in December 1964 he recommended to the Minister for Defence that the pentropic division be abandoned.[19] It just did not make sense, he thought, for Australia to be on a different organisation and doctrine from those of her major allies. The recommendation was accepted with alacrity—the fastest response Wilton could remember —and Cabinet approval was given sixteen days later. Sir Wilfrid Kent Hughes, MP, was close to the mark when he wrote to the Minister for Defence: 'Personally I would like to congratulate you on having changed back and having abolished something that never should have been set up. We brought it in at the time America had abandoned it and Britain never adopted it...'[20]

Indeed, the pentropic division may well have been an organisation that should never have been set up. But once adopted it required a judgment backed by professionalism, confidence, and foresight to change direction. Soon after the reorganisation commenced Australia received a request from the United States to provide combat troops for Vietnam. In June 1965 a battalion on the new Tropical Warfare establishment was fitted neatly into the US 173rd Airborne Brigade (Separate) at Bien Hoa.

The strategic situation which had spurred Wilton's decision concerning the reorganisation of the division also rendered more critical the Army's manpower shortage. Concerned that the manpower target of 33 000 men could not be met by normal recruitment, Wilton, supported by the Chiefs of Staff Committee, recommended in March 1964 to the Minister for Defence that pay and conditions of service be improved in order to attract more volunteers in this time of full employment. The introduction of conscription was eschewed as being necessary only if Australia was under direct threat, whereupon the Citizen Military Forces would be mobilised as well.[21] The government nevertheless decided ultimately in favour of selective National Service—twenty-year-olds picked by a birthday ballot for two years' service which could be overseas. This introduced further turmoil into the Army which had to find and train more staff, provide more equipment, more buildings, and support a sudden increase in strength when the first intake appeared. Once adjustments were made, however, selective National Service answered the Army's needs to the letter. By 1965 with a battalion in

Vietnam, Special Air Service, infantry and artillery units in Borneo in reaction to President Soekarno's policy of confrontation, and the introduction of National Service, the Army under Wilton, but no other element of Australian society, had begun to put itself on a war footing.

One month after the despatch of a battalion to Vietnam Wilton directed his staff to consider ways in which the battalion might be increased to a task force (brigade group) should the government so require. In January 1966, just as alternative proposals were being considered at the highest political level, the Prime Minister, Sir Robert Menzies resigned and was replaced by Harold Holt. The Minister for Defence Shane Paltridge, died after a short illness and his position was taken by Sir Allen Fairhall, and the Minister for the Army, Alexander (Jim) Forbes, was replaced by Malcolm Fraser. After these disruptions, Cabinet resumed consideration of the deployment of further forces to Vietnam, and on 8 March 1966 the new Prime Minister announced the decision to send a two-battalion task force, leaving Wilton with the problem of getting the force away in the next two months. Wilton left immediately for Vietnam to negotiate a military working agreement with the United States and Vietnamese commands, and to argue for the allotment of an independent area of operations in Vietnam with which Australia could be identified. He was successful on both counts; the working agreement became the directive to the Australian force commander and was repeated with little amendment over the succeeding years.

The stresses to which the Army was exposed during this period highlighted the weaknesses in the system of control at the top level. The faults were common to the three services and, Wilton believed, stemmed from two main sources: an awkward and ill-defined relationship between the Service minister, the secretary or permanent head, and the Service chief, and the differing interpretations of such ideas as 'supremacy of the civil power' and 'civilian control of the Services'. Some public servants seemed to understand these terms to mean control of the defence forces through or by the public service; to the defence forces they meant control by Parliament through the minister.[22] The permanent heads of government departments supervised their departments by statutory authority subject only to their minister. The Service departments, however, were administered under the Defence Act by the Service board—in the case of the Army by the Military Board. The chairman of the Service board was the Service chief, while the secretary of the department was one of its members. Yet outside the Service board the secretary had the same statutory authority over his department as had permanent heads of non-service departments. One result was that proposals by the Service board to the Defence or Service minister had to be submitted through the secretary and be signed by him, rather than by the chairman of the board.

Another factor which complicated the line of authority arose because Service ministers were junior in the ministerial hierarchy and were anxious not to make mistakes which might impede their promotion. The safest way to operate, they seemed to believe, was through the secretary as was done in other departments. This appeared to place the secretary between the minister and the service chief in the chain of command. Furthermore, the secretary tended to become bogged down in professional and technical matters which he was not equipped to handle. Throughout his period as Chief of the General Staff Wilton endeavoured to change this arrangement but made little headway—mainly, he thought, because the ministers concerned did not want to upset a system which appeared to suit them, and which they little understood.[23] In Wilton's view it was only because he had such an able permanent head as Mr Bruce White that serious friction did not occur. Towards the end of his period as Army chief, Wilton, with support from White, had almost completed a document proposing a clarification of the statutory responsibilities of the chief and the secretary. But the matter was never finalised. On learning of his appointment as Chairman, Chiefs of Staff Committee Wilton's thoughts turned to the reorganisation of the Defence Group of Departments as a whole; if his ideas were accepted the problem in the Services would also be overcome. Wilton was promoted to general and took up his new position on 19 May 1966. He was succeeded as Chief of the General Staff by Lieutenant-General Sir Thomas Daly.

*A joint Service approach*

Wilton's move to the post of Chairman, Chiefs of Staff Committee occurred after the preparation and despatch of the task force to Vietnam but before its establishment in Phuoc Tuy province. Because the Australian force in Vietnam was a joint force with elements of all three Services, the commander in Saigon was a joint commander. His title became 'Commander, Australian Force Vietnam' instead of the former 'Commander, Australian Army Force Vietnam' which was appropriate when the force was a battalion group. The new position was raised to major-general.[24] While the Commander, Australian Army Force Vietnam had been clearly subject to the Chief of the General Staff, the Commander, Australian Force Vietnam needed to be under some form of joint control from the Defence Department. But the only machinery which could provide joint control was a committee of the Chiefs of Staff headed not by a commander but by a chairman. Wilton, as Chairman, Chiefs of Staff Committee had no statutory responsibility for or control over the joint force commander in Vietnam. He was the adviser to the Minister for Defence and government, and he was responsible to execute the will of the Defence and Chiefs of Staff Committees, but for the latter responsibility he had no authority. He would stretch his directive to its limits to achieve a desired result, but in reality he was akin to the position of a staff officer with no commander. Knowing Wilton's character and his experience to this point, his unreserved determination to replace this structure with one which would allow the defence forces to execute efficiently the will of the government can be understood.

Wilton was driven by two philosophies: that war was a joint Service venture and the concept of the military ethos. He believed strongly that no operational activity was undertaken by one Service which did not in some way affect the others, and which should thus be considered under joint control. 'You can't have your three services fighting three separate wars', he would say. 'It's as simple as that.'[25] His ideas on the need for a formal system for joint Service management had developed when he was posted to the Australian Military Mission, Washington, a tri-service organisation. There he saw the Air Force officer sending to Australia Air Force reports of strategic interest by Air Force channels, while he sent back independent Army reports by different channels. 'That', recalled Wilton, 'is where my first interest came in the necessity to have a proper system of joint service control as opposed to cooperation which is so dependent on personalities...'[26] Wilton never veered from that resolve. Although Wilton spoke of the military ethos largely in Army terms, he applied it to all three Services. He believed that few civilians in high places in the Defence Department had any notion or understanding of the military ethos; this had contributed to the lack of empathy and even hostility which existed between the defence forces and the public service.

Weaknesses of organisation and structure in the Defence Group of Departments, although in some instances known for many years, were exacerbated by the pressures to satisfy the Vietnam commitment. Excessive independence of the Services and lack of centralised control was seen by Wilton as one of the major difficulties in readying the force for Vietnam, and would present a serious obstacle should Australia have to defend herself. Internecine rivalry had always clawed back the Services, and in the first few months of the task force's arrival in Vietnam bitter quarrels occurred between the Army and Air Force over the use of RAAF helicopters.[27] Difficulties between the services were made worse by the dissonant civilian–military relationship and the absence of a single line of authority linking the minister, the public service, and the Defence Force. Wilton tackled the issue at three levels: the integration of institutions, the conferring of statutory authority on the position of Chairman, Chiefs of Staff Committee, and the integration of the Defence Group of Departments.

Integration of Service academies was not popular among the Services in the early 1960s. Wilton had been unable to combine Duntroon and Point Cook in 1957, and a study in 1960 had revealed 'unusual difficulties' in integrating the three cadet colleges and had recommended against it.[28] He therefore felt some satisfaction in being invited in 1968 as Chairman, Chiefs of Staff Committee to join the committee headed by Professor Sir Leslie Martin to enquire into the question of a tri-service academy. There were many opponents to such a step, but Wilton himself believed in it strongly. While he considered tertiary education was vital to keep officers alert to modern technology and the human side of service management, this had to be carried out in a military environment where cadets could be inculcated with the military ethos.[29] His experience as Commandant at the Royal Military College, his personal convictions, and his logical, incisive style of argument, ensured that Wilton exerted a strong influence on the committee which recommended in favour of a joint Service academy. This was accepted and the decision to establish the Australian Defence Force Academy was announced by the government in 1974.

The idea of a joint Services staff college put forward in 1960 and 1961 also foundered initially because of lack of Service support.[30] The matter gained renewed impetus when the Chiefs of Staff Committee under Wilton's chairmanship resolved in favour of such an establishment in 1967 and directed a working party to examine it.[31] Some dissent existed among the chiefs on proposed locations and Wilton visited several places which were possibilities. Not until he visited the Schreiner Complex on the Cotter Road, Canberra, did he make up his mind. 'This is it,' he said to his staff officer, Commander G.J.H. Woolrych. 'This will scrub up beautifully.'[32] Thus the Schreiner Complex became the immediate temporary home for the Joint Services Staff College which opened in 1969 and included visiting students from thirteen countries.

A further area which required rationalising was that of Defence intelligence. Although in 1961 the Defence Department considered there to be no overlap between the intelligence functions of the three services and the Joint Intelligence Bureau (JIB) in Melbourne,[33] in the late 1960s this position was reviewed. Wilton, who thought that separating the intelligence role of the three services was dysfunctional for both the intelligence process and Service integration, and also that there was insufficient military influence in the JIB, led a committee to investigate the intelligence system serving the Defence Department. As a result the JIB and the overseas functions of the Service intelligence directorates were abolished. The Joint Intelligence Organization (JIO) was established within the Defence Department, incorporating all the former agencies together with a means of input from the Department of Foreign Affairs. Wilton fought hard and successfully to ensure adequate Service representation at the top level. The new organisation was vastly superior to what had existed, and provided a full and timely service to Defence.

Important as these changes were in the move towards integration, they were always overshadowed in Wilton's mind by the issue of Defence reorganisation. This held the prospect of combining the activities of the Service and civilian staffs who had hitherto operated in parallel if not in conflict, facilitating inter-service coordination by the establishment of a central authority, increasing Service representation at the decision-making levels of the Defence Department, and an abandonment of the separate Service departments which had the effect of fragmenting the defence effort. Any attempt at creating an organisation which would appear capable of fulfilling these aspirations, however, would threaten long-held and established interests. It could be expected to face opposition and delay, if not outright rejection. The contribution made by Wilton towards achieving such a reorganisation, even though meeting only part of his objectives, is a measure of his clarity of vision, tenacity, and boldness.

A single Defence Department had been proposed by the Morshead Committee in 1958. Although its recommendations were supported by the Chiefs of Staff, they were in the main rejected by the government of the day. The Prime Minister attempted to make some changes administratively, but without statutory support these were of little avail. In April 1965 the Chiefs of Staff Committee had recommended a far-reaching examination of the defence structure but nothing had eventuated by the time Wilton became chairman. Wilton had already given much thought to the question, and had had discussions in the UK with Lord Louis Mountbatten, when Chief of the Defence Staff, on the British experience.[34] Thus on taking up his appointment Wilton soon opened discussions with the Secretary and Minister for Defence. While the Morshead report had claimed administrative efficiency and economy as the grounds for reorganisation, Wilton stressed the military and political requirement. Verbal discussions with the Minister and Secretary were not productive, and in September 1967 Wilton set out his views formally to the Minister. 'A Defence organisation will fail in its basic purpose,' he wrote,

> unless in addition to its other important functions, it provides effectively for the direction, command and management of our armed forces and ensures that they can be employed with maximum operational effectiveness with adequate logistic and other support to carry out roles as directed by the Government.[35]

The necessity for reorganisation and a joint Service approach to warfare transcended in Wilton's mind all but the Vietnam war itself.

Some small improvements followed, such as the provision of a joint staff to supplement the one overworked staff officer, but no fundamental decisions were made. Again in 1970 Wilton put forward a similar proposal for Defence reorganisation to the new minister, Malcolm Fraser.[36] But once more nothing transpired, Fraser having required more time to understand the Department. This was often the reason given for procrastination on the issue; either the secretary or the minister was usually new.

Wilton retired in October 1970. Little had been accomplished concerning an integrated Defence Department and no substantial changes were made to strengthen the position of the Chairman, Chiefs of Staff Committee. Certainly he had not succeeded in obtaining what was most needed, statutory authority for the chairman over the three Services. Some minor changes had occurred, and institutions were either being integrated or integration was being planned, but the fundamental reorganisation of the Defence Group of Departments seemed as far away as when Wilton became Chief of the General Staff almost eight years earlier. Yet Wilton had by no means lost interest. His chance came when the Labor opposition in the 1972 Federal election year, announced its support for the integration of the Service departments. Wilton contacted Lance Barnard, the Shadow Minister for Defence, who indicated he would appreciate further discussion on Wilton's views. During the election campaign in November 1972 the two men met in Melbourne.

Wilton had a complete paper prepared suggesting to Barnard how the integration of the departments might be accomplished.

Soon after winning the election, Barnard announced publicly on 19 December 1972 the abolition of the Defence Group of Departments and outlined his plans for an immediate study with a view to restructuring Defence. It was a gratifying moment for Wilton who saw some of the more fundamental proposals he had made to Barnard already agreed, and others awaiting the outcome of the full enquiry. The Defence Reorganization Act was passed in 1975 with little debate or opposition and the reorganisation was carried out the following year. Although Wilton could not agree with all the details of the new administration he believed the fundamental principles were sound. When it was pointed out that it took about 30 years from his resolution in Washington during the war to the final integration of Defence, Wilton replied, 'Oh yes, I'm an incredibly patient man'.[37]

## 'Happy Jack'

Outside the conference room Wilton's reserve showed itself in many ways. When visiting the Australian forces in Vietnam he preferred to talk in the mess with the commanders—Commander of the 1st Australian Task Force and Commander, Australian Force Vietnam—than talk with subalterns. He had little militarily to discuss with lieutenants. His manner sometimes created awkward social moments making it difficult for his juniors to feel at ease. On one occasion he was expected at a battalion officers' mess in Vietnam. The officers were assembled inside while the Commanding Officer, Lieutenant-Colonel P.H. Bennett (later Chief of the General Staff) kept an eye on the entrance. When Wilton appeared, Bennett welcomed him, receiving in reply a thin smile. As officers do in these situations, they deferred to the senior to begin the conversation. No one spoke. Then Bennett, breaking the silence, commented on the trying heat and the change in climate from the Australian winter. The general answered with a 'humph'. The discomfort passed when the conversation immediately settled on military topics. A small incident, but of a kind often repeated, earning Wilton the nickname 'Happy Jack'.

His emotionless exterior and his rejection of small-talk may have given some the impression that he lacked compassion. Indeed this was not so, as is evident from his reaction to enemy slain in Korea as well as his concern for the comfort and well-being of his own soldiers. It is apparent too from the tone of his private memoirs where he refers to the 'harrowing' task of confronting staff cadets who had made insufficient progress, or deciding on promotions or appointments in the interests of the Army but which would hurt old colleagues and friends.[38]

## The leader

Wilton's passing from the Defence Force had coincided with the end of an era in Australian strategic thinking. At the beginning there had been a commitment to forward defence, SEATO, and reliance on support from the United Kingdom and the United States in the provision of bases in South-East Asia. By 1970 SEATO had been discredited and first Britain, then America, had begun their withdrawal from Asia. In their consideration of the strategic basis for Australian defence policy only months after Wilton had departed, the Defence Committee agreed that not again would Australia be likely to support counter-insurgency operations unless her own safety were directly affected. The ideological basis for international confrontation had greatly receded, and the term 'continental defence' was used for the first time.[39] Australian attitudes towards regional defence and cooperation in South-East Asia began taking on a new perspective.

Wilton had led first the Army then the Services through the long years of the Vietnam war. Before he retired the first Australian withdrawals had been announced, following the wake of the Americans. It had been a strife-torn period, fraught with political perils. Conscription, the effect of civil disobedience and dissent on the forces in training, the questioning of the commitment, and the constant readiness of the press and sections of the public to carp at the military effort could have created restlessness and even disunity in the ranks. Wilton, believing implicitly in the separation of the military role from the political role, saw his duty unequivocally as carrying out the will of government and getting on with the job. That the Services survived this period with cohesion and an undeterred sense of direction is a tribute in part to the effect of his leadership.

He was not flamboyant, nor did he try to sell an image. He could not be described as ambitious for himself, but rather for what he thought should be done. His leadership stemmed from the respect that all who knew him felt for him—he was admired for his keen intellect, his tenacity, and his professionalism. He was respected also as a man, for the integrity he displayed and the loyalty he gave to his subordinates and to the idea of service.

His aim was always to achieve consensus, but when that was not possible his decisions were firm and unyielding. His keen mind gave him a strength and incisiveness in conference. His ability for deep concentration and appreciation of key points in a document or in conversation enabled him to absorb complex situations readily. Wilton's Service experience had been broad and his appointments and opportunities gave him a depth of comprehension seldom possible in the Australian forces. In many of his postings he developed a close knowledge of matters which were to be highly relevant to him as head of the Army or Chairman Chiefs of Staff Committee. He fervently believed in the integration of Defence, a conviction held for most of his career and which he did much to achieve.

Realising that his strength did not lie in the more overt, traditional model of military leadership, Wilton, consciously or unconsciously, appears to have emphasised those traits and developed the skills he already had. It is said that societies spawn the kind of leaders they need. Wilton was appointed to the pinnacle of his profession in the tempestuous 1960s, when orthodoxy was challenged and the armed forces were sometimes seen as the apparatus of repression. With his quiet demeanour, brilliant and unruffled mind, Wilton steered the Services through these years with dignity and honour. He was indeed a commander for his time.

# Notes

## 1 Introduction

1. S. Encel 'The Study of Militarism in Australia', in J. Van Doorn (ed.), *Armed Forces and Society* The Hague: Mouton & Co., 1968, p. 134.
2. B.H. Liddell Hart *Thoughts on War* London: Faber, 1943, p. 219.
3. Robert D. Heinl, Jr *Dictionary of Military and Naval Quotations* Annapolis: U.S. Naval Institute, 1966, p. 62.
4. ibid., p. 61.
5. E.G. Sinclair-MacLagan was not an Australian, but had lived in Australia for some years before the First World War.
6. Sir Archibald Wavell *Generals and Generalship* London: Times Publishing Co., 1941, p. 2. Montgomery of Alamein *A History of Warfare* London: Collins, 1968, p. 23. Napoleon's Maxims, quoted in T.R. Phillips, (ed.) *Roots of Strategy* Harrisburg, Penn.: Military Service Publishing Co., 1943, p. 235.
7. J.F.C. Fuller *The Foundations of the Science of War* London: Hutchinson, 1925, p. 126.
8. Sir Francis Tuker *Approach to Battle* London: Cassell, 1963, p. 391.
9. J. Connell *Auchinleck* London: Cassell, 1959, p. 237.
10. Montgomery *A History of Warfare* p. 25.
11. J.F.C. Fuller *Generalship, Its Diseases and Their Cure* London: Faber, 1933, pp. 32, 87, 88.
12. Liddell Hart *Thoughts on War* p. 222.
13. Carl von Clausewitz *On War* vol. I, London: Routledge & Kegan Paul, 1968, p. 68.
14. For more details on these issues see D.M. Horner, 'Staff Corps versus Militia, The Australian Experience in World War II' in *Defence Force Journal*, No. 26, January/February 1981.
15. Barbara W. Tuchman *Practicing History* New York: Alfred A. Knopf, 1981, p. 276.
16. Liddell Hart *Thoughts on War* p. 218.

## 2 Major-General Sir William Bridges

1. *Age* (Melbourne), 8 August 1914.
2. He was not, however, the first Australian to command a truly 'Australian' (that is, representative of more than one colony) force, this being a distinction that belonged to Colonel J.C. Hoad—later Australia's first local major-general—who led the 'Australian Regiment' during the Boer War.
3. *Commonwealth Gazette* 1915, vol. 2, p. 2312, carried the delegation of AIF powers to Birdwood dated 15 September 1915. His appointment as GOC of the force appeared in mid-September 1916 in *Commonwealth Gazette* 1916, vol. 2, p. 2582, but was backdated to have effect from 18 September 1915. The appointment was terminated with effect from 10 September 1920 (*Commonwealth Gazette* 1920, vol. 2, p. 1323).
4. *Report upon the Department of Defence from the First of July, 1914, until the Thirtieth of June, 1917*, Part I, Melbourne, 1917, p. 16.
5. Sir Ronald Munro-Ferguson (?), Governor-General of Australia, to General Sir Ian Hamilton, 25 December 1914, Australian War Memorial (hereafter AWM).
6. C.E.W. Bean, vol. I of *The Official History of Australia in the War of 1914–1918* Sydney: Angus & Robertson, 1942, p. 65.
7. ibid., p. 66.
8. C.E.W. Bean *Two Men I Knew* Sydney: Angus & Robertson, 1957, p. 75.
9. Hamilton to General Sir John French, 10 March 1914, Hamilton Papers, Liddell Hart Centre for Military Archives, King's College, London, 7/8/5.
10. Birdwood to Hutton, 30 September 1916, Hutton Papers, British Library, add MS 50089, pp. 1–2.
11. C.M.H. Clark *The People Make Laws, 1888–1915* vol. V of *A History of Australia* Melbourne: Melbourne University Press, 1981, pp. 377–8.
12. R.A. Preston *Canada's RMC* Toronto: University of Toronto Press, 1969, p. 61.
13. Deakin to Sir George Clarke, Secretary of the Committee of Imperial Defence, 8 January 1906, cited in L.D. Atkinson 'Australian Defence Policy: a Study of Empire and Nation (1897–1910)' unpublished PhD thesis, Australian National University, 1964, p. 276.
14. Bridges to Senator G.F. Pearce, Minister for Defence, 20 April 1910, Pearce Papers, AWM file 419/80/2. Bundle 6.
15. C.D. Coulthard-Clark *A Heritage of Spirit* Melbourne: Melbourne University Press, 1979, p. 90.
16. Bridges to Colonel H.G. Chauvel, 20 August 1914, Chauvel Papers. The interpretation given here as 'most probable' is based on the fact that Bridges was offered and accepted command on 7 August and discussions with the War Office to determine the composition of the force took place in the subsequent few days. See *A Heritage of Spirit*, pp. 117–18.

17. Bean, vol. I of *The Official History* p. 36, and also *Two Men I Knew* p. 33.
18. N.K. Meaney *The Search for Security in the Pacific, 1901–14* Sydney: Sydney University Press, 1976, p. 63n.
19. Bridges to Munro-Ferguson, 31 January 1915, Novar Papers, National Library of Australia MS 696/3525-6.
20. A.G. Butler, vol. I of *The Official History of the Australian Army Medical Services in the War of 1914–1918* Melbourne: Australian War Memorial, 1930, p. 33.
21. Bean, vol. I of *The Official History*, pp. 135–6.
22. For a listing of the powers of GOC, AIF, see *Commonwealth Gazette* 1914, vol. 2, p. 2227.
23. W.L. Gammage *The Broken Years* Canberra: Australian National University Press, 1974, p. 269.
24. This is demonstrated nowhere so graphically as in *The Anzac Book*, written and illustrated on Gallipoli by the men of the ANZAC. Published in London in 1916, the book carries a frontispiece featuring the King's testimony that 'The Australian and New Zealand troops have indeed proved themselves worthy sons of the Empire'.
25. *Military Board Papers of Historical Interest* vol. 2, Australian Archives, CRS A2657/T1.
26. R.A. Preston *Canada and 'Imperial Defence'* Durham, NC: Duke University Press, 1967, p. 472.
27. J. Hetherington, *Blamey: Controversial Soldier* Canberra: Australian War Memorial, 1973, p. 33.
28. Bean, *Two Men I Knew* p. 72.
29. Bean, vol. I of *The Official History* p. 550.
30. ibid., p. 547.
31. R.R. James *Gallipoli* London: Batsford, 1965, p. 91.
32. ibid., p. 129n.
33. Bridges to Hutton, 14 March 1915, Hutton Papers.
34. Hutton's note dated August 1915, Hutton Papers.
35. Copies of both Operation Orders are held in the AWM.
36. Bean, vol. I of *The Official History* pp. 259, 261.

## 3 General Sir Brudenell White

1. Diary of C.E.W. Bean, 18 June 1918, No. 116, Papers of Dr C.E.W. Bean (AWM). C.E.W. Bean, *Two Men I Knew* Sydney: Angus & Robertson, 1957, pp. 222–3.
2. G. Serle, *John Monash* Melbourne: Melbourne University Press, 1982, p. 397. Field Marshal Lord Birdwood 'General Sir Brudenell White', *Australian Quarterly* vol. XII, no. 4, December 1940, pp. 5–7. Diary of C.E.W. Bean, 16 June 1918, No. 115. Bean *Two Men I Knew* p. 222; the stories of Birdwood swimming in the nude at Gallipoli within the range of Turkish sniper fire are legendary; see his autobiography, *Khaki and Gown* London: Ward, Lock & Co., 1941. Opinion of Birdwood by Fihelly, Acting Premier of Queensland in the Papers of Major-General H.E. Elliott, AWM.
3. Diary of C.E.W. Bean, 16 May 1918, No. 111. Dyson also added that 'the Jew will always get there'; R. McMullin 'Will Dyson: War Artist' *Sabre Tache* vol. XXI, no. 4, October–December 1980, pp. 19–26.
4. Private Memoranda by Brigadier-General H.E. Elliott, 25 May 1918 in Papers of Major-General H.E. Elliott, AWM. For a complete documentation of his views see boxes 36–38 of his papers.
5. N. Pixley 'John Warren White and Family' *Journal of the Royal Historical Society of Queensland* vol. 10(1), 1975–76, pp. 11–23; Before Federation, White saw service in the Wide Bay Infantry Regiment, a volunteer unit, and the Queensland Permanent Artillery. His initial interest in military matters was influenced by his friend, T.W. Glasgow (later a major-general in the AIF, Senator, and Minister for Defence) who worked in the Queensland National Bank in Gympie and encouraged him to join the Wide Bay Infantry Regiment.
6. *Herald* (Melbourne), 16 March 1940.
7. Commonwealth of Australia, Parliamentary Papers, 1901–1902 Session, Vol. II, 'Military Forces of the Commonwealth—Minute Upon the Defence of Australia by Major-General Hutton, 7 April 1902'; Letter from Hutton to Sir M. Ommaney, 24 February 1902, Papers of Major-General Sir E. Hutton, British Museum, add. MS 50089.
8. 'Minute Paper' by Major-General Sir E. Hutton, 7 November 1904, AA MP 84/1 file no. 2002/4/21; 'Reports by Lieutenant C.B.B. White from Staff College, 1906–1908', AA MP 84 file no. 2002/4/18.
9. Gellibrand and White wrote to each other in endearing terms: White was called, 'Boomerang', and Gellibrand was called 'Jolly Frog'. Evidence from Lady Derham, University of Melbourne.
10. See the monthly report by White for March, 1907 p. 4, AA MP 84, file 2002/4/18.
11. 'Observations' by Lieutenant C.B.B. White, Papers of General Sir Brudenell White in the possession of Lady Derham, University of Melbourne.
12. Major-General Sir Brudenell White, 'Australia in the Great War and After', *Australia Today*, 22 November 1919, pp. 53–71.
13. 'Report on Lieutenant C.B.B. White, 21 December 1907', by H. Wilson, Staff College, AA MP 84, file 2002-4-32. For a report on his activities in Great Britain, 1908–11, see AA MP 84/1, file 1894/6/104. There is also a glowing report by Colonel Adye, July 1910, on White's performance at the War Office. White was private secretary to Senator Pearce, Defence Minister, when he visited London for the Imperial Conference in 1911.
14. Minute from Chief of the Commonwealth Section, Imperial General Staff to the Secretary for Defence, 4 October 1912, AA MP 84/1, file 1856/1/33.

15 Minute, Chief of the Commonwealth Section, Imperial General Staff to the Secretary, Department of Defence, 4 October 1912. Proceedings of the Conference between Major-General A.J. Godley and Brigadier-General J.M. Gordon, 18 November 1912, AA MP 84/1 file 1856/1/33; for background on the Australian concern about Japan and the Pacific see N.K. Meaney *A History of Australian Defence and Foreign Policy, 1901–1923: Volume I—The Search for Security in the Pacific, 1901–1914* Sydney: Sydney University Press, 1976, and 'Secret—Australia and New Zealand: Strategic Situation in the Event of the Anglo–Japanese Alliance being Determined', 3 May 1911, CAB 8/5, 422M, PRO.
16 Letter, Colonel J.G. Legge to Major C.B.B. White, 25 July 1913, AA MP 826, box I.
17 Minute, Gordon to Secretary, Department of Defence. 2 July 1913, AA MP 84/1 file 1856/1/23. Bean *Two Men I Knew* p. 90.
18 ibid., pp. 91–2.
19 'Origin of the A.I.F.'—Answer by Brudenell White to C.E.W. Bean on 1 October 1919, Papers of Dr C.E.W. Bean, file 419/8/1, 153–9/12. Bean *Two Men I Knew* p. 93.
20 Letter, White to Bridges, 22 January 1909, Papers of General Sir W.T. Bridges, Bridges Memorial Library, RMC Duntroon. See also letter, White to Hutton, 4 March 1913, Hutton Papers.
21 C.E.W. Bean *The Official History of Australia in the War of 1914–1918. Vol. I. The Story of Anzac* Sydney: Angus & Robertson, 1942, p. 37, and Bean *Two Men I Knew* p. 94.
22 Bean *Two Men I knew* p. 47.
23 Letter, White to Hutton, 19 June 1915, Hutton Papers.
24 This chapter does not include a discussion of White's brief association with General Walker on Gallipoli.
25 Letter, Birdwood to Hutton, 30 September 1916, Hutton Papers.
26 Diary, 24 May 1918, p. 7, Papers of General Lord Rawlinson, Churchill College, Cambridge, MS 1/11.
27 Major-General Sir Brudenell White, 'Australia in the Great War and After', pp. 53–67.
28 Report on the Military Defence of Australia, 6 February 1920, Papers of Sir G.F. Pearce, National Library of Australia, MS 1827, item 14.
29 Letter, Field Marshal Lord Birdwood to White, 22 November 1922, Papers of General Sir Brudenell White.
30 Bean *Two Men I Knew* p. 206.
31 White had a high regard for Blamey's professional ability—evidence from Lady Derham.
32 'White cannot now be asked what he said or what he meant, but the natural inference of his hearers would be that he was thinking in terms of 1918, and that his mind lacked its former elasticity.'—Bean *Two Men I Knew*, p. 215. Bean makes the point that White's advice was 'possibly true', and the May offensive by Hitler may have occupied a number of months.
33 Letter, Sir Alfred Kemsley to D.M. Horner, 21 October 1982. Kemsley, a staff officer on HQ Australian Corps, based his views on the comments of Major-General Harington from HQ 2nd Army, when visiting the Australian Corps. R.G. Menzies *Afternoon Light* Melbourne: Cassell 1969, p. 18.

4 Vice-Admiral Sir William Creswell

1 Sir William Creswell *Close to the Wind: The Early Memoirs (1866–1879) of Admiral Sir William Creswell, K.C.M.G., K.B.E.* edited by Paul Thompson, London: Heinemann, 1965.
2 Armstrong subsequently married Nellie Melba. See John Hetherington *Melba* Melbourne: Cheshire, 1967, p. 30.
3 Stephen D. Webster, 'Creswell, the Australian Navalist: A Career Biography of Vice Admiral Sir William Rooke Creswell, K.C.M.G., K.B.E. (1852–1933)', unpublished PhD thesis, Monash University, 1976, p. 43.
4 See Arthur J. Marder, *The Anatomy of Seapower: A History of British Naval Policy in the Pre-Dreadnought Era, 1880–1905*, New York: Octagon Books, 1940.
5 Webster, 'Creswell, the Australian Navalist', pp. 111–66.
6 The Anglo–Japanese Alliance had been first concluded in 1903. Australia had shown considerable concern about it from the start.
7 Forty-one years earlier Creswell and Henderson had served together in the screw frigate HMS *Phoebe* during a round the world training cruise that had called at Australian ports. See Log of HMS *Phoebe*, ADM 53/9670, Public Record Office (PRO).
8 Commonwealth of Australia, Parliamentary Papers 1911, vol. II.
9 MP 504, S8, 2310/7, Australian Archives.
10 Commonwealth of Australia, Parliamentary Papers 1911, vol. II.
11 'Strategical Report (Part II)', Royal Australian Naval Archives 185i.
12 Senator G.F. Pearce, 29 April 1910 to 24 June 1913, and Senator E.D. Millen, 24 June 1913 to 17 September 1914.
13 Thring to Bazley, 22 September 1938, Bazley Papers, AWM.
14 Memorandum of 8 August 1914, ADM 1/8388/235, PRO.
15 A.W. Jose, *The Royal Australian Navy*, Vol. IX of *The Official History of Australia During the War of 1914–1918* Sydney: Angus & Robertson, 1939, Chapters I–V.
16 After a brief, but spectacular raiding career in the Indian Ocean SMS *Emden* was sunk by HMAS *Sydney* at the Cocos Islands on 9 November 1914.
17 Munro-Ferguson to Harcourt, 29 September 1914, Novar Papers, MS 696/604, National Library of Australia (NLA).
18 Creswell to Allen, 13 December 1914, Allen Papers, New Zealand National Archives.
19 MP 1049, S 1, 14/307, Australian Archives.

20 Colonial Office to Governor-General, 14 October 1914, MP 1049, S 1, 14/055.
21 Colonial Office to Governor-General, 23 November 1914, MP 1049, S 1, 14/0455.
22 ibid., Colonial Office to Governor-General, 3 December 1914.
23 ibid.
24 Although a Council of Defence had been established as early as 1905 it had fallen into disuse during the war. Creswell pressed hard for the re-establishment of an effective Council of Defence. See Webster, 'Creswell, the Australian Navalist', Chapter XI.
25 Latham, who later became Chief Justice of the High Court, was appointed to undertake counter-espionage activities.
26 ADM 1/8520/103, PRO.
27 ibid.
28 MP 1049, S 1, 17/0115A.
29 Jose *The Royal Australian Navy* p. 8.
30 MP 1049, S 1, 17/0153.
31 ibid.
32 Munro-Ferguson to Bonar Law, 13 July 1915, Novar Papers, MS 696/728, NLA.
33 *Reports of the Royal Commission on Navy and Defence Administration 1917–1919*, Commonwealth of Australia, Parliamentary Papers, 1917–1919, vol. IV.
34 Minutes of the Naval Board Meeting of 29 July 1916, A 2585, Australian Archives.
35 Minutes of the Naval Board Meeting of 29 July 1916. Prior to the confrontation he had frequently attended.
36 Retirement age for rear admirals in the Royal Australian Navy had not been set; hence, the reference to the Royal Navy rule. Creswell was promoted to vice-admiral in September 1922, some three years after retirement.
37 Creswell to Latham, 15 July 1918, Latham Papers, MS 1009/1/743-744, NLA.
38 Creswell to Latham, 25 June 1918, Latham Papers, MS 1009/1/707, NLA.

5 *General Sir Harry Chauvel*

1 For details of Chauvel's life I have drawn largely on my biography *Chauvel of the Light Horse* Melbourne: Melbourne University Press, 1978.
2 The Desert Column included two mounted divisions and, at times, an infantry division.
3 Hill, *Chauvel* pp. 35–7.
4 C.E.W. Bean *The Story of Anzac* Sydney: Angus & Robertson, 1942, vol. I, p. 138.
5 Captain P.A. Pedersen, 'The Development of Sir John Monash as a Military Commander', unpublished PhD thesis, University of New South Wales, pp. 142–52.
6 Geoffrey Serle, *John Monash* Melbourne: Melbourne University Press, 1982, p. 222.
7 Hill, *Chauvel* pp. 54–5.
8 Chauvel to his wife, 20.9.1915 in Hill, *Chauvel* pp. 57–8.
9 Quoted *ibid.*, p. 57.
10 Pedersen 'The Development of Sir John Monash' pp. 170–1.
11 ibid., p. 172 but see Bean, *The Story of Anzac*, vol. II, pp. 218–20.
12 Pedersen, 'The Development of Sir John Monash' p. 172.
13 Hill *Chauvel* p. 58.
14 ibid., p. 66.
15 ibid., p. 72.
16 No. 2 Section Canal Defence was responsible to Murray, not Lawrence, the corps commander.
17 Hill *Chauvel* p. 83.
18 ibid., p. 81.
19 ibid., p. 83; H.S. Gullett, *Sinai and Palestine* Sydney: Angus & Robertson, 1941, pp. 190–1.
20 The Imperial Camel Corps Brigade was recruited largely from the AIF and NZEF.
21 John Connell *Wavell: Scholar and Soldier*, London: Collins, 1964, p. 265.
22 For the background to this incident see Hill, *Chauvel* pp. 94–5.
23 Hill, *Chauvel* pp. 106–7.
24 An army troops yeomanry brigade also usually operated under Chauvel.
25 Cyril Falls, *Armageddon 1918* London: Weidenfeld & Nicolson, 1964, p. 41. Falls was good enough to add: 'British commanders and staffs were inclined to be too patronizing in this respect, to the annoyance of Australians and Canadians.'
26 H.S. Gullett *Sinai and Palestine* p. 403.
27 Field Marshal Liman von Sanders, *Five Years in Turkey* London: Ballière, Tindall & Cox, 1927, p. 221.
28 Hill *Chauvel*, p. 145.
29 Liddell Hart, *Through the Fog of War* London: Faber, 1938, p. 100, asserts that the cause of the failure 'lay rather with some of the subordinates on whom Allenby had to depend than with himself. Moreover, he learnt from each of these mistakes [the two Trans-Jordan operations]—not least as to the men on whom he could rely.' The innuendo is unwarranted, so much so that one wonders whether Liddell Hart ever studied the battle. Certainly Grant's dispositions were faulty until Chauvel moved him back; also it may be argued that Chauvel should have acceded to his request for another regiment. Allenby himself acknowledged Grant's achievement in extricating his brigade. Apart from this, Chauvel and his divisional commanders, Shea (60th) Chaytor and Hodgson put up an admirable performance but neither they nor their troops could compensate for the fumbling of Allenby and his staff. It may well be that Allenby learnt from this battle as he clearly did from Third Gaza. Chetwode disliked the operation at the time and later wrote to Wavell: 'These two expeditions of Allenby's across the Jordon were the stupidest thing he ever did, I always thought, and very risky'. Hill *Chauvel* pp. 151–2.
30 The capture of Damascus and Chauvel's relations with Lawrence are discussed at length in Hill *Chauvel* Ch. 11. Lawrence himself warned Liddell Hart of the danger of following his account of

Damascus in *Seven Pillars of Wisdom*. The latter did not seek Chauvel's views on Damascus when writing *T.E. Lawrence In Arabia and After*.
31 Hill p. 243, note 1.
32 ibid., p. 158.
33 In December 1918 a party of New Zealanders and Australians attacked and killed or wounded 30 or more Bedouin in retaliation for the murder of a New Zealander.
34 John Hetherington *Blamey: Controversial Soldier* Canberra: Australian War Memorial, 1973, p. 183.

6 *General Sir John Monash*

1 W.S. Churchill *The World Crisis 1916–1918* 2 Parts, Part 1, London: Butterworth 1927, p. 348. A.G. Serle *John Monash: A Biography* Melbourne: Melbourne University Press, 1982. J. Monash *The Australian Victories in France in 1918* London: Hutchinson, 1920. C.E.W. Bean *Official History of Australia in the War* 6 vols, Sydney: Angus & Robertson, 1920–42.
2 J. Lewis' Notes in Serle *John Monash*, p. 59.
3 In this period Monash also qualified as a Municipal Surveyor, Engineer of Water Supply and Patent Attorney.
4 R.G. Menzies, 'Sir John Monash' *Australian Jewish Historical Society* VI, 1966, pp. 81–2.
5 Serle *John Monash*, p. 140.
6 ibid., pp. 176–7.
7 H. Essame *The Battle for Europe—1918* London: Batsford 1972, p. 117. S.H.E. Barraclough, 'Sir John Monash—Engineer' *Journal of the Institute of Engineers of Australia* III, 1931, p. 363.
8 M. Janowitz *The Professional Soldier* New York: Free Press, 1968, p. 9.
9 J. Monash, 'Leadership in War', address to the Beefsteak Club, 30 March 1926, p. 7. (Box 126, Monash Papers. At the time of writing these were about to be returned to the National Library of Australia for public access. As this will involve a complete recataloguing, references to them will be by box number, followed by the designation MP.)
10 After working as a schoolteacher and insurance salesman, Currie turned to real estate, capitalising on the boom in British Columbia. He was almost ruined by the crash of 1913 and was burdened by his debts throughout the war.
11 B.H. Liddell Hart *Thoughts on War* London: Faber, 1944, p. 122.
12 Monash to Lieutenant E. Matthews, 6 January 1910 (Box 78, MP).
13 Notes by Colonel G. Farlow, June 1933 (Box 76, MP).
14 A. Hollingsworth to Monash, 20 December 1907 (Box 77, MP).
15 Corps District Order V2/08, 2 October 1908 (Box 78, MP).
16 Monash to M'Cay, 17 February 1910; Monash to HQ 3MD, 18 November 1912 (Box 78, MP). 'Report on the Annual Continuous Training—Seymour Camp, 10–17 January 1910, Part 1—Training', (Box 77, MP).
17 Monash 'Staff Duties in Operations' (Notes for two lectures to the USI), 18 June 1911 (Box 77, MP).
18 Notes for 11 October, 'Syllabus for 1909 Course' (Box 77, MP).
19 Liddell Hart *Thoughts on War* p. 223.
20 Monash, 'Notes for Conferences with OCs', 12 August 1913; 'The Development of a Soldierly Spirit', 3 January 1914 (Box 79, MP). Lecture by Monash to officers and senior NCOs, 11 February 1914 (Box 77, MP). Hamilton to E.D. Millen (Defence Minister), 17 February 1914 (Item 7/8/106, Hamilton Papers, King's College, London. Hereafter KC.)
21 Points 7–9, '100 Hints for Company Commanders', November 1914. (Copy in Book 1, Box 79, Monash Collection, Australian War Memorial. Hereafter MC.) Brigade Routine Order 53, 1 February 1915 [File 707/9(2), Australian War Records Section, AWM. Hereafter AWRS]. Letter, 18 February 1915 in F.M. Cutback *War Letters of General Monash* Sydney: Angus & Robertson, 1934, p. 19.
22 4 Bde Orders, 27 April 1915 (Book 3, Box 80, MC). Entry for 3 May 1915, Bean Diary 7 (Bean Collection, AWM. Hereafter BC).
23 Entry for 3 May 1915, Bean Diary 7.
24 Entry for 8 and 17 May 1915, Monash Diary, Box 87, MP.
25 Bean *Official History* vol. 4, p. 488, fn 181.
26 Allanson in R. Rhodes James *Gallipoli* London: Batsford, 1965, p. 272.
27 Monash, Lecture to 3rd Division Officers, 25 October 1916 (Book 14, Box 86, MC).
28 Birdwood to Munro-Ferguson, 26 November 1915 (Box 215, Birdwood Collection, AWM). Cox to Birdwood, 3 May 1916 (Box 212, ibid).
29 Monash, Notes for First Conference, 26 July 1916 (Book 13, Box 85, MC).
30 Monash to Durrant, 2 October 1916 (Box 39, MP).
31 Monash to Plumer, 8 February 1917 (Box 39, MP).
32 Letter, 1 June 1917 in War Letters—Original, p. 302 (Book 1, Box 79, MC).
33 Monash *Australian Victories* p. 96.
34 Monash to Grimwade, 9 December 1916 (Box 39, MP).
35 Entry for 24 May 1917, Haig Diary (Acc 3155, Papers of Field Marshal Sir Douglas Haig, Scottish National Library. Hereafter Haig Diary). Jobson to Monash, 7 May 1917 (Book 15, Box 86, MC).
36 See, for example, A.J.P. Taylor *The First World War* Harmondsworth: Penguin, 1972, p. 190.
37 OAD 291/27, 'Proceedings of Army Commanders' Conference, 14 June', dated 15 June, Haig Diary.
38 Monash to Walter Rosenhain, 14 June 1917 (Box 39, MP).
39 See Monash to Godley, 10.40 am, 4 October 1917 (Item 4, Box 250, 'Operational Records', AWM.

Hereafter, OR). Monash to F. Swinburne, 20 July 1925 (kindly sent to author by Dr A.G. Serle).
40 Monash to F. Swinburne, 20 July 1925.
41 Jackson to Liddell Hart, 4 October 1935 (Item I/516/Ib, Liddell Hart Papers, KC). Harington to Edmonds, 27 January 1932 (Official Historian's Correspondence, BC).
42 Bean *Official History* vol. 5, p. 177.
43 Monash to Birdwood, 28 March 1918 (Box 40, MP). Bean, 'Monash the Soldier' *Reveille* 31 October 1931, p. 2.
44 Letter, 14 May 1918 in War Letters—Original, p. 404.
45 Monash *Australian Victories* p. 296.
46 Monash, 'The History and Constitution of the Australian Army Corps', 3 October 1918 (Notes prepared for Dr R.W. Springthorpe, Box 92, MC). Monash, 'Notes for Visits to 5-6 Bdes', 22-23 June; '11 Bde', 9 July 1918 (Book 19, Box 88, MC). Monash, 'Personal Notes', 4 October 1918 (Springthorpe Papers, Box 92, MC).
47 Elles to GHQ, 3 January 1918 (Item 8, Box 481, OR). Major-General H.J. Elles commanded the Tank Corps.
48 Monash to Fourth Army, 26 June 1918 (Item 2, Box 361, OR).
49 Flavoured smoke consisted of smoke and gas to force the Germans to wear gas masks. Before the battle only smoke was fired, enabling the assaulting troops to attack without gas masks and catching the Germans in theirs. Monash was the first to use it in the BEF.
50 Monash to Rawlinson, 5 July 1918 (Box 40, MP).
51 Monash *Australian Victories* p. 56.
52 C.R.M. Cruttwell *A History of the Great War* Oxford: Clarendon, 1964, p. 532. 'GHQ Notes on Recent Fighting', No. 10, 5 August 1918 (Book 20, Box 89, MC). Courage to Fourth Army, 13 July 1918 (Item 17, Box 358, OR). See, for example, Geoffrey Blainey in *Sydney Morning Herald* 5 May 1973.
53 Monash *Australian Victories* p. 44.
54 Monash to Bruche, 10 October 1919. Sent to author by Dr A.G. Serle, 3 November 1980. Serle *John Monash* p. 340.
55 Bean to White, 30 June 1935 (Box 178A, White Collection, AWM).
56 The exclusion of the gun line from the first objective at Hamel had been soundly criticised by the attacking troops.
57 Monash *Australian Victories* p. 74.
58 Monash 'Allotment of Infantry to Objectives', c. 23 July 1918 (Book 20, Box 89, MC). Monash *Australian Victories* p. 95.
59 Bean, 'The Battle of August 8', *Sydney Sun* 8 October 1931; Bean in *Reveille* 31 October 1931, p. 2.
60 Monash, 'Allotment of Infantry to Objectives'. Monash *Australian Victories* p. 84.
61 Draft dated 7 August 1918 (Book 20, Box 89, MC).
62 G.S. Cook to Sir J. Cook, 2 September 1918 (Box 40, MP).
63 E. Ludendorff *My War Memories, 1914-18* 2 vols, London: Hutchinson, 1919, 2, pp. 679, 684.
64 Entry for 6 August 1918, Bean Diary 116.
65 Monash, 'Leadership in War'.
66 'Battle Instruction One, Series B', 8 August 1918 (Box 94A, MC).
67 Bean *Official History* vol. 6, p. 685. Entry for 11 August 1918, Bean Diary 116.
68 Monash *Australian Victories* p. 145.
69 Liddell Hart to Major-General Dorman Smith, 26 November 1942 (Item I/242/47, Liddell Hart Collection, KC).
70 Bean, 'Narrative of Australian Corps Operations to September 1918' (Box 42, MC). Monash *Australian Victories* p. 159.
71 Fourth Army G38 to Corps, 24 August 1918 (Item 73, Montgomery-Massingberd Collection, KC). Monash *Australian Victories* pp. 166-7. Battle Instruction 10, Series C, 26 August 1918 (Box 94A, MC).
72 Monash *Australian Victories* p. 176.
73 Cited in Bean *Official History* vol. 6, p. 822.
74 ibid., p. 872.
75 A.J. Smithers, *Sir John Monash* London: Leo Cooper, 1973, pp. 247-8. Monash *Australian Victories* pp. 194, 177.
76 ibid., pp. 222, 223. Entry for 18 September 1918, Fourth Army War Diary (Item 4, Box 473, OR). Entry for 18 September 1918, Bean Diary 116.
77 Monash *Australian Victories* p. 202. Bean *Official History* vol. 6, p. 940. Cited in entry for 8 September 1918, Bean Diary 116.
78 Monash *Australian Victories* p. 242.
79 Monash, 'Plan for Beaurevoir Offensive', 21 September 1918 (Book 21, Box 89, MC).
80 Gellibrand, 'The Action of the Third Australian Division at Boy, 29 September–1 October 1918' (Box 190A, Gellibrand Collection, AWM).
81 Monash, 'Leadership in War'.
82 Monash *Australian Victories* p. 244.
83 Entry for 28 September 1918, Haig Diary.
84 A. Bazley, 'Sir John Monash' *Reveille* 1 May 1937, p. 15.
85 J. Terraine *Douglas Haig: The Educated Soldier* London: Hutchinson, 1963, p. 215. Serle *John Monash* p. 375. Gellibrand, 'Comments on Australian Victories' (Box 90, Gellibrand Collection, AWM). B.H. Perry (14 BN) to author, November 1980. Bean *Official History* vol. 6, p. 1042. Monash *Australian Victories* pp. 284-6.
86 Serle *John Monash* p. 375.

*7 Lieutenant-General Sir John Lavarack*

1 J.D. Lavarack 'The Defence of the British Empire, with Special Reference to the Far East and Australia', *Army Quarterly*, vol. XXV, no. 2, January 1933. See Admiral Sir H. Richmond 'An Outline of Imperial Defence' *Army Quarterly*, July 1932.
2 Lavarack, 'Defence of the British Empire' pp. 209-10. G. Long *To Benghazi* Canberra: Australian War Memorial, 1952, pp. 7-9.
3 Lavarack, 'Defence of the British Empire' pp. 210-13, 216.

4 E.M. Andrews 'The Broken Promise—Britain's Failure to Consult its Commonwealth on Defence in 1934, and the Implications for Australian Foreign and Defence Policy' *Australian Journal of Defence Studies*, vol. 2, no. 2, November 1978, pp. 103, 108-9. N.H. Gibbs *Grand Strategy*, vol. 1, London: HMSO, 1976, p. 375.
5 Andrews, 'The Broken Promise' p. 110.
6 Minutes of Defence Committee Meeting, 21 March 1935, Agenda 4/1935, Appendix B, 'Memorandum by Colonel J.D. Lavarack, CMG, DSO, Chief of the General Staff Designate, on Report on Certain Aspects of Australian Defence by Sir Maurice Hankey, GCB, GCMG, GCVO, Secretary, Committee of Imperial Defence', CRS A2031, vol. 3. (Bruche's memorandum is contained in Appendix A.) Memorandum, Lavarack to CGS, 6 March 1930, Shedden Papers, Australian Archives Accession, MP 1217, item Box No. 39.
7 Minutes of Defence Committee Meeting, 19 July 1935, Annex 8, 'Australian Coast Defences. Reconsideration of Revise of CID Paper 249C (DC Paper 35/34). Memorandum by the Chief of the General Staff', CRS A2031, Volume 3.
8 Minutes of Defence Committee Meeting, 19 July 1935, CRS A2031, Volume 3, pp. 8-9. See also Minutes of Council of Defence Meeting, 19 June 1935, AA1971/216.
9 Council of Defence Meeting, 24 August 1936, Summary of Proceedings, p. 8, AA1971/216.
10 Council of Defence Meeting, 17 December 1937, Agenda No. 4/1937, 'Imperial Conference 1937. Questions raised by Australian Delegation on Empire and Australian Defence Policy', pp. 15-16, AA1971/216.
11 Council of Defence Meeting, 24 February 1938, Summary of Proceedings, pp. 2-3, AA1971/216. D.M. Horner *Crisis of Command, Australian Generalship and the Japanese Threat 1941-1943* Canberra: Australian National University Press, 1978, p. 2.
12 Council of Defence, 24 February 1938, Summary of Proceedings, pp. 13, 17, AA1971/216.
13 Herring to Hetherington, Hetherington Papers, Box 2, 3DRL 6224, AWM. J. Hetherington *Blamey: Controversial Soldier* Canberra: Australian War Memorial and AGPS 1973, pp. 83-4. Minutes of Defence Committee Meeting, 16 August 1938, Agenda No. 45/1938, CRS A2031, vol. 5. Letter, Lavarack to Piesse, 7 August 1935, Piesse Papers, MS 882, Series 9, NLA. Gavin Long Diaries, D 11/22-23, AWM.
14 Letter, Lavarack to Richmond, 27 March 1936, Lavarack Papers (private collection).
15 Letter, Lavarack to Dill, 9 June 1936, Lavarack Papers.
16 War Diary 7 Div G Branch, 3, 6 April 1941, 1/5/14, AWM. Lavarack diary, Lavarack Papers 3, 6 April 1941. Wilson diary, (Lavarack's ADC), Wilson Papers (private collection), 3 April 1941. I.S.O. Playfair *History of the Second World War, The Mediterranean and Middle East, Volume II, The Germans Come to the Help of Their Ally (1941)* London: HMSO, 1956, p. 33. G.M. Long *The Six Years War* Canberra: Australian War Memorial and AGPS 1973, p. 75. B. Maughan *Tobruk and El Alamein* Canberra: Australian War Memorial 1966, p. 115. 'Narrative of Operations in Cyrenaica During Period April 7-14, both inclusive, by Lt-Gen Sir John Lavarack', p. 3, Lavarack Papers.
17 'Narrative of Operations in Cyrenaica', p. 3. Lavarack diary, 7 April 1941. War Diary 7 Div G Branch, 7 April 1941, 1/5/14, AWM. Anthony Eden *The Reckoning* Boston: Houghton Mifflin, 1965, pp. 278-279. Maughan *Tobruk and El Alamein* pp. 118-19.
18 'Narrative of Operations in Cyrenaica', pp. 4-5.
19 See Maughan, *Tobruk and El Alamein* pp. 143, 148-55. Signal, DCGS to Lavarack, 12 April 1941, HQ Cyrcom War Diary 12 April 1941, Appendix B, WO 169/1240, PRO. Lavarack diary, 12 April 1941. Wilson diary, 12 April 1941. Signal, Wavell to Lavarack, 14 April 1941, War Diary HQ Cyrcom, 14 April 1941, Appendix C, WO 169/1240, PRO.
20 'Narrative of Operations in Cyrenaica', p. 6. War Diary 7 Div G Branch, 8 April 1941 (refers to 9 April), 1/5/14, AWM. Maughan *Tobruk and El Alamein* pp. 121-2, 137. Signals, Mideast to Cyrcom, 11, 12 April 1941, and Lavarack to DCGS, 12 April 1941, War Diary HQ Cyrcom 11, 12 April 1941, Appendices C & D, WO 169/1240, PRO. Lavarack diary, 11 April 1941.
21 See A.B. Lodge 'A Share of Honour: Lavarack and Tobruk—April 1941', *Journal of the Royal United Services Institute of Australia*, vol. 5, no. 1, April 1982, pp. 56-9.
22 War Diary 7 Div G Branch, 22 May 1941, 1/5/14, AWM. Lavarack diary, 22 May 1941. G. Long *Greece, Crete and Syria* Canberra: Australian War Memorial, 1953, p. 336.
23 Lavarack diary, 28 May, 5 & 8 June 1941. 'Notes on Chapter 16', 26 September 1952, pp. 1-2, Gavin Long Correspondence, J.D. Lavarack, Part 1, AWM.
24 Lavarack diary, 12-15 June 1941. Letter, Lavarack to Long, 1 October 1952, p. 3, Gavin Long Correspondence. Long *Greece, Crete and Syria* pp. 386-8, 395-401. Berryman diary, 15 June 1941, Berryman Papers (private collection).
25 'Report on Operations—1st Australian Corps—In the Campaign in Syria, June-July 1941', Appendix A, "Summary of Operations—Syrian Campaign 1941 up to 18 Jun 41', Blamey Papers 51, 3DRL 6643, AWM. Long *Greece, Crete and Syria* pp. 513, 527.
26 'Notes on Chapter 22', pp. 5, 7, Gavin Long Correspondence. '6 [UK] Division Report on Syrian Operations June-July 1941', p. 1, Blamey Papers 51, 3DRL 6643, AWM.
27 Letter, Lavarack to Long, 1 October 1952, Gavin Long Correspondence. Long *Greece, Crete and Syria* p. 527.
28 '7 Aust Div Report on Ops in Syria', p. 38, Blamey Papers.
29 Long *Greece, Crete and Syria* pp. 515-22.

30 Lavarack diary, 26–28 January 1942. L. Wigmore *The Japanese Thrust* Canberra: Australian War Memorial, 1957, pp. 442–3.
31 Lavarack diary, 10 February 1942.
32 Lavarack diary, 12–13 February 1942. Wigmore *The Japanese Thrust* pp. 444–6.
33 Wigmore *The Japanese Thrust* p. 449.
34 Lavarack diary, 18 February 1942. Wigmore, *The Japanese Thrust*, p. 450.
35 Lavarack diary, 14–15 February 1942. Wigmore, *The Japanese Thrust* pp. 453–4.
36 Lavarack diary, 16–19 February 1942. Wigmore, *The Japanese Thrust* pp. 446–7, 454–7. C.E.M. Lloyd to Long, Gavin Long Notebooks, N1/22–23, AWM. Letter, Lavarack to Long, 10 October 1952, Gavin Long Correspondence.
37 'Notes of discussions With Commander-in-Chief Australian Military Forces 27th September 1943', Shedden Papers, item Box No. 4. Wigmore *The Japanese Thrust* pp. 464–5, 506.
38 D.M. Horner *High Command Australia and Allied Strategy, 1939–1945* Sydney: George Allen & Unwin, 1982, p. 164. See C.E.M. Lloyd to Long, Gavin Long Notebooks, N1/27, AWM.
39 'Notes of Discussions with Commander-in-Chief Australian Military Forces 27th September 1943', Shedden Papers. Letter, Lavarack to Long, 14 February 1945, Gavin Long Correspondence, Part 2.

## 8 Lieutenant-General Sir-Vernon Sturdee

1 I.D. Rae 'Outline History of the Australian Army Organization', prepared for the Army Review Committee, 1970.
2 Military Board Agendum 92/1935 and Army War Book, July 1939, with revisions to September 1939, both in AWM 577/7/33.
3 Interview with W.M. Hughes, 20 February 1944, Gavin Long Notes, No. 45, AWM. War Cabinet Minute 345, Melbourne, 18 June 1940, in H. Kenway, H.J.W. Stokes, P.G. Edwards (eds) *Documents in Australian Foreign Policy 1937–49, Volume III: January–June 1940* Canberra: Australian Government Publishing Service, 1979, pp. 451–3.
4 C.E.W. Bean *Two Men I Knew* Sydney: Angus & Robertson, 1957, p. 220.
5 Letter, Bennett to Sturdee, no date, MP 727/7, item 17/421/72, Australian Archives (Melbourne).
6 Lieutenant-General Sir Sydney Rowell, 'General Sturdee and the Australian Army' *Australian Army Journal*, August 1966.
7 Wilton to author, 15 November 1974. Hopkins to author, 12 August 1974. Also interviews with Rogers (25 June 1974), Barham (11 December 1974), B.J.F. Wright (12 October 1978), Buckley (22 January 1979). S.F. Rowell, *Full Circle* Melbourne: Melbourne University Press, 1974, p. 31.
8 Lecture on the Plan of Concentration by Lieutenant-Colonel V.A.H. Sturdee, Senior Officers' Course 1933, AWM 243/6/150.

9 Sir Arthur Fadden, 'Forty Days and Forty Nights, Memoir of a War-time Prime Minister', *Australian Outlook* no. 27, 1973, p. 6. AHQ Operation Instruction No. 12, 24 February 1941, AWM 243/6/42. An amendment to the Defence Act had provided that the militia might serve in Papua, but not in New Guinea.
10 War Cabinet Agendum 146/1941. War Cabinet Minute No. 1146, 10 June 1941, MP 729/6, item 2/401/40.
11 Instructions to GOC-in-C Home Forces, MP 729/6, item 2/401/40.
12 Rowell 'General Sturdee and the Australian Army', p. 8.
13 Defence of Australia and Adjacent Areas—Appreciation by Chiefs of Staff, 11 December 1941, CRS A2671, item 14/301/227, Australian Archives (Canberra). War Cabinet Minute 1529 quoted in P. Hasluck, *The Government and the People, 1939–1941* Canberra: Australian War Memorial, 1952, p. 401.
14 War Cabinet Agendum No. 418/1941, 12 December 1941, and Chiefs of Staff Appreciation, 15 December 1941, War Cabinet Agendum, No 418/1941, 18 December 1941, CRS A2671, item 14/501/227.
15 Chiefs of Staff Appreciation, 15 December 1941, War Cabinet Agendum No 418/1941, 18 December 1941, CRS A2671, item 14/301/227. E.G. Keogh *South West Pacific, 1941–45* Melbourne: Grayflower, 1965, p. 109. Rowell to author 26 June 1974.
16 L. Wigmore *The Japanese Thrust* Canberra: Australian War Memorial, 1957, pp. 422, 423n.
17 Hasluck *The Government and the People* pp. 673, 674.
18 Wigmore *The Japanese Thrust* p. 258.
19 Notes by CGS (Sturdee) in Memorandum for the Minister by Shedden, 20 February 1947, CRS A816, item 37/301/330.
20 Letter, Curtin to Forde, 23 February 1942, with attached extracts dated 13 February 1942, MP 729/6, file 15/401/649.
21 Letter, Sturdee to Secretary, Department of the Army, 4 March 1942.
22 Stanner to author, 4 June 1979.
23 Memoranda, Sinclair to Forde, 9 December 1941, Sturdee to Minister 30 January 1942, Blamey Papers DRL 6643, item 'Secretary and Public Service Miscellaneous'.
24 His sense of propriety explains his cold attitude towards General Bennett after the latter's escape from Singapore. He thought Bennett's action was wrong. It was not a matter of personal animosity. On 9 May 1940, while GOC Eastern Command, Sturdee wrote to the Adjutant-General that Bennett was 'probably the most efficient militia officer in the Commonwealth... He is looked up to as an excellent leader and organizer and one who maintains the dignity of a senior officer... like all energetic public spirited men, he has some critics, especially amongst the less efficient'. Military Board Proceedings 65, 40.
25 Defence of Australia, 4 February 1942, CRS

A2684, item 905.
26 ibid., Appendix II.
27 Paper by the Chief of the General Staff on Future Employment of AIF. 15 February 1942, AWM 541/1/4.
28 Interview with B.J.F. Wright QC, 12 October 1978. Wright was Sturdee's ADC and later MA. Wright was not present at the Chiefs of Staff meeting but was told these details by the late Douglas Menzies (later the Right Honourable Sir Douglas Menzies), then Secretary to the Chiefs of Staff Committee.
29 Curtin to Gavin Long, Sydney, 2 October 1943, Gavin Long Diary No. 1, AWM. Rowell, 'General Sturdee and the Australian Army', p. 9. Letter from Colonel J.P. Buckley, 20 August 1974, and interview, 22 January 1979. Buckley is Sturdee's son-in-law and was a First Assistant Secretary in the Department of Defence until he retired. War Cabinet Minute 1896, Sydney, 18 February 1942, CRS A816, item 52/302/142.
30 Buckley to author, 8 April 1983. For a draft of the cable see MP 1217, Box 573.
31 Buckley to author, 23 September 1982.
32 Memorandum, 'Redesignation of Formations', Sturdee to Minister, February 1942, MP 729/6, item 14/401/151.
33 Letter, Vasey to his wife, 10 August 1942, Vasey Papers 2, 6. Rowell, 'General Sturdee and the Australian Army' p. 9.
34 Transcript of interview with F.M. Forde, 4 March 1971, TRC 121/8, NLA.

## 9 · Lieutenant-General Henry Gordon Bennett

1 Percival, from comments written on F. Owen *The Fall of Singapore* London: Pan, 1962, March 1959, Percival Papers 44, p. 8.
2 Quoted in L. Wigmore *The Japanese Thrust* Canberra: Australian War Memorial, 1957, p. 65.
3 Bennett Diary, Gavin Long Correspondence, H.G. Bennett, AWM, 9 August 1941. Wigmore *The Japanese Thrust* p. 96.
4 Bennett Diary, 28-31 December 1941.
5 War Diary, HQ 8 Div, G Br, 1/5/17, 14 February 1942, AWM.
6 Letter, Bennett to Secretary, Military Board, 27 March 1942, CRS A 2671, item 192/1942; Blamey Papers, AWM, 33.21; Bennett Papers, Mitchell Library, ML MSS 807/5.
7 Letter, Bennett to Long, undated, Bennett Papers, item 183. See also Bennett Diary, 3 April 1941.
8 Letter, Bennett to Forde, 19 December 1941, Forde Papers AA1974/398; Bennett Diary, 30 January 1942; letter, Bennett to Long, 15 January 1954, Bennett Papers.
9 S.W. Kirby *Singapore: Chain of Disaster* London: Cassell, 1971, p. 94. See also letter, Bennett to Forde, 27 January 1942, Forde Papers; 'Report by Major-General H. Gordon Bennett, CB, CMG, DSO, VD, GOC AIF Malaya on Malayan Campaign, 7th December 1941 to 15th February 1942', WO 106/2569. p. 27.
10 H.G. Bennett, *Why Singapore Fell* Sydney: Angus & Robertson, 1944, pp. 20, 21. Letter, Bennett to Forde, 27 January 1942, Forde Papers, AA 1974/398. B. Bond (ed.) *Chief of Staff: The Diaries of Lieutenant-General Sir Henry Pownall* vol. 2, 1940-1944, London: Leo Cooper, 1974, p. 76. See Wigmore *The Japanese Thrust* p. 204.
11 'Comments by Lieut-General A.E. Percival CB DSO OBE MC', 1 December 1953, Percival Papers 48, Imperial War Museum. Letter, Bennett to Forde, 27 January 1942, Forde Papers.
12 Bond *Chief of Staff* pp. 70, 75-6.
13 'Comments by Lieut-General A.E. Percival CB DSO OBE MC on the Official Historians' Comments on the Fighting in Johore', 30 November 1953, Percival Papers 42 [subsequently referred to as 'Percival, "Comments" 30 November 1953'].
14 'Notes by Lieut-General A.E. Percival CB DSO OBE MC on Certain Senior Commanders and other matters', 8 January 1954, Percival Papers 42 [subsequently referred to as 'Percival, "Notes" 8 January 1954'].
15 Sir John Smyth *Percival and the Tragedy of Singapore* London: MacDonald, 1971, p. 256.
16 Bennett Diary, 30 January 1942. Cable, PM(Aust) to PM(UK), 24 January 1942, CAB 66/21. 'Note by Lt. Gen. A.E. Percival CB, DSO, OBE, MC, on the "Brief Outline Narrative of the Mainland Operations in Malaya 8th Dec. 1941 to 31st Jan. 1942" prepared by the Combined Inter-Service Historical Section (India)', p. 7, Percival Papers 43. Bond *Chief of Staff* p. 70. Lieut-General A.E. Percival 'Despatch on Operations of Malaya Command, 8th December 1941—15th February 1942', Second Supplement to *London Gazette*, 20 February 1948, p. 1300 [subsequently referred to as 'Percival, "Despatch"']; see also letter, Percival to Kirby, 4 September 1954, Percival Papers 48. Bennett Diary, 14 January 1942.
17 Kirby *Singapore: Chain of Disaster* p. 96.
18 Letter, Percival to Simpson, 29 October 1945, Percival Papers 48.
19 Percival, 'Notes', 8 January 1954.
20 ibid.
21 Letter, Kirby to Percival, 2 February 1954, Percival Papers 48.
22 Cable, PM (Aust) to PM(UK), 24 January 1942. See also Bennett Diary, 27 December 1941.
23 Bennett Diary, 16 August 1941. A.E. Percival *The War in Malaya* London: Eyre & Spottiswoode, 1949, p. 284.
24 Percival, comments on Owen, p. 7, Percival Papers 44. Letter, Percival to Denis Russell-Roberts, 24 August 1964, Percival Papers 44. 'Comments by Lieut-General A.E. Percival CB DSO OBE MC', 2 December 1953, Percival Papers 42.
25 Letter, Percival to Kirby, 4 September 1954, Percival Papers 48. 'Plans for the Defence of Johore, Additional Comments by Lieut-General A.E. Percival, C.B., D.S.O., O.B.E., M.C.', p. 1, Percival Papers 48.
26 Percival, 'Notes', 8 January 1954.

27 Kirby *Singapore: Chain of Disaster* p. 223. Percival, 'Comments', 30 November 1953.
28 Wigmore *The Japanese Thrust* p. 97. J.H. Thyer and C.H. Kappe '8th Australian Division, Report on Operations in Malaya, December 1941–February 1942', CAB 106/162, p. 191 The report was compiled by Thyer from a narrative prepared by Kappe. Taylor to Long in Gavin Long Notebooks and Diaries, AWM, N109/78. Thyer, 'Report', p. 191. Letter, Percival to Denis Russell Roberts, 24 August 1944, Percival Papers 44.
29 Gavin Long Notebooks and Diaries, AWM, N109/14,21. See E.G. Keogh *Malaya 1941–42* Melbourne: Printmaster, 1962, p. 179.
30 T.W. Mitchell, interview, Canberra, 16 December 1980.
31 Bennett Diary, 27 December 1942.
32 Percival, 'Notes' 8 January 1954.
33 Thyer and Kappe, '8th Australian Division', p.(v).
34 Bennett Diary, 12 April 1941. Thyer and Kappe, '8th Australian Division' pp. 191–2. Bennett Diary, 18 April 1941.
35 See J. Leasor *Singapore: The Battle That Changed the World* London: Hodder & Stoughton, 1968, pp. 159–60.
36 Quoted in Wigmore *The Japanese Thrust* p. 155.
37 Quoted in I. Simson *Too Little Too Late* London: Leo Cooper, 1970, p. 43; see also Louis Allen *Singapore 1941–1942* London: Davis Poynter, 1977, p. 206.
38 Leasor *Singapore* pp. 159–60. H. Belloc *The Tactics and Strategy of The Great Duke of Marlborough* London: Arrowsmith, 1933, p. 8. War Diary 2/30th Battalion, January 1942, 'Narrative 1: Sungei Gemencheh and Gemas', p. 3, 8/3/30, AWM.
39 See Wigmore *The Japanese Thrust* p. 273; Bennett Diary, 28 January 1942; J.W.C. Wyett, interview, Corryong, 22 February 1980.
40 Letter, Bennett to Forde, 13 December 1941, MP 729/6, item 13/401/402. Bennett Diary, 5 January 1942. Letter, Bennett to Forde, 27 January 1942, Forde Papers. Letter, Colonel J.P. Buckley to author, 6 September 1982.
41 C.G.W. Anderson, interview, Canberra, 22 August 1980; 14 and 20 July 1982. Gavin Long Notebooks and Diaries, N109/23.
42 Quoted in Smyth *Percival* p. 256. Percival 'Notes' 8 January 1954.
43 G. Long, *To Benghazi* Canberra: Australian War Memorial, 1952, pp. 4–5. Commonwealth Parliamentary Debates, 10 August 1926, vol. 114, p. 5181. Directorate of Military Training, AHQ, 'The Basis for Expansion for War', *Australian Army Journal*, May 1950, p. 7. Long *To Benghazi* p. 30.
44 C. Neumann 'Australia's Citizen Soldiers, 1919–1939: A Study of Organization, Command, Recruiting, Training and Equipment', unpublished MA thesis, Faculty of Military Studies, University of New South Wales, 1978, pp. 214–15. J.F.C. Fuller *Generalship: Its Diseases and Their Cure* London: Faber & Faber, 1933, p. 50. D.M. Horner, *Crisis of Command* Canberra: Australian National University Press, 1978, p. xvii.
45 R. Lewin *The Chief* New York: Farrar-Straus-Giroux, 1980, p. 94.
46 I. Morrison *Malayan Postscript* Sydney: Angus & Robertson, 1943, p. 157.
47 C.E.W. Bean *The AIF in France: May 1918–Armistice* Sydney: Angus & Robertson, 1942 p. 404.

10  Lieutenant-General Sir Leslie Morshead

1 C.E.W. Bean *The Story of Anzac* vol. 1, Sydney: Angus & Robertson, 1921, p. 515.
2 C.E.W. Bean *The AIF in France in 1918* vol. V, Sydney: Angus & Robertson, 1937, pp. 302–9.
3 Barton Maughan, *Tobruk and El Alamein*, Canberra: Australian War Memorial, 1966, p. 8.
4 Morshead Diary, 16 March, 29 March, 2 April, 3 April 1941, Morshead Collection 98/8 AWM.
5 ibid., 16, 17 March 1941; Wavell's comment on Neame is in John Connell, *Wavell: Scholar and Soldier* London: Collins, 1964, pp. 385–6. Blame for the confusion in the early stages of Rommel's advance must also go to Wavell who, having lost confidence in Neame before the battle, sent O'Connor forward during the battle, but not to command, and interfered directly with Neame's orders to his armour.
6 'Credit must be given to Morshead and his staff for this feat of organization which included an extensive mining and demolition programme as well as the extrication of the troops without loss.' General Sir W.G.F. Jackson, *The Battle for North Africa 1940–43* London: Batsford, 1975, p. 105.
7 Maughan *Tobruk and El Alamein* pp. 69–76.
8 Morshead to Sir John Lavarack 11 August 1943, Morshead Collection, 99/18.
9 Connell *Wavell* p. 403. The account is quoted with approval in Ronald Lewin's study of Wavell, *The Chief* London: Hutchinson, 1980.
10 Maughan's account is confirmed in a letter, Morshead to Lavarack, 11 August 1943, Morshead Collection, 99/18. I thank Mr Brett Lodge for drawing my attention to General Lavarack's Summary of Events 4 April–16 April 41 among the Lavarack Papers on loan to him.
11 Maughan *Tobruk and El Alamein* p. 159.
12 GS War Diary, 9th Australian Division, April 1941, AWM.
13 Chester Wilmot *Tobruk* Sydney: Angus & Robertson, 1944, p. 111.
14 Maughan *Tobruk and El Alamein* pp. 338–9.
15 ibid., pp. 195–202.
16 ibid., p. 217.
17 B.H. Liddell Hart (ed.) *The Rommel Papers* London: Collins, 1953, p. 132.
18 Maughan *Tobruk and El Alamein* pp. 234, 316–27, 332–3.

19 C.E.W. Bean, *The AIF in France in 1918* vol. V, p. 301. Recollection of the author.
20 Maughan *Tobruk and El Alamein* p. 353.
21 Morshead Diary, 19 May 41.
22 Quoted in Maughan *Tobruk and El Alamein* pp. 400-1.
23 Quoted in J.F.C. Fuller *The Decisive Battles of the Western World* vol. 3, London: Eyre & Spottiswoode, 1963, p. 88.
24 Morshead Diary, 9 February, 1942.
25 Maughan *Tobruk and El Alamein* p. 546.
26 ibid., pp. 551-2; Jackson *The Battle for North Africa* p. 242. See Field Marshal Lord Harding's comments in N. Nicolson *Alex* London: Weidenfeld & Nicolson, 1973, p. 155.
27 Liddell Hart (ed.) *The Rommel Papers* pp. 250, 253-4, 262, 264-70.
28 Maughan *Tobruk and El Alamein* p. 554.
29 Morshead Diary, 21 July 1942.
30 ibid., quoted in Maughan *Tobruk and El Alamein* p. 579.
31 Corelli Barnett, *The Desert Generals* London: Kimber, 1960, omits this. Liddell Hart, *The Tanks* vol. 2, London: Cassell, 1952, and J. Connell, *Auchinleck* London: Cassell, 1959, have the Australians driven back to their original positions.
32 Maughan *Tobruk and El Alamein* p. 598.
33 Hugh Paterson was serving on HQ 20th Brigade.
34 Maughan *Tobruk and El Alamein* p. 598.
35 ibid., pp. 610, 613-4, 616.
36 C.E.L. Phillips *Alamein* London: Heinemann, 1962, pp. 107, 55.
37 N. Hamilton *Monty: The Making of a General* London: Hamish Hamilton, 1981, pp. 718-28, 732-41.
38 Maughan *Tobruk and El Alamein* p. 662. GS War Diary 9 Australian Division, 22 October 1942. Author's recollection.
39 Maughan *Tobruk and El Alamein* p. 663. The slit trenches were to be dug and occupied in no man's land the night before.
40 Hamilton *Monty* p. 801.
41 Notes by Morshead on articles by DGPR, LHQ, 29 August 1944, Morshead Collection, 99/14.
42 Maughan *Tobruk and El Alamein* pp. 709-20.
43 ibid., pp. 720-8.
44 Lieutenant-General Sir Brian Horrocks *A Full Life* London: Leo Cooper, 1960, p. 140.
45 Maughan *Tobruk and El Alamein* p. 746.
46 GS War Diary 9 Australian Division, 17 November 1942; also Maughan *Tobruk and El Alamein* p. 747.
47 Maughan *Tobruk and El Alamein* p. 750.
48 D. Dexter *The New Guinea Offensives* Canberra: Australian War Memorial, 1961, pp. 557-8.
49 J. Hetherington *Blamey: Controversial Soldier* Canberra: Australian War Memorial, 1973, p. 329.
50 G. Long *The Final Campaigns* Canberra: Australian War Memorial, 1963, pp. 457-8.
51 ibid., p. 506.
52 Hamilton *Monty* p. 834.
53 Maughan *Tobruk and El Alamein* p. 754.
54 ibid., p. 748.

*11 Field Marshal Sir Thomas Blamey*

1 A.J. Sweeting 'The War in Papua' *Stand To.* vol. 6, no. 6, November 1958-January 1959.
2 Shedden Manuscript, Book 4, Box 4, Ch 55, p. 2.
3 John Hetherington, *Blamey: Controversial Soldier* Canberra: Australian War Memorial, 1973, p. 80. L.E. Beavis, review of G. Long *The Final Campaigns*, in *Stand To*, vol. 9, no. 1. G. Long *Six Years War* Canberra: Australian War Memorial, 1973, p. 22.
4 Notes by BGS for Military History Section, 25 January 1941, Blamey Papers 67.
5 Letter, Rowell to Long, 20 January 1947, Gavin Long Correspondence-Rowell, Australian War Memorial. Hetherington *Blamey* p. 155.
6 J. Field to Gavin Long, 14 February 1945, G. Long Notes, AWM 577/7/32. Hetherington *Blamey*, p. 216.
7 Letter, Blamey to Curtin, 29 June 1942, CRS A2670, Item 231/1942.
8 Letter, Wynter to Berryman, 29 March 1944, AWM 225/1/16.
9 Letters, Blamey to Curtin, 24 September, Blamey Papers 23:81; 8 October 1942, Hetherington *Blamey*, p. 299.
10 S.F. Rowell *Full Circle* Melbourne: Melbourne University Press, 1975, p. 110. Jay Luvaas (ed.) *Dear Miss Em, General Eichelberger's War in the Pacific, 1942-1945* Westport, Connecticut: Greenwood Press, 1972, pp. 27, 21.
11 Letter, Blamey to Curtin, 8 October 1942, Blamey Papers 23.81; and letter, MacArthur to Curtin, 10 October 1942, RG4, MacArthur Memorial. See also account in Chapter 12.
12 Letter, Sturdee to Smart, 20 July 1942, AWM 425/11/12. Commander-in-Chief's Diary, Blamey Papers DRL 6643, Item 144. He spent 28 of the 53 days (1 August to 23 September) in Brisbane. Letter, Vasey to Rowell, 1 September 1942, Rowell Papers, AWM.
13 Wilkinson Diary, 19 October 1942, Churchill College, Cambridge. Notes of Discussions (by Shedden) with Commander-in-Chief, Southwest Pacific Area, Brisbane, 20-26 October 1942, MP 1217, Box 2.
14 Interview with Berryman, 26 June 1974. D. MacArthur *Reminiscences* Greenwich, Connecticut: Fawcett, 1964, p. 168.
15 Minutes of Prime Minister's War Conference, Canberra, 17 July 1942, MP 1217, Box 1. F.T. Smith Reports, No. 13, 23 July 1942. Wilkinson Diary quoted in C. Thorne 'MacArthur's British Liaison Officer (Part II)', *Australian Outlook*, vol. 20, no. 2, August 1975. Blamey's Memoirs.
16 Radio, MacArthur to Marshall, 22 September 1942, RG 165, OPD Exec 10, Item 236, National Archives.
17 Advisory War Council Minute No. 1067, 17 September 1942, CRS A2682, Item, vol. V. Notes of

Secraphone Conversation between the Prime Minister and the Commander-in-Chief, Southwest Pacific Area, 17 September 1942, MP 1217, Box 532. 'C-in-C's Press Conference', Perth, 9 July 1945, Blamey Papers 138.3. Letter, Dunstan to Rowell, 29 September 1942, Rowell Papers.
18 Letters, Blamey to MacArthur, 27 December 1942, MacArthur to Blamey, 28 December 1942, Blamey Papers 43.361. Memoirs of Air Marshal Sir George Jones.
19 Blamey Memoirs.
20 Signal, MacArthur to Marshall, 11 January 1943, RG4, MacArthur Memorial. W. Krueger *From Down Under to Nippon* Washington: Combat Forces Press, 1953, p. 10.
21 Dewing to Wilkinson, 10 February 1943, quoted in C. Thorne *Allies of a Kind* London: Hamish Hamilton, 1978, p. 263. Interview tapes, Papers of General George H. Decker, US Military History Institute, Carlyle, Pennsylvania. Luvaas *Dear Miss Em* p. 67.
22 E.G. Keogh *The South West Pacific Area*, Melbourne: Grayflower, 1965, p. 473. Memorandum by Shedden, 19 November 1942 and letter, Curtin to the Minister for Air, 24 December 1942, CRS A 816, Item 31/301/196A. War Cabinet Agendum 107/1943, Sup 1, 15 April 1943, CRS A816, Item 31/301/196A.
23 Keogh *The South West Pacific Area* p. 474.
24 Letter, Krueger to General Orlando Ward, 12 September 1951, RG 319, National Archives. Long *The Final Campaigns* p. 599. Letter, Diller to MacArthur, 17 May 1943, RG4, MacArthur Memorial. Transcript of interview with F.M. Forde, 4 March 1971, NLA TRC 121/8.
25 Letter, Gairdner to Ismay, 30 May 1945, WO, 216/317.
26 D. McCarthy *South-West Pacific Area, First Year* Canberra: Australian War Memorial, 1959, p. 82.
27 'The Defence of Port Moresby' by F.G. Shedden, 30 September 1942, MP 1217, Box 587. Notes of Discussions with C-in-C, SWPA, 20–26 October, 1942, MP 1217, Box 2. Rowell to author, 26 June 1974. G. Long Diary No 81, Sydney 1 August 1942, p. 76.
28 'The Defence of Port Moresby' and memorandum by Shedden, 3 October 1942, MP 1217, Box 587.
29 Appreciation on Operations of the AMF in New Guinea, New Britain and the Solomon Islands, 18 May 1945, Berryman Papers.
30 Appreciation in Wills Papers, Folder 11, AWM.
31 Changes in Machinery for Higher Direction of War, 14 April 1942, signed by Curtin, Military Board Minutes, misc 41/1942. Minutes of Prime Minister's War Conference, Melbourne, 8 April 1942, MP 1217, Box 1.
32 Letter, Blamey to Curtin, 2 June 1943 and following correspondence in MP 1217, Box 306.
33 Shedden Diary, MP 1217, Box 16.
34 Notes of Discussion with C-in-C, SWPA, 27 June 1944, MP 1217 Box 3.
35 Interviews with Major-Generals Barham and Hopkins, and letter, Northcott to Blamey, 13 June 1945, Blamey Papers 23.12.
36 Letter, Blamey to Northcott, 28 May 1945, Blamey Papers 27.

*12 Lieutenant-General Sir Sydney Rowell*

1 Rowell to author, 26 June 1974. Letter, Rowell to Brigadier G.A. Vasey, 19 May 1941, Vasey Papers.
2 Letter, Blamey to Lieutenant-General V.A.H. Sturdee, 26 June 1941, Blamey Papers 5A.
3 Letter, DO/78/CGS, Sturdee to Blamey, 1 August 1941, Blamey Papers 5A. Letter, Vasey to his wife, 16 July 1941, Vasey Papers 2/5.
4 Letter, Lavarack to Forde, 23 March 1942, AWM 33/1/4.
5 Advisory War Council Minute 1013, 6 August 1942, CRS A2682, item vol 5. *Bulletin*, 12 August 1942.
6 Letter, Rowell to Vasey, 27 August 1942, AWM 225/2/5.
7 S. Milner, *Victory in Papua* Washington: OCMH, 1957, p. 71.
8 Signal, Landops to NGF, 26 August 1942; Signal, NGF to Milne Force, 26 August 1942, AWM 579/6/2.
9 Letter, Vasey to Blamey, 26 August 1942, Vasey Papers 2/9.
10 Letter, Vasey to Rowell, evening 28 August 1942, Rowell Papers.
11 Radio, MacArthur to Marshall, 30 August 1942, RG 165, OPD Exec 10 Item 23.a. National Archives, Washington.
12 Letters, Rowell to Vasey, 28, 30 August 1942; AWM 225/2/5.
13 Notes for War Diary, 8 September 1942, Allen Papers, File 1, AWM 419/3/9. Spry to author 8 August 1974. Spry still believes that it would have been the best solution.
14 Letter, Rowell to Vasey, 8 September 1942, AWM 225/2/5. Report on Operations New Guinea Force, AWM 58/7/35.
15 'Gen Willoughby's Dispositions', GHQ File, written possibly 25 September 1942. AWM 923/1/7. S.F. Rowell *Full Circle* Melbourne: Melbourne University Press, 1974, p. 174. Letter, Rowell to Vasey 28 August 1942, AWM 225/2/5.
16 Brigadier R.R. Vial, to author 25 August 1974.
17 'Denial of Resources to the Enemy' War Diary, NGF, 9 September 1942, AWM 1/5/51.
18 Note for War Diary, 13 September 1942, Allen Papers, file 1, AWM, 410/3/9.
19 NGF to 7 Div, 16 September 1942, War Diary 7 Div, September 1942, AWM 1/5/14.
20 Letter, Rowell to Clowes, 14 September 1942, Rowell Papers.
21 G.C. Kenney *General Kenney Reports* New York: Duell, Sloan and Pearce, 1949, p. 94. Minute 1067, Advisory War Council Meeting, Canberra, 17 September 1942, CRS A2682, Vol 5.
22 Colonel W.L. Ritchie to Marshall, 21 September

1942, RG 165, OPD, Exec 10, Item 236, Nat Archives (Wash). Notes of Secraphone Conversations between the Prime Minister and the Commander in-Chief, SWPA, 17 September 1942, MP 1217, Box 532.
23  Minute No 1067, Advisory War Council Meeting, Canberra, 17 September 1942, CRS A2682, item vol. 5.
24  Notes of Secraphone Conversation. R.A. Paull *Retreat from Kokoda*, Melbourne: Heinemann, 1958, p. 247.
25  'C-in-C's Press Conference', Perth 9 July 1945, Blamey Papers 138.3.
26  J. Hetherington *Blamey: Controversial Soldier* Canberra: Australian War Memorial, 1973, p. 239.
27  ibid., pp. 241, 242.
28  Letter, Rowell to Clowes, 22 September 1942, Rowell Papers.
29  Hetherington *Blamey* p. 245.
30  Rowell *Full Circle* p. 128. Letters, Rowell to Clowes, 14 September 1942, Rowell Papers.
31  Kenney *General Kenney Reports* p. 118, my emphasis.
32  Letter, Rowell to Clowes, 27 September 1942, Rowell Papers.
33  Letter, Blamey to MacArthur, 25 September 1942, Blamey Papers 43.631 and AWM 557/3/1.
34  Rowell to author, 26 June 1974.
35  'Report on N.G. Operations 23 September 1942–22 January 1943', by Blamey, AWM 519/6/58.
36  Hopkins to author, 13 July 1974, Letter, Rowell to Clowes, 14 September 1942, Rowell Papers.
37  Letter, Vasey to his wife, 25 September 1942, Vasey Papers 2/8.
38  Letter, Vasey to his wife, 28 September 1942, Vasey Papers 2/8.
39  Rowell *Full Circle* pp. 137, 138.
40  Letter, Dunstan to Rowell, 29 September 1942, Rowell Papers.
41  D. McCarthy *South West Pacific Area-First Year* Canberra: Australian War Memorial, 1959, p. 237. Letter, Rowell to Clowes, 22 September 1942, Rowell Papers.
42  Letter, Rowell to Vial, 28 September 1942, Allen Papers, File 7, AWM 419/3/9.
43  Letter, Rowell to Allen, 28 September 1942, Allen Papers, File 7.
44  Letter, Rowell to Clowes, 27 September 1942, Rowell Papers. McCarthy to author, 23 February 1974.
45  Willoughby to Gavin Long, Brisbane 16 June 1944, Gavin Long Notes, AWM 577/7/32. 'Crete' should read 'Greece'.
46  Major-General R.N.L. Hopkins to author, March 1973. Allen to Gavin Long, Sydney 27 September 1944, Gavin Long Notes.
47  Brigadier J.D. Rogers to author, 25 June 1974. Rogers visited Rowell during this time.
48  Letter, Rowell to Dunstan, 24 September 1942, Rowell Papers.
49  Letter, Blamey to Shedden, 10 August 1943, Blamey Paper 163.1. Letter, Vasey to his wife, 8 October 1982, Vasey Papers 2/8.

## 13 Lieutenant-General the Honourable Sir Edmund Herring

1  D. McCarthy *South West Pacific Area—First Year* Canberra: Australian War Memorial 1959, pp. 72-3, 140 note 6, 238-9. D.M. Horner *Crisis of Command, Australian Generalship and the Japanese Threat 1941-1943* Canberra: ANU Press, 1978, Appendix 16, pp. 315-16.
2  GOC 1st Corps Diary, 1 October 1942. Letters, Herring to his wife, 30 September, 3 October 1942. Gavin Long *The Final Campaigns* Canberra: Australian War Memorial, 1963, p. 594.
3  Letter, Herring to his wife, 3 October 1942. Horner *Crisis of Command* p. 189.
4  Horner *Crisis of Command* p. 316.
5  Stuart Sayers *Ned Herring: A Life of Sir Edmund Herring* Melbourne: Hyland House, Canberra: Australian War Memorial, 1980, pp. 218-19.
6  Letter, Blamey to Herring, 25 August 1945; Lieutenant-General R. Bierwirth to author, 4 January 1983. Letter, Colonel J. Buckley to author, 17 February 1983. Horner *Crisis of Command* pp. 199-200, p. 183.
7  Letters, Herring to his wife, 29 May, 1 September, 6 September 1942. Letter, Herring to Dexter, 19 March 1954.
8  David McNicoll in the *Bulletin*, 20 May 1980, p. 74. Sayers *Ned Herring* pp. 9, 131.
9  Sayers *Ned Herring* p. 127.
10  Bierwirth to author, 4 January 1983.
11  Horner *Crisis of Command* p. 189. Sayers *Ned Herring* p. 225. Bierwirth to author, 4 January 1983.
12  Sayers *Ned Herring* p. 218. Letter, Herring to his wife, 19 October 1942. Letter, Herring to Dexter, 19 March 1954. Horner *Crisis of Command* p. 200.
13  Horner *Crisis of Command* pp. 188-203. Sayers *Ned Herring* pp. 223-5.
14  Horner *Crisis of Command* pp. 200. Letter, Herring to his wife, 25 October 1942. Bierwirth to author, 4 January 1983.
15  Horner, *Crisis of Command*, pp. 230-64. Sayers *Ned Herring* pp. 227, 229. Letter, Herring to his wife, 17 November 1942.
16  Sayers *Ned Herring* p. 222.
17  ibid., p. 243. GOC 1st Corps Diary, 25 December 1942. Letter, Herring to his wife, 28 December 1942.
18  Sayers *Ned Herring* pp. 225, 227, 244. Letters, Herring to his wife, 11, 14 December 1942; 8, 19 January 1943.
19  D. Dexter *The New Guinea Offensives* Canberra: Australian War Memorial, 1961, pp. 55-60. D.M. Horner *High Command, Australia and Allied Strategy, 1939-1945* Sydney: George Allen & Unwin, 1982, pp. 267-72. Sayers *Ned Herring* pp. 249-50.

20 Dexter, *The New Guinea Offensives* pp. 58–60 Sayers *New Herring* pp. 253–4.
21 Dexter *The New Guinea Offensives* pp. 84–106 137–40, 164. Horner *High Command* pp. 201–5 Sayers *Ned Herring* pp. 254–7. Bierwirth to author, 4 January 1983. John Hetherington *Blamey: Controversial Soldier* Canberra: Australian War Memorial, 1980, pp. 310–13.
22 Sayers *Ned Herring* pp. 257–8. Dexter *The New Guinea Offensives* pp. 197–200, 216.
23 Sayers *Ned Herring* pp. 259–66. Dexter *The New Guinea Offensives* pp. 264–392.
24 Sayers *Ned Herring* pp. 267–76. Dexter *The New Guinea Offensives* pp. 279–81, 444–79. Letters, Herring to Dexter, 19 March 1954, 21 January 1955.
25 Sayers *Ned Herring* pp. 271–3. Horner *High Command* pp. 297–300. Ivan Chapman *Iven G. Mackay, Citizen and Soldier* Melbourne: Melway, 1975, pp. 227–8. Dexter *The New Guinea Offensives* pp. 480–3. Letter, Herring to Dexter, 21 January 1955.
26 Sayers *Ned Herring* pp. 273–4. Horner *High Command* pp. 300–1. Chapman *Iven G. Mackay* pp. 278–81. Dexter *The New Guinea Offensives* p. 488. Letter, Herring to Dexter, 21 January 1955.
27 Sayers *Ned Herring* pp. 270–8. Dexter *The New Guinea Offensives* pp. 480–500. Letters, Herring to Dexter, 7 August 1950, 20 August 1954, 21 January 1955, 23 July 1956. Horner *High Command* pp. 297–301. Hetherington *Blamey* p. 312.
28 Sayers *Ned Herring* pp. 277–8. Letters, Herring to Dexter, 7 August 1956, 20 August 1954, 21 January 1955.
29 Robert L. Eichelberger *Jungle Road to Tokyo* London: Odhams, 1951, pp. 54–5. George C. Kenney *General Kenney Reports* New York: Duell, Sloan and Pearce, 1949, p. 138. Letter and private communication, Colonel J. Buckley to author, 23 February 1983. Sayers *Ned Herring* p. 286. Chapman *Iven G. Mackay* p. 280.

## 14  Major-General George Alan Vasey

1 Spry to author, 8 August 1974. J. Hetherington *Blamey: Controversial Soldier* Canberra: Australian War Memorial, 1973, p. 265. For a detailed discussion of this incident see D.M. Horner *Crisis of Command, Australian Generalship and the Japanese Threat 1941–1943* Canberra: Australian University Press, 1978, Ch. 9.
2 F.K. Norris *No Memory for Pain* Melbourne: Heinemann, 1970, p. 169.
3 Vasey to his wife, 6 November 1942, Vasey Papers 2/6. Norris *No Memory for Pain* p. 170.
4 Eather to author, 29 April 1974. Memorandum to all commanders, 23 September 1942, War Diary HQ 6 Division, G Branch AWM 1/5/12. Lieutenant-Colonel M.S. Alexander to H.J. Manning, 14 July 1961, Vasey Papers 1/6. J. Hetherington *Australians, Nine Profiles* Melbourne: Cheshire, 1960, p. 17.
5 Signal Advanced HQ 7 Australian Division to 16 and 25 Brigades, 1 November 1942, Vasey Papers 2/8.
6 D. McCarthy *South West Pacific Area—First Year* Canberra: Australian War Memorial, 1959, p. 321. Eather to author, 29 April 1974. Spry to author, 8 August 1974.
7 Letter, Blamey to Shedden, 14 November 1942, Blamey Papers 136.1. Address, 'The Problems of a Commander', 1944, Vasey Papers 6/5.
8 Sir Reginald Pollard to H.J. Manning, 29 June 1962.
9 Blamey to Herring, 13 December 1942, Blamey Papers 170.2. Hetherington, *Blamey*, pp. 269–70. Vasey to his wife, 2 January 1943, Vasey Papers 2/6.
10 R.L. Eichelberger *Jungle Road to Tokyo* London: Odhams, 1951, p. 42. Herring to author, 25 June 1974. Vasey to his wife, 30 November 1942, Vasey Papers 2/6. J. Burns, *The Brown and Blue Diamond at War* Adelaide: 2/27 Battalion Ex-Servicemen's Association, 1960, p. 147.
11 Vasey to his wife, 2 January 1943, Vasey Papers 2/6.
12 Eichelberger to Manning, 13 March 1961, Vasey Papers 1/6.
13 Notes on Jungle Warfare pamphlet, 13 March 1943, War Diary 7 Division G Branch AWM 1/5/14.
14 Canet to H.J. Manning, 7 September 1961.
15 Letter, Vasey to Herring, 22 September 1943, Vasey Papers.
16 Letter, Mackay to Blamey, 7 October 1943, Blamey Papers. Letter, Morshead to Vasey, 18 October 1943, Vasey Papers.
17 Letter, Vasey to Major-General H.C.H. Robertson, 12 January 1944, Vasey Papers 2/6. Chilton to author, 3 May 1974.
18 Hetherington *Australians* p. 16. Canet to H.J. Manning, 7 September 1961. Address, 'The Problems of a Commander'. Robertson to author, 24 November 1982.
19 André Beaufre *An Introduction to Strategy* London: Faber & Faber, 1965, p. 57.
20 Address, 'The Problems of a Commander'.
21 Address by Chaplain D.L. Redding, 11 March 1945, Vasey Papers 6/6.
22 Shelford Bidwell *Modern Warfare, A Study of Men, Weapons and Theories* London: Allen Lane, 1973, p. 78. Notes by Father Richard Scarfe, 6 June 1962, Manning Papers.
23 Hetherington *Australians* pp. 16, 17.
24 ibid., pp. 3, 18. Interviews with senior AIF doctors who knew Vasey well. Disher to H.J. Manning, 3, 4 February 1962.
25 Blamey recommended him for a KBE after Sanananda and it was initially approved by Curtin. Apparently it was eventually rejected on advice from London after it was pointed out that Lieutenant-Generals Eichelberger, Herring and

Kenney—all one rank higher—were to receive the same award. (See cablegrams in Bruce Papers, Monthly War Files, December 1942, January 1943, CRS M100.) Vasey was the first graduate from RMC Duntroon to receive the American DSC. Notes on Broadcast by C-in-C, 15 April 1945, Blamey Papers 136.61. G.C. Kenney

## 15 Lieutenant-General Sir Horace Robertson

1. C.E.W. Bean *ANZAC to Amiens* Canberra: Australian War Memorial, 1961, p. 153.
2. H.S. Gullett *The A.I.F. in Sinai and Palestine 1914-18* Sydney: Angus & Robertson, 1923, Chapter XIV.
3. G.D. Solomon *A Poor Sort of Memory* Canberra: Roebuck, 1978, p. 85.
4. Gavin Long *To Benghazi* Canberra: Australian War Memorial, 1952, p. 83.
5. *Bulletin* (Sydney), 27 July 1960.
6. The Italian warship *San Giorgio* was aground in Tobruk harbour.
7. Ken Moses, ex 2/4 Battalion; *Sun* (Melbourne), 20 April 1946.
8. Chester Wilmot *Tobruk: 1941: Capture—Siege—Relief* Sydney: Angus & Robertson, 1945, p. 14.
9. John Hetherington *Blamey: Controversial Soldier* Canberra: Australian War Memorial, 1977, p. 125.
10. *Sydney Morning Herald*, 29 May 1960.
11. Hetherington *Blamey* pp. 143-4.
12. For example, in a letter to Hetherington on 29 November 1970, Brigadier C.M.L. Elliott said that he was present during the visit when Robertson was alleged to have made the statement, and he recalled no such statement (Hetherington Papers). See also Peter Charlton *The Thirty Niners* Melbourne: Macmillan, 1981, p. 152.
13. Letter, Blamey to Sturdee, 5 December 1940, Blamey Papers, DRL 6643, Item 5A, Australian War Memorial.
14. Gavin Long Papers No. 27, December 1943. Australian War Memorial.
15. Hetherington *Blamey* p. 213.
16. Information from Robertson's niece and her husband who served in the 1st Armoured Division in WA.
17. Memorandum, Sturdee to Forde, 2 April 1946, and teleprinter message, Coleman to Shedden, 3 April 1946, CRS A816, item 98/301/186.
18. Robertson Papers. Untitled History of BCOF, p. 93.
19. JCOSA had by then been dissolved since the British, Indian and New Zealand contingents had been withdrawn. Robertson reported directly to the Australian Chiefs of Staff.
20. Robert O'Neill *Australia in the Korean War, 1950-53, Vol. 1 Strategy and Diplomacy* Canberra: Australian War Memorial and AGPS, 1981, p. 253. I was astonished to read the criticism that Robertson occasionally displayed vulgarity. The statement, naturally, would not have been made without good cause but, in my own experience, I cannot recall a single instance to support the suggestion.
21. *Sun* (Melbourne), 24 April 1960.
22. R.G. Hodge in *Bulletin* (Sydney), 27 May 1960.

## 16 Air Chief Marshal Sir Frederick Scherger

1. ABC television interview, 26 May 1962.
2. National Library interview, 20 January 1975.
3. *Aircraft*, April 1957.
4. Extract from RAAF Form PP29 (Confidential Report).
5. ibid.
6. Scherger to author, 16 December 1980.
7. National Library interview, 13 November 1973.
8. Lowe Report transcripts, CRS A816, item 37/301/293, pp. 221, 222.
9. Barry Report summary, CRS A816 item 37/301/307, paragraph 291.
10. G.C. Kenney *General Kenney Reports* New York: Duell, Sloan and Pearce, 1949, p. 563.
11. Barry Report, paragraph 295.
12. Lewis H. Brereton *The Brereton Diaries* New York: William Morrow, 1946, p. 25.
13. Letter to author.
14. Letter to author.
15. Extract from RAAF Form PP29.
16. The RSL magazine *Stand-To* saw the appointment as 'another step in a career which has moved steadily towards the highest post available to the Australian regular Air Force officer'.
17. Letter to author.
18. Letter to author.
19. Letter to author.
20. *Aircraft*, April 1957.
21. National Library interview, 20 January 1975.
22. Mr J. Corrigan, former Secretary, Chiefs of Staff Committee to author.
23. In 1971 Air Marshal Sir Richard Williams was moved to comment, 'Scherger's too much a politician. He favoured his friends—still does, even in TAA and CAC' (comment to author).
24. National Library interview, 20 January 1975.
25. 1963 *Defence Report*, Tables 1 and 2, pp. 58-9.
26. *Aircraft*, April 1957.
27. 'Scherger knew politics, and he had the ear of Menzies, Townley and Hicks'—Vice-Admiral Sir Henry Burrell. 'Scherger was a politician in uniform'—Navy Minister (later Sir) John Gorton. Both in comments to author.
28. Bruce White, 'Defence and Its Environment' (APSA Conference, CCAE, 12 July 1975).
29. National Aviation Press Club, July 1975.
30. Defence Committee Minute No 15/1965.
31. Comment to author, 17 May 1980.
32. Sir Edwin Hicks, to author.
33. Comment to author.

## 17 General Sir John Wilton

1. General Sir John Wilton, Diary and Memoirs, p. 38, Wilton papers, held by author.

2   Major-General Sir Douglas Kendrew to Joseph Eisenberg and Janis Wilton (JE JW), 26 September 1977, Wilton papers.
3   Lieutenant-General Sir Sydney Rowell *Full Circle* Melbourne: Melbourne University Press, 1974, p. 180.
4   Molly Wilton, interview by author, 27 December 1982, Tape Revolutions (Revs) 60–70. Maurice Wilton, interview by author, 9 July 1982, Revs 85–100.
5   General Sir John Wilton, interviews by JE JW, 1978–80, p. 8, Wilton papers. Lieutenant-General Sir Thomas Daly, interview by author, 21 January 1983, Revs 17–50. Confidential Report, John Gordon Noel Wilton, 9 December 1930, Army Office (MS) file Wilton J.G.N.
6   Wilton, Diary and Memoirs, p. 21.
7   ibid., p. 26.
8   ibid., p. 142.
9   Major-General S.G. Savige to Major-General J.S. Whitelaw, AHQ Melbourne, 13 December 1942, Wilton papers.
10  Daly, interview by author, Revs 320–325. General Daly was the General Staff Officer Grade 1 of the 5th Division which relieved the 3rd Division at Salamaua.
11  Foreign Liaison Officer US War Department, to Lieutenant-General Sir John Lavarack, 22 February 1945. Lieutenant-General F.H. Berryman, Confidential Report on Colonel J.G.N. Wilton, Advance Land Headquarters Morotai and Manila, 1 February 1946. Major-General S.F. Rowell, Confidential Report on Colonel J.G.N. Wilton, 19 July 1949. Major-General R.N.L. Hopkins, Confidential Report Colonel J.G.N. Wilton, 10 August 1950. Lieutenant-General S.F. Rowell, Army Office (MS) file Wilton J.G.N.
12  Extract from Charter of the Royal Military College, reprinted in annual *Royal Military College Handbook*.
13  Wilton, interview by JE JW, p. 131.
14  Joint Planning Committee, Report No. 80/59, 18 November 1959, Defence file 249-10-13.
15  Defence Committee Minute (DCM) No. 122/1959 26 November 1959, Defence file 249-10-13.
16  Defence file 292-2-220.
17  Lieutenant-General Sir Reginald Pollard to J.O. Cramer, Minister for the Army, 20 February 1962, Defence file 19-4-10. Minister for the Army to Athol Townley, Minister for Defence, 28 February 1962, Defence file 19-4-10.
18  Wilton, interview by author, p. 4.
19  Secretary for Defence to Minister for Defence, 15 December 1964, Defence file 206-3-25.
20  Sir Wilfred Kent Hughes to Minister for Defence, 23 December 1964, Defence file 206-3-25.
21  DCM No. 56/1964, 6 April 1964.
22  Wilton, Diary and Memoirs, p. 86.
23  ibid., p. 87.
24  Because the Army component was the largest of the three services, the force commander remained an Army officer but his deputy was provided from the Air Force.
25  Wilton, interview by JE JW, p. 115.
26  ibid., p. 114.
27  Group Captain P.G. Latham, Staff Officer to General Wilton 1966-68, interview by author, 11 January 1983, Revs 604 ff.
28  Secretary for Defence to Chairman Public Service Board, 14 March 1980, Defence file 194-1-4 (29).
29  Rear Admiral G.J.H. Woolrych, Staff Officer to General Wilton 1968-70, interview by author, 27 January 1983.
30  Secretary for Defence to Chairman Public Service Board, 12 May 1961, Defence file 194-1-4 (29).
31  DCM No. 19/1967, 2 February 1967.
32  Woolrych, interview by author, Revs 135–172.
33  Secretary for Defence, 12 May 1961.
34  Wilton, interview by JE JW, p. 114.
35  Wilton, Minute to Minister for Defence, 5 September 1967, Wilton papers.
36  Wilton, Minute to Minister for Defence, 27 January 1970, Wilton papers.
37  Wilton, interview by JE JW, p. 115.
38  Wilton, letters to Lady Wilton from Korea, Wilton papers. Wilton, Diary and Memoirs, pp. 46, 84.
39  DCM No. 2/71, 5 March 1971.

# Index

Aboukir, 24
Adair, Lieutenant-Commander C.C., 257
Admiralty, 49, 54, 55, 56
Afrika Corps, 195, 200
AIF, 4, 7, 9, 13, 15, 16, 17, 18–19, 20, 21, 22, 24, 26, 27, 28, 32, 33, 34, 35, 36, 37, 38, 39, 40, 41, 66, 72, 82, 92, 96, 122, 140, 141, 145, 151, 154, 156, 159, 160, 161, 162, 164, 165, 167, 174, 175, 176, 178, 204, 205, 207, 208, 217, 225, 226, 228, 231, 246, 247, 248, 261, 262, 288, 289
Air Board, 301, 302, 307
Alamo Force, 214, 215
Alexander, General Sir Harold, 189, 190, 197, 200
Ali Muntar, 72–3
Allanson, Major C.J.L., 95
Allen, Major-General A.S., 4, 51, 227, 228, 231, 233, 234, 240, 241, 243, 245, 248, 250, 251, 253, 263, 265, 266, 268, 275
Allenby, Sir Edmund, 66, 68, 73, 76, 78, 79, 81, 82, 83, 84, 284
Allied Air Forces, 302, 304
Allied Land Forces, 210, 211, 214, 215, 219, 220, 244, 245, 250, 257, 261
Amiens, 108, 109
Anderson, J.T. Noble, 86
Anderson, Major S.M., 22, 172
Anderson, Air Vice-Marshal, 300
Angaur, 52, 54
Anglo-Japanese Alliance, 46, 54
ANZAC Corps, 13, 23, 26, 34, 35, 36, 38, 61–6, 68, 70, 98, 101, 205, 282, 283, 304
Anzac Cove, 61–6, 92–6, 175
Anzac Light Railways, 37
Anzac Mounted Corps, 72, 73, 74–5, 78, 79, 83
ANZUS Treaty, 325
Arabs, 78, 79, 81
Armstrong, Charles, 44
Auchinleck, General, 183, 184, 185, 186, 188, 200, 205, 289
*Australia*, 50, 51
Australian Air Squadron, 5
Australian Corps, 3, 38, 39, 41, 42, 103, 105, 114, 122, 123, 129, 140, 141, 221, 226, 228, 239, 261
Australian Council of Defence, 131

Australian Defence Force Academy, 330
Australian Fleet Unit, 46, 47, 49, 50, 51, 55
Australian Imperial Forces *see* AIF
Australian Intelligence Corps, 90
Australian Military Board, 143, 148, 157, 161, 207, 208, 222, 327
Australian Military Forces, 202, 204, 207, 210, 284
Australian Military Mission, 320, 321, 329
Australian Mounted Division, 74, 79, 81
Australian Naval Board, 2, 44, 47–8, 49, 51, 53, 55, 56, 57, 58
Australian Reinforcement Training Centre, 37
Australian Task Force, 5

Baby 700, 62, 63, 65, 94, 175, 282
Bangkok, 324
Barbey, Rear-Admiral Daniel, 257, 259
Barnard, Lance, 331, 332
Barry, J.V., 303
Barstow, Major-General, 163
BCFESR, 323, 325
Bean, C.E.W., 16, 22, 26, 27, 34, 38, 40, 61, 65, 85, 93, 94, 102, 103, 109, 110, 113, 115, 117, 119, 174, 184, 282
Beaufre, General André, 275
Beavis, Major-General L.E., 205, 207
Beersheba, 72–6
Belgium, 28
Bennett, Lieutenant-General Henry Gordon, 4, 6, 142, 145, 159–74, 212, 236, 289
Bennett, Lieutenant-Colonel P.H., 332
Berryman, Lieutenant-General Sir Frank, 3, 4, 8, 9, 138, 141, 146, 173, 198, 208, 211, 218, 247, 248, 250, 253, 255, 256, 259, 260, 278, 319, 320, 322
Bidwell, Brigadier Shelford, 276
Bierwirth, Brigadier R., 251, 256
Birdwood, Sir William, 13, 17, 19, 22, 23, 24, 26–7, 33, 34, 35, 36, 37, 38, 39, 40, 61, 63, 64, 66, 68, 82, 83, 84, 92, 93, 96, 98, 101, 102, 103, 105, 109
Bladin, Air Vice-Marshal F.M., 305
Blamey, Field Marshal Sir Thomas, 4, 6, 7, 21, 41, 61, 84, 104, 105, 119, 133, 135, 138, 141, 142, 146, 156, 157, 159, 161, 166, 172, 176, 184, 186, 190, 198, 200, 202–24, 225, 226, 227, 228, 229, 230, 231, 234, 235, 236, 237, 239, 240, 242, 243, 244, 245, 246, 247, 248, 250, 251, 253, 254, 257, 259, 260, 261, 262, 263, 264, 266, 268, 272, 277, 278, 288, 289, 290, 319, 320, 321
Boer War, 5, 28, 33, 60–1
Bogadjim, 272
Borneo, 4, 198, 215, 221, 303, 304, 312
Bostock, Air Vice-Marshal William, 5, 302, 303
Bougainville, 3, 218
Bourne, Major, 66
Boxer Rebellion, 46
Braund, G.F., 175
Bridges, Major-General Sir William, 3, 6, 13–25, 32, 33, 34, 61, 92
British Army Council, 21
British Commonwealth Occupation Force, 2, 281, 290–3, 294, 295
British Expeditionary Force, 103, 146
Broodseinde, 101, 110
Brooke-Popham, Air Chief Marshal, 165
Bruche, Major-General J.H., 108, 131, 132
Buna, 266–70
Burma, 318, 319
Burnett, Air Chief Marshal, Sir Charles, 149, 156, 300, 301
Butler, Lieutenant-General R.H.K., 109, 111

Canada, 108, 109, 111, 118
Canet, Major-General L.G. 270, 275
Cannan, 207
Cape Helles, 23, 33
Carlyon, Norman, 208
Carpender, Vice-Admiral A.S., 259, 260
Casey, R.G., 204
Chamberlin, Brigadier-General S.J., 257
Chapman, I., 3
Chauvel, General Sir Harry, 3, 40, 60–84, 94, 145, 282
Chetwode, Lieutenant-General Sir Philip, 60, 66, 71, 72, 73, 74, 76
Chiefs of Staff Committee, 311–13, 328, 329, 330, 331
Chifley, G., 157
Chilton, Brigadier, 270
China, 316, 323
China Fleet, 46
China Force, 3

Chipilly Spur, 109-11
Chunuk Bair, 96
Churchill, Sir Winston, 49, 85, 140, 156, 163, 165, 189, 190
Citizen Military Forces, 176, 284, 325, 326
Clarkson, Captain, 48, 56, 58
Clowes, Major-General Cyril, 4, 8, 151, 228, 229-31, 234, 236, 237, 240, 242
Cobby, Air Commodore A.H., 303
Cold War, 323
Collins, Vice-Admiral Sir John, 3, 5
Colonial Office, 29, 52
Colvin, Sir Ragnar, 133
Committee of Imperial Defence, 29
Connell, John, 179
Cook, Sir Joseph, 31, 32, 53, 54, 111
Coral Sea, Battle of the, 5, 153, 216, 225
Coronel, 51
Coulthard-Clark, C., 3, 6, 13-25
Courage, Brigadier-General A., 105, 106, 107, 108
Cox, C.F., 72, 175
Cox, Brigadier-General H.V., 95, 96
Crace, Admiral Sir John, 5
Creswell, Vice-Admiral Sir William, 2, 44-59
*Cumberland*, 56
Currie, Sir Arthur, 88, 89, 113, 114
Curtin, John, 140, 142, 148, 153, 156, 163, 165, 190, 198, 208, 211, 213, 214, 215, 216, 219, 220, 221, 222, 235, 236, 239, 241, 242, 243, 244, 245, 262
Cutlack, F.M., 27
Cyrenaica, 134, 135, 178, 179, 180, 185, 274, 288

Daly, General Sir Thomas, 4, 316, 317, 318, 328
Damascus, 81-2, 138
Dardanelles, 16, 23, 33
Darwin, 42, 48, 147, 151, 202, 245, 246, 285, 301, 307
Deakin, Alfred, 18
Debeney, General, 111, 113
Defence Act, 38, 40
Defence Committee, 312, 324, 333
Defence Reorganization Act, 332
Depression, 318
Desert Column, 60, 71, 73
Desert Mounted Corps, 3, 73-4, 76, 81, 82
Dewing, General, 214
'Digger' tradition, 21
Dill, Field Marshal, 7
Diller, Colonel L.A., 215
Disher, Brigadier Clive, 278
Dobell, Lieutenant-General Sir C. Dobell, 72, 73
Dorman-Smith, Major-General, 188
Dougherty, Brigadier, 3
Downes, Colonel R.M., 74
Dunkirk, 41, 181, 213
Dunstan, Lieutenant-General Sir Donald, 4

Dunstan, William, 213, 236, 241
Duntroon, 8, 15, 23, 61, 129, 226, 247, 263, 281, 282, 285, 297, 298, 299, 311, 313, 316, 318, 323, 324, 329
Dutch East Indies, 42
Dyson, W., 27

Eather, Brigadier, 234, 265, 266
Eden, Sir Anthony, 134
Egypt, 13, 19, 21, 24, 28, 32, 33, 37, 61, 66, 68, 71, 73, 82, 92, 96, 98, 134, 176, 184, 246, 282, 284
Egyptian Expeditionary Force, 73
Eichelberger, Lieutenant-General Robert L., 210-11, 214, 253, 261, 268, 270
Eisenhower, General, 320
El Alamein, 4, 185-97, 200, 201
El Arish, 71-2
Ellington, Air Chief Marshal Sir Edward, 300, 301
Elliott, Brigadier-General Pompey, 3, 27, 37, 38, 113, 117, 175
*Emden*, 50
*Emerald*, 57
Empire Air Training Scheme, 5, 300
Encel, Professor S., 1
Es Salt, 76-9
Essame, Major-General H., 88
Evans, Brigadier B., 198
Fadden, Arthur, 147, 148
Fairhall, Sir Allen, 327
Farlow, George, 89
Farncomb, Commodore, 5
Federation, 46
Feisal, 81
Field, Brigadier, 208
Finschhafen, 256-61, 272
First World War, 2, 8, 26, 27, 28, 31, 32, 39, 41, 42, 44, 82, 89, 92, 129, 162, 172, 173, 204, 205, 246, 248, 281, 282-4, 297, 298, 303
Fisher, Andrew, 30, 53, 54
Fisher, Sir John, 49
Foch, Marshal, 2, 36, 109, 111
Foote, Brigadier-General C.H., 38
Forbes, Alexander, 327
Forde, F.M., 154, 155, 158, 162, 172, 208, 215, 222, 227, 289
Foster, Colonel H.J., 91
France, 41, 50, 66, 72, 74, 226
Fraser, Malcolm, 327, 331
Free French, 247
Freyberg, General, 190
Fuller, Major-General Horace H., 225
Fuller, Colonel J.F.C., 6, 7, 105, 173
Fusion Party, 32

Gaba Tepe, 23, 33
Gabo Island, 56
Gallipoli, 13, 16, 21, 22, 32, 34, 36, 37, 39, 61-6, 92-6, 175, 226, 282
Garrett, Sir Ragnar, 325
Gaza, 66, 68, 72-6, 82, 197
Gellibrand, Brigadier-General John, 3, 27, 28, 37, 38, 104, 116, 117, 122, 123

Germany, 41, 46, 47, 48, 49, 50, 52, 53, 103, 105-6, 108-13, 118-22, 131, 173, 184, 186, 188, 193, 195, 200, 205, 264, 276, 282
German East Asia Squadron, 49
Ghurka troops, 306
Gilbert, W.S., 174
Glasgow, Major-General T.W., 3, 104, 113, 114, 115, 175
*Gneisenau*, 50
Godfrey, Brigadier, A.H.L., 180, 198
Godley, Major-General Sir A.J., 30, 34, 61, 62, 63, 92, 94, 96, 98
Gona, 266-70
*Good Hope*, 51
Gordon, Brigadier-General J.M., 30, 31
Gough, General, 37
Grant, William, 75, 79
Greece, 205, 207, 213, 226, 247, 248, 264, 266, 274, 288, 289
Guadalcanal, 234, 268
Guam Conference, 312
Gullett, H.S., 66, 71, 75, 204

Haig, Field Marshal Sir Douglas, 37, 38, 82, 85, 89, 99, 103, 107, 109, 115, 119, 121, 122, 124, 146
Hamel, 105-8, 109, 115, 124
Hamilton, General Sir Ian, 16, 23, 33, 34, 66, 91
Hammer, H.H., 256
Hancock, Air Marshal, 310
Hankey, Sir Maurice, 130, 131, 132
Harding, Major-General E.F., 253
Harding, Brigadier John, 178, 179
Hardman, Air Marshal Sir Donald, 305
Harington, Sir Charles, 98, 102, 104
Hart, Sir Basil Liddell, 1, 8, 9, 91, 114
Hasluck, Sir Paul, 1
Haworth-Booth, Admiral, 55
Heath, Lieutenant-General, 160, 162, 163, 169, 170, 172
Heliopolis, 61
Henderson, Admiral Sir Reginald, 47, 58
Henderson Report, 47, 48, 58
Herring, Lieutenant-General Sir Edmund, 4, 6, 7, 197, 198, 207, 217, 218, 232, 241, 244-62, 263, 268, 289, 296
Hetherington, John, 235, 236, 237, 275, 277, 288, 289
Hewitt, Air Commodore J.E., 5, 305
Hicks, Sir Edwin, 309, 313
Hill, A.J., 3, 4, 60-84, 175-201
Hindenberg Line, 118-22, 123
Hitler, Adolf, 147
Hoad, Major-General Sir John, 32
Hobbs, Major-General J.T.T., 3, 22, 40, 104, 116, 117, 118, 124, 175
Hodgson, General, 75
Holt, Harold, 327
Hopkins, Major-General R.N.L., 2, 6, 146, 239, 241, 250, 253, 281-97, 322
Horii, Major-General, 266
Horner, Major D.M., 143-58, 202-24,

225-43, 245, 251, 263-78
Horrocks, General, 195
Howard-Vyse, Brigadier-General R.G., 74
Hughes, Sir Wilfred Kent, 326
Hughes, William, 26, 54, 58, 118, 145
Hughes-Onslow, Captain Constantine, 48, 49, 52, 53, 54, 55
Hutton, Major-General Sir Edward, 16, 17, 18, 24, 27, 28, 32

Imamura, General, 157
Imperial Camel Corps Brigade, 72
Imperial Conference 1923, 129
Imperial Defence College, 322
Imperial Defence Conference 1909, 46, 47
Imperial Mounted Division, 72-3
Imperial Navy, 54
Imperial War Conference 1918, 54
India, 264, 318, 324
Indian Ocean, 53
Indonesia, 323
Irvine, Sir William, 31

Jacka, A., 123
Jackson, Colonel G.H.N., 102
Janowitz, Morris, 88
Japan, Japanese, 9, 30, 40, 41, 42, 48, 50, 52, 53, 54, 129-33, 139, 140, 141, 143, 144, 145, 146, 147, 148, 149, 151, 152, 153, 154, 157, 163, 168, 169, 171, 173, 174, 202, 204, 205, 208, 210, 217, 218, 225, 228, 229, 231, 232, 233, 234, 237, 239, 242, 244, 245, 248, 251, 255, 256, 260, 261, 263, 265, 266, 267, 268, 270, 272, 275, 289, 290-3, 295, 296, 297, 301, 304, 320, 321
Java, 139-41
Jellicoe, Lord, 58
Jensen, J.A., 53, 56, 57, 58
Jobson, Brigadier-General, 99
Johnston, Brigadier W.W.S., 233
Johore, 162, 163, 166, 169, 171, 172, 174
Joint Intelligence Bureau, 330
Joint Planning Committee, 42
Joint Services Staff College, 330
Jones, Air Marshal Sir George, 5, 214, 302, 303, 305, 308
Jordan River, 76-9
Jose, A.W., 55

Kemal, Mustapha, 82
Kemsley, Sir Alfred, 42
Kenney, Major-General George, 5, 234, 237, 252, 259, 261, 278, 305
Keogh, Colonel, 152, 215
Kirby, General, 162, 163, 164, 165, 166
Kitchener, Field Marshal Lord, 18
Kokoda Track, 4, 7, 217, 225, 228, 229, 231-4, 239, 240, 241, 242, 245, 250, 251, 252, 263, 264-6, 270
Korean War, 262, 293-5, 296, 313, 316, 321, 322
Kota Bharu, 149
Krueger, General Walter, 214, 215

Kuala Lumpur, 306
Kumusi River, 265-6
Kusaka, Admiral, 157

Lae, 253, 254, 255, 256-61, 270, 276
Land Headquarters, 207
Latham, Lieutenant-Commander J.G., 54
Lavarack, Lieutenant-General Sir John, 4, 129-42, 144, 145, 146, 157, 179, 181, 212, 218, 227, 228, 236, 237, 241, 246, 247, 285, 319, 320
Lawrence, Major-General H.A., 66, 68, 70, 72, 81
Leese, Sir Oliver, 189, 190, 191, 196, 200
Legacy, 28
Legge, Major-General J.G., 3, 15, 30, 32, 37, 39, 40
Lewin, 174
Lloyd, Major-General C.E.M., 180, 183, 222, 241
Lodge, A.B., 4, 6, 129-42, 159-74
Lone Pine, 175, 191
Long, Gavin, 1, 167, 205, 215, 217, 286, 289
Lowe, Mr Justice, 301
Ludendorff, 108, 109, 111, 118
Luftwaffe, 184
Lyons, Joseph, 130, 204

Maadi, 61
MacArthur, General, 7, 153, 154, 198, 200, 202, 204, 210, 211, 212, 213, 214, 215, 216, 217, 218, 219, 220, 221, 222, 223, 224, 228, 229, 230, 232, 234, 235, 236, 237, 239, 242, 243, 244, 245, 247, 250, 251, 252, 253, 257, 259, 260, 268, 278, 292, 293, 294, 295, 302
MacDonald, General Sir Arthur, 4
Mackay, Lieutenant-General Sir Iven, 3, 148, 151, 155, 156, 157, 173, 175, 198, 218, 244, 245, 247, 248, 255, 257, 259, 260, 262, 264, 272, 286, 287, 288
MacKechnie, Colonel Archibald R., 255
Madras, 50
Magdhaba, 72, 283
Mahan, Alfred Thayer, 50
Malaysia, 323, 325
Malaya, 4, 42, 147, 153, 159, 160, 161, 162, 163, 164, 165, 167, 168, 169, 170, 172, 173, 174, 205, 235, 302, 305-7, 321, 325
Malaya Command, 161, 164-5, 167, 306
Mangin, General, 108
Manisty, H.W.E., 48
Manpower, 205
Marshall, General, 213, 214, 230
Marshall, Norman, 175
Martin, Professor Sir Leslie, 329
Maughan, Barton, 180, 189, 201
Maxwell, Lieutenant-General Sir John, 20, 166, 167, 172
McArthy, Dudley, 236, 241
M'Cay, Major-General J.M., 3, 39, 40
McCauley, Air Vice-Marshal J.P.J., 305

McFarlane, A.B., 309
McNamara, Wing Commander F., 300
McNeill, I., 2, 316-34
McNicoll, Brigadier-General W.R., 101, 114
McSharry, T.P., 175
Meggido, Battle of, 79-81
*Melbourne*, 50, 51
Menzies, Sir Robert, 40, 43, 87, 133, 157, 204, 288, 299, 305, 311, 327
Merdjayoun, 4, 137-8
Messines, 101, 110, 113
Milford, Major-General E.J., 4, 199, 256, 286, 319
Military Board, 14-15, 40, 41
Millen, Senator Edward, 31
Milne Bay, 4, 228, 229-31, 236, 237, 239, 242, 259
Mitchell, D., 87
Monaghan, R.F., 256
Monash, General Sir John, 3, 6, 7, 8, 37, 38, 39, 40, 42, 61, 62, 63, 65, 85-125, 175, 204
*Monmouth*, 51
Montgomery, Field Marshal, 6, 7, 108, 113, 189, 190, 191, 193, 195, 276, 320
Mont St Quentin, 116-18, 124
Morlancourt, 105
Morotai, 320
Morrison, Ian, 174
Morshead, Lieutenant-General Sir Leslie, 4, 6, 134, 175-201, 210, 215, 218, 241, 253, 259, 260, 262, 272, 311, 331
Moten, Brigadier Murray, 255, 256
Mountbatten, Lord Louis, 301, 331
Munro-Ferguson, Sir Ronald, 50, 55, 56
Murdoch, Sir Keith, 38, 103, 241
Murray, General Sir Archibald, 37, 68, 71, 72, 73, 84, 151, 175, 180

Nadzab, 270-4, 275, 278
National Service, 295, 313, 321, 326, 327
National Service Act 1951, 40-1
Navy Office, 47, 48, 49, 50, 51, 52, 55, 56, 57, 58
Neame, Lieutenant-General Sir Philip, 178, 179, 186, 201
New Caledonia, 42, 149, 153
New Guinea, 3, 4, 7, 50-4, 148, 149, 153, 154, 173, 197, 198, 202, 207, 210, 211, 213, 214, 215, 217, 218, 225, 228, 229, 230, 231, 232, 233, 235, 236, 237, 239, 244, 245, 246, 247, 248, 250, 251, 252, 255, 256, 260, 261, 270, 274, 275, 276, 290, 302, 304, 305, 319, 320
New Guinea Force, 3, 4, 198, 233, 244, 245, 250, 251, 253, 255, 257, 259, 260, 261, 263, 270, 271, 272, 278
New Zealand, 30, 31, 47, 54
Norris, Sir Kingsley, 264, 265
Northcott, Major-General Sir John, 4, 148, 151, 176, 210, 222, 241, 289, 290, 293

O'Connor, General R.N., 178
Oliver, Major-General Sir William, 307
Operation OBOE, 303
Operation RECKLESS, 302
Operations Directorate, 31
*Orcades*, 141

Pacific, 30, 31, 39, 40, 46, 47, 49, 51, 52, 53, 54, 90, 130, 157, 202, 203, 218, 219, 220, 221, 229, 259, 302-5, 319
Pakistan, 324
Palestine, 3, 5, 74, 79, 82, 134, 176, 197, 226, 246, 288, 289, 297
Papua, 228, 229, 237, 239, 244, 245, 246, 248-53, 261, 270, 271
Park, Sir Keith, 298
Parkhill, Sir Archdale, 132
Passchendaele, 101, 102, 117
Patey, Rear Admiral, 50, 51, 52
Pearce, Senator, 30, 31, 38, 39, 52, 53
Pearl Harbour, 149
Pederson, P., 3, 8, 65, 85-125
Penang, 50
Percival, Lieutenant-General A.E., 139, 159, 160, 161, 162, 163, 164, 165, 166, 167, 168, 169, 170, 171, 172
Perowne, Major-General Lance, 306
Pershing, General J.J., 106
Pétain, 98, 99
Phipps, Rear-Admiral Sir Peter, 312
Plumer, General Sir Hubert, 98, 99, 104
Point Cook, 308, 329
Pollard, Lieutenant-General Sir Reginald, 268, 324, 325
Port Moresby, 225, 229, 231, 232, 233, 234, 235, 236-40, 242, 244, 248, 250, 251, 253, 257, 259, 263, 264, 268, 270, 320
Porter, Brigadier, 231, 233
Potts, Brigadier, 228, 231, 232, 233, 234
Pozieres, 37
Preston, R.A., 21
Prime Minister's War Conference, 219
*Protector*, 44, 46

RAAF, 298-315, 323
Rabaul, 151, 152-3, 157, 229
RAF, 298, 300, 305-7
Ramsay, Brigadier A.H., 193
Ramsden, Major-General W.H.C., 186, 188, 189
Ramu Valley 270-4
RAN, 5, 44, 47, 48, 49, 56, 58, 300
Ranks, Guide to, xv
Rawlinson, General Sir Henry, 36, 105, 107, 108, 109, 113, 115, 117, 118, 119, 120, 121, 122, 124
Rayner, H.R., 2, 298-315
Redding, D.L., 276
Red Line, 180-5
Reid, Sir George, 20
Richmond, Admiral Sir Henry, 129
Robertson, Lieutenant-General Sir

Horace, 2, 3, 6, 7, 8, 142, 151, 212, 248, 281-97
Robertson, Colonel W.T., 275
Rogers, Brigadier, 242
Romani, Battle of, 68, 69, 71, 82
Rommel, General, 134, 135, 179, 181, 183, 185, 186, 188, 195, 200
Rosenthal, Major-General C., 3, 104, 105, 113, 114, 116, 117, 118, 124, 175
Rowell, Lieutenant-General Sir Sydney, 4, 6, 7, 8, 9, 66, 142, 146, 148, 151, 152, 157, 208, 211, 213, 214, 217, 225-43, 244, 245, 246, 247, 248, 250, 251, 261, 278, 296, 317, 322
Royal Artillery, 318
Royal Australian Artillery, 319
Royal Commission on Navy and Defence Administration, 58
Royal Navy, 17, 29, 41, 42, 44, 46, 51, 55, 58, 59, 130, 131, 133, 300
Royle, Admiral, 149, 156
Royston, Brigadier-General, 70
Russell, W.B., 4
Russia, 50, 60
Russo-Japanese War, 28
Ryrie, Sir Granville, 173

Sailly-Laurette, 103, 105
Salamaua, 3, 250, 252, 253-6, 260, 261, 270, 320
Samoa, 50, 51
Sanananda, 266-70
Sari Bair, 96
Sattleberg, 3
Savige, Lieutenant-General Sir Stanley, 3, 4, 151, 207, 248, 254, 255, 256, 319, 320
Sayers, S., 4, 6, 244-62
Scharnhorst, 50
Scherger, Air Chief Marshal Sir Frederick, 2, 3, 298-315
Scott, Humphrey, 175
SEATO, 323, 324, 325, 333
Second World War, 2, 8, 9, 40, 41, 42, 129, 173, 202, 204, 205, 224, 226, 246, 261, 263, 264, 266, 281, 285-90, 297, 317, 319-21, 323
Sellheim, Colonel, 37
Serle, A.G., 26, 85, 87
Sharp, Admiral, 312
Shedden, Sir Frederick, 131, 132, 156, 204, 211, 212, 214, 215, 217, 219, 220, 222, 262
Sinai, 66, 68, 71, 74, 82, 282, 283, 297
Sinclair, F.R., 208
Sinclair-MacLagan, Major-General E.G., 3, 23, 88, 104, 105, 106, 111, 119, 124
Singapore, 41, 42, 54, 129-33, 139-41, 145, 151, 154, 155, 159, 162, 163, 166, 168, 171, 174, 202, 305
Smith, Major-General W.E.B., 38, 68, 70
Smithers, A.J., 117
Solomon, Brigadier Geoffrey, 285

South Africa, 17, 32, 60-1, 82; *see also* Boer War
South-East Asia, 323, 325, 333
Spender, Sir Percy, 148
Spry, Colonel C.C.F., 231, 263, 266
Squires, Lieutenant-General E.K., 41, 133, 144
Staff College, Camberley, 28, 29, 60, 162, 176, 226, 284
Staff Corps, 8, 9, 173, 285
Stansfield, Lieutenant-General W., 74
Steele, General, 289
Stevens, General, 4
Steward, Major George, 56
Stewart, Alex, 169, 305
Sturdee, Lieutenant-General Sir Vernon, 2, 7, 143-58, 167, 207, 208, 210, 211, 222, 226, 262, 289, 290, 319, 320
Sudan, 17
Suez Canal, 68, 282
Sutherland, Brigadier R., 229, 251, 256, 260
Sweeting, A.J., 202
Swinbourne, G., 39
*Sydney*, 50
Syria, 4, 79, 135-9, 226, 247

Taylor, Brigadier, 167, 168, 171
Tedder, Air Chief Marshall, 184
Tell es Sabe, 75
Templer, General Sir Gerald, 306, 307
Tet offensive, 4, 5
Thorby, H.V.C., 132
Thring, Commander Walter, 48, 49, 52
Thyer, Colonel, 167, 168, 172
Timor, 202
Tobruk, 4, 134-5, 136, 137, 148, 176-85, 186, 198, 199, 200, 201, 205, 286, 287, 288, 296
Tovell, R.W., 180
Tropical Force, 52
Trotman, Brigadier-General, 94
Tuchman, B., 10
Turkey, Turks, 61-6, 67, 68, 70, 71, 72, 73, 74, 79, 81, 82, 92, 93, 96

United Nations, 294
United States Airforce, 228, 233, 252, 257, 272
United States Navy, 42, 257
Universal Training Scheme, 91

Vasey, Major-General George, 4, 6, 7, 8, 142, 151, 157, 207, 211, 226, 228, 229, 230, 232, 240, 242, 244, 245, 248, 251, 253, 257, 259, 263-78, 288, 289
Verney, G., 2, 26-43
Versailles Peace Conference, 54, 58
Vial, Lieutenant-Colonel R.R., 233
Vichy French, 135-9
Vietnam War, 3, 4, 9, 306, 312, 313, 325-8, 331, 332, 333
von Kressenstein, Kress, 68, 71

*Index*

von Moltke, Field Marshal H., 2
von Sanders, Liman, 79
von Spee, 49, 50, 51

Wackett, Air Vice-Marshal E.C., 305, 314
Walcot, Commander John, 44, 46
War Book, 31
War Cabinet, 41, 148, 151, 156
Ward, Eddie, 154
Wark, Blair, 175
War Office, 18, 19, 21, 30, 37, 68, 293
War Railway Council, 28
Washington Conference, 40
Wavell, Field Marshal Lord, 6, 7, 88, 134, 138, 138, 140, 141, 162, 163, 169, 170, 178, 179, 180
Webster, S.D., 2, 44-59
Wells, Lieutenant-General Sir Henry, 311, 323
Wewak, 218
White, Bruce, 311, 328
White, General Sir Brudenell, 2, 7, 16, 22, 23, 26-43, 82, 103, 105, 109, 124, 144, 145
Whitehead, Brigadier-General Ennis C., 233, 252, 257, 272, 302, 304

Wigmore, Lionel, 153, 161
Williams, Air Marshal Sir Richard, 300
Willoughby, Major-General C.A., 233, 240, 251
Wills, Brigadier Sir Kenneth, 218, 288
Wilmot, Chester, 181
Wilson, Air Commodore D.E.L., 301
Wilson, Major-General Sir Henry, 26, 29, 135, 136, 138, 175, 185
Wilton, General Sir John, 2, 3, 146, 310, 313, 316-34
Wimberley, D.N., 191
Windeyer, Brigadier W.J.V., 3, 189, 257, 259, 260
Woolrych, Commander G.J.H., 330
Wootten, Major-General Sir George, 4, 179, 180, 197, 198, 250, 257, 259, 260
Wynter, Major-General H.D., 145, 176, 207, 208

Xenophon, 8

Yap, 52, 54
Young, Captain Hubert, 81
Ypres, 99-100

Lieutenant-General Sir Horace Robertson

Air Chief Marshal Sir Frederick Scherger

General Sir John Wilton

Major-General George Alan Vasey

For Product Safety Concerns and Information please contact our EU
representative GPSR@taylorandfrancis.com
Taylor & Francis Verlag GmbH, Kaufingerstraße 24, 80331 München, Germany

www.ingramcontent.com/pod-product-compliance
Lightning Source LLC
Chambersburg PA
CBHW081148290426
44108CB00018B/2479